Advances in
PARASITOLOGY

Remote Sensing and Geographical Information Systems in Epidemiology

VOLUME 47

Advances in PARASITOLOGY

Remote Sensing and Geographical Information Systems in Epidemiology

Series editors

J.R. BAKER, R. MULLER and D. ROLLINSON

Guest editors

S.I. HAY, S.E. RANDOLPH and D.J. ROGERS

Department of Zoology, University of Oxford, Oxford, UK

VOLUME 47

ACADEMIC PRESS

An imprint of Elsevier Science

Amsterdam Boston London New York Oxford Paris
San Diego San Francisco Singapore Sydney Tokyo

CONTRIBUTORS TO VOLUME 47

P.M. ATKINSON, *Department of Geography, University of Southampton, Highfield, Southampton SO17 1BJ, UK*

L.R. BECK, *California State University, Monterey Bay, NASA Ames Research Center, Moffett Field, CA 94035, USA*

M.R. BOBO, *JCWS, NASA Ames Research Center, Moffett Field, CA 94035, USA*

S. BROOKER, *Wellcome Trust Centre for the Epidemiology of Infectious Disease, Department of Zoology, University of Oxford, South Parks Road, Oxford OX1 3FY, UK*

J. COX, *Department of Infectious and Tropical Diseases, London School of Hygiene and Tropical Medicine, Keppel Street, London WC1E 7HT, UK*

M.H. CRAIG, *MARA/ARMA Investigating Centre, Medical Research Council, 771 Umbilo Road, PO Box 17120, Congella, 4013, South Africa*

P.J. CURRAN, *Department of Geography, University of Southampton, Highfield, Southampton SO17 1BJ, UK*

A. FLAHAULT, *Institut National de la Santé et de la Recherche Médicale (INSERM) Unité 444, WHO Collaborating Centre for Electronic Disease Surveillance, Faculté de Médecine Saint-Antoine, 27 Rue Chaligny, F-75571 Paris Cedex 12, France*

G.M. FOODY, *Department of Geography, University of Southampton, Highfield, Southampton SO17 1BJ, UK*

J. GOETZ, *Department of Geography, University of Maryland, College Park, Maryland, MD 2074–8225,USA.*

S.I. HAY, *Trypanosomiasis and Land Use in Africa (TALA) Research Group, Department of Zoology, University of Oxford, South Parks Road, Oxford OX1 3PS, UK*

B.M. LOBITZ, *JCWS, NASA Ames Research Center, Moffett Field, CA 94035, USA*

E. MICHAEL, *Wellcome Trust Centre for the Epidemiology of Infectious Disease, Department of Zoology, University of Oxford, South Parks Road, Oxford OX1 3FY, UK*

E.J. MILTON, *Department of Geography, University of Southampton, Highfield, Southampton SO17 1BJ, UK*

M.F. MYERS, *Human Health Initiative, NASA—Goddard Space Flight Center, Code 902, Bldg 32/S130E, Greenbelt, Maryland, MD 20771, USA*

J.A. OMUMBO, *KEMRI/Wellcome Trust Collaborative Programme, PO Box 43640, Nairobi, Kenya*

S.D. PRINCE, *Department of Geography, University of Maryland, College Park, Maryland, MD 2074–8225,USA*

S.E. RANDOLPH, *Department of Zoology, South Parks Road, Oxford OX1 3PS, UK*

T.P. ROBINSON, *International Livestock Research Institute, PO Box 30709, Nairobi, Kenya*

D.J. ROGERS, *Trypanosomiasis and Land Use in Africa (TALA) Research Group, Department of Zoology, University of Oxford, South Parks Road, Oxford OX1 3PS, UK*

J. SMALL, *Department of Geography, University of Maryland, College Park, Maryland, MD 2074–8225,USA*

R.W. SNOW, *KEMRI/Wellcome Trust Collaborative Programme, PO Box 43640, Nairobi, Kenya*

B.L. WOOD, *NASA Ames Research Center, Moffett Field, CA 94035, USA*

Front cover: This shows a global image of processed Normalized Difference Vegetation Index (NDVI) data for 1986. The original 10-day composite data, produced by the Goddard Distributed Active Archive Center (Agbu, P.A., Smith, P.M., James, M.E., Kailuri, S.N.V. and Wolf, D.L. (1994) *NOAA/ NASA Pathfinder AVHRR Land Data Set*, Vol. 1, *Ten-day Composite Subset—1986*, NASA Goddard Distributed Active Archive Center Science Data Series, published on CD-ROM by NASA) was temporal Fourier processed (for a brief description, see Rogers, this volume) to extract the mean annual NDVI (displayed in the red channel of the output screen), the amplitude of the annual cycle of vegetation growth (in the blue channel) and the phase or timing of this cycle (in the green channel). Processing thus captured features of annual seasonality that are often important in the transmission of vector-borne diseases. In the image, areas of evergreen forest appear mostly red (when the annual cycle of vegetation activity—which is often unpronounced in such habitats—peaks early in the year) or yellow (when this peak occurs later in the year); areas with a strong blue coloration show strong seasonality (deciduous woodlands in the tropics or seasonally active vegetation at mid- to high latitudes), and areas that are mostly dark (or light) green have sparse or no vegetation cover, and little annual seasonality, peaking early (or late) in the year (deserts and desert edges). (© David Rogers, TALA Research Group, University of Oxford, UK.)

FOREWORD

In the mere 42 years since the United States first ventured into outer space, our astronauts, space scientists, and a supporting cast of government and private workers have achieved truly amazing things. We have orbited the world, analysed global weather patterns and surface features, detected previously hidden archaeological ruins, transformed human communications, and in the process profoundly influenced social and political institutions. We have actually lived in space, gazed as the Earth rose over the lunar horizon, set foot (and skipped and golfed) on the Moon, and dispatched our technological handmaidens to the surface of other celestial bodies. We have experienced the elation and suffered the tragedies that all pioneers share. As we stride boldly into the new millennium, however, a question that sometimes arises is 'why spend scarce money on spacecraft operations when people on Earth are in need?'

On one hand, the benefits seem so obvious. You might as well challenge the value of the development of tools or the exploration of the globe. On the other hand, such high-flying confidence may elude feverish patients, potential victims of worldwide epidemics, or tribes that depend on livestock decimated by disease. NASA scientists, however, have focused on such concerns since researchers at Ames first applied Landsat data to the study of malaria 20 years ago—and the latest advances in remote sensing and geographical information systems (GISs) give medical doctors, epidemiologists, and veterinary scientists a powerful new weapon.

I am therefore truly pleased to introduce this collection of cutting-edge essays on the theoretical and practical application of remote sensing and GIS technologies to medical and epidemiological issues. After decades of diligent and wide-ranging study, we can now combine remote sensing and GIS data from earth observing satellites, to battle significant and widespread diseases and health hazards such as tsetse and trypanosomiasis (Rogers, this volume), mosquitoes and malaria (Hay *et al.*, this volume), ticks and tick-borne diseases (Randolph, 2000, this volume), and helminth diseases (Brooker and Michael, this volume). We can analyse similarities in disease ecology across regions, document changes in the environment and land use patterns during disease outbreaks, and perhaps even predict outbreaks. For each disease, acknowledged experts review progress to date (and any problems), real options to improve public health policies, the historic implication of environmental and meteorological factors in the disease epidemiology, and future expectations for technologies, techniques, and communications. The book also summarizes the current state of the art in relevant satellite technologies, associated software, sophisticated sampling and analytical techniques, and internet access to relevant data.

In a broader sense, the book further solidifies the partnership between the medical and earth science communities through the reciprocal application of remote sensing and GIS capabilities to controlling many of the diseases that threaten the global community and individual survival.

The grandmother of a colleague once remarked that she travelled to Kansas in a covered wagon and lived to see her grandson launch rockets. I can only begin to imagine the accomplishments of space pioneers in my life and in the lifetimes of my grandchildren—but the frontier of this millennium must be marked by exciting concepts such as treating and preventing disease and global epidemics by applying remote sensing and GIS technologies. This book ranks as an important landmark in medical sciences and epidemiology, earth sciences, and human potential.

Compton J. Tucker
NASA—Goddard Space Flight Center

SERIES EDITORS' PREFACE

This is the second of the occasional special volumes of *Advances in Parasitology*, the first of which—on opportunistic protozoa—was published in 1998 as volume 40. Remote sensing, geographical information systems and geodesy represent some of the many exciting new tools for the study of epidemiology, and their recent application to parasitology is indeed an advance.

We are therefore very fortunate in having Simon Hay, Sarah Randolph and David Rogers as the guest editors of this volume, which originated as a result of initial discussions with Simon, and has been planned, assembled and edited by him and his colleagues at the University of Oxford, UK. We are sure the volume will serve as a valuable resource for those who wish to apply these new techniques to the study of the epidemiology of parasitic diseases, and also to others who want only to keep up to date with the important developments in the field.

We sincerely thank the guest editors and all who have contributed to the planning and writing of this volume.

J.R. Baker
R. Muller
D. Rollinson

GUEST EDITORS' PREFACE

Maps have always been recognized as a source of power and an inspiration for adventure. The information they contain, arrayed in spatially explicit patterns, empowers map owners to direct their multifarious activities and to assess the consequences of their actions within a predictive framework. Only if you know where you stand, physically and metaphorically, can you plan where to go. Historically, map-makers set the foundations of empires. From the medieval *Mappa Mundi* to the electronic atlases of today, traditional cartography has focused on geographical patterns. Increasingly important, however, are the underlying processes that give rise to these patterns. Processes are especially important in biology, and can occur at many different scales of both space and time. We believe that only satellite sensor data can give biologists appropriate and timely information on our natural environments and the animals and plants living within them. This book concentrates on a fascinating and important subgroup of this biodiversity—'vector-borne' diseases of animals in which part of the life cycle of the parasite is spent within invertebrate hosts (snails, ticks or insects).

Today, epidemiologists are empowered as never before, but the inherent complexity of biological systems—spanning the range of phenomena from those so fine that they operate at the level of the molecule to those so extensive that they can be studied only from space—warns against any expectation of simple solutions. We now have the technologies to match this range of scales from DNA sequencing to satellite imagery. Vector-borne disease systems illustrate this range well. From microscopic, genetically diverse transmitted pathogens, via invertebrate vectors and vertebrate hosts, these diseases spin patterns across the world, enmeshing many poor communities in the tropics in a cycle of poverty, ill-health and despair. We believe that satellite technology can break this cycle through increasing both our understanding of disease transmission and our ability to map and monitor the impacts our interventions might have on transmission.

In the first three chapters of this book, the theoretical and practical background of remote sensing and its integration into geographical information systems (GISs) are presented. Technology alone is not enough. The first problem is one of biophysics: how does the stream of electromagnetic radiation detected by sensors on board satellites relate to geophysical conditions on the ground? And which of those conditions are relevant to vector-borne disease systems? Next, the flood of data must be archived, processed and analysed in ways both rigorous and meaningful, while retaining its biological relevance. The burgeoning field of spatial statistics, now embraced by ecologists of all sorts, plays a vital role in allowing correct interpretations of the newly revealed patterns.

The next four chapters present case studies in the epidemiology of contrasting indirectly-transmitted disease systems, the African trypanosomiases, malaria, tick-borne pathogens, and helminths. Epidemiology means different things to different people. In its most general, medical usage, epidemiology involves matching patterns in the distribution of ill-health with patterns of extrinsic or intrinsic factors. Any statistical correlation between these patterns is used as a springboard from which to launch more intensive studies to establish causality. Biologists put more emphasis on the processes underlying observable patterns of infections. This is a more demanding procedure, but has the virtue of yielding a more versatile product, models that can predict future patterns born of complex dynamic interactions between parasites and their hosts, and the important biotic and abiotic factors extrinsic to these interactions. Vector-borne diseases are prime examples of systems requiring the process-based approach, although many of them are too complex for us yet to be able to produce robust and accurate biological models for them. We must begin our journey slowly, using the statistical approach to refine the questions to be asked in our biological studies.

The eventual aim of all types of models is to make predictions about other times and other places, whether it is the effects on transmission of global climate change, or the occurrence of epidemic conditions in an area not yet surveyed. By identifying environmental determinants of the rates of the demographic processes that drive these systems, we can extend our understanding from small-scale intensive studies into uncharted territories, and from the present into the future. With data from the right sorts of satellite, the temporal dynamics of natural systems are rendered as transparent as the spatial patterns. As the former very often determine the latter, we can achieve an understanding of interest extending beyond public health, to ecologists in general.

This is a discipline that is moving faster than is comfortable for many, but not fast enough for those who are straining for ever better, more detailed environmental information with which to test new ideas. The final section of the book looks forward to future, increasingly sophisticated technological developments for the retrieval of geophysical data from every part of the Earth's surface. Whatever the epidemiological question, if there is an element of environmental input, satellites of one sort or another offer the potential for developing surveillance and early-warning systems to address it on a global scale.

To be informed is to be fore-armed. To cope with the health problems of the future we need not only data, but instruction on how to acquire and interpret those data. The final chapter is a practical guide to encourage more scientists in public health to overcome their reluctance to enter this new field. We hope it also dispels the myth that remote sensing is too expensive for biologists to use. With both free software and free imagery (of certain sorts) available over the internet, there is little to limit the exploitation of the information content of remotely-sensed data, except our own imagination. With fewer than 20 papers per year published in the 1990s on the subject of remote sensing and vector-borne diseases, this is not yet an over-subscribed area of research.

Finally, we want to share with you our enthusiasm for remote sensing applications in biology. The sheer beauty of many of the images that now appear routinely on office walls is matched only by their unrealized power to provide information that will help us face the many environmental problems that lie ahead.

We thank the editors of *Advances in Parasitology* for their vision in recognizing this topic as one of increasing importance and already mature enough for a series of reviews to merit publication as a book.

S.I. Hay
S.E. Randolph
D.J. Rogers

CONTENTS

An Overview of Remote Sensing and Geodesy for Epidemiology and Public Health Application

S.I. Hay

Linking Remote Sensing, Land Cover and Disease

P.J. Curran, P.M. Atkinson, G.M. Foody and E.J. Milton

Spatial Statistics and Geographical Information Systems in Epidemiology and Public Health

T.P. Robinson

Satellites, Space, Time and the African Trypanosomiases

D.J. Rogers

Earth Observation, Geographic Information Systems and *Plasmodium falciparum* Malaria in Sub-Saharan Africa

S.I. Hay, J.A. Omumbo, M.H. Craig and
R.W. Snow

Ticks and Tick-borne Disease Systems in Space and from Space

S.E. Randolph

The Potential of Geographical Information Systems and Remote Sensing in the Epidemiology and Control of Human Helminth Infections

S. Brooker and E. Michael

Advances in Satellite Remote Sensing of Environmental Variables for Epidemiological Applications

S.J. Goetz, S.D. Prince and J. Small

Forecasting Disease Risk for Increased Epidemic Preparedness in Public Health

M.F. Myers, D.J. Rogers, J. Cox, A. Flahault and S.I. Hay

Education, Outreach and the Future of Remote Sensing in Human Health

B.L. Wood, L.R. Beck, B.M. Lobitz and M.R. Bobo

Plates 1–4 between pp. 78 and 79; Plates 5–10 between pp. 174 and 175; Plates 11–13 between pp. 270 and 271; Plates 14–18 between pp. 334 and 355

An Overview of Remote Sensing and Geodesy for Epidemiology and Public Health Application

S.I. Hay

Trypanosomiasis and Land Use in Africa (TALA) Research Group, Department of Zoology, University of Oxford, South Parks Road, Oxford OX1 3PS, UK

ADVANCES IN PARASITOLOGY VOL 47
0065-308-X $30.00

ABSTRACT

The techniques of remote sensing (RS) and geodesy have the potential to revolutionize the discipline of epidemiology and its application in human health. As a new departure from conventional epidemiological methods, these techniques require some detailed explanation. This review provides the theoretical background to RS including (i) its physical basis, (ii) an explanation of the orbital characteristics and specifications of common satellite sensor systems, (iii) details of image acquisition and procedures adopted to overcome inherent sources of data degradation, and (iv) a background to geophysical data preparation. This information allows RS applications in epidemiology to be readily interpreted. Some of the techniques used in geodesy, to locate features precisely on Earth so that they can be registered to satellite sensor-derived images, are also included. While the basic principles relevant to public health are presented here, inevitably many of the details must be left to specialist texts.

1. INTRODUCTION

1.1. Remote Sensing Definition

Remote sensing (RS) is the process of acquiring information about an object, area or phenomenon from a distance. This broad definition covers everything from the eyes reading this page to radio telescope installations that receive data that are processed to yield information from distant galaxies. The diversity of RS systems, however, can be usefully categorized as active or passive, differing simply in the source of the energy from which information is gathered. Active systems generate their own energy and passive systems rely on ambient energy from an external source which on Earth arises mainly from the Sun. In this chapter I shall consider both active (briefly) and passive (in depth) RS systems that measure the amount of radiant energy, i.e. the magnitude of electromagnetic radiation (EMR) reflected and radiated from the Earth's surface and atmosphere, with a view to deriving information about surface conditions.

A recent review of the evolution of RS in the last two decades can be found in Cracknell (1999). There are numerous books devoted to RS and for general background to RS and its applications see, among others, Swain and Davis, 1978; Colwell, 1983; Curran, 1985; Asrar,1989; Cracknell and Hayes, 1991; Cracknell, 1997a; Morain and Budge, 1997; Quattrochi and Goodchild, 1997; Sabins, 1997; Henderson and Lewis, 1998; Barrett and

Curtis, 1999; Rencz, 1999; Richards and Jia, 1999; Gibson and Power, 2000a; Lillesand and Kiefer, 2000. Two useful books and tutorial CD-ROMs that provide image processing software, specimen data and examples of many of the procedures outlined in this chapter, can be found in Mather (1999) and Gibson and Power (2000b). The Idrisi32 geographic information system and image processing software also provides useful tutorials (Eastman, 1999a,b).

1.2. Electromagnetic Radiation

Electromagnetic radiation is emitted by all objects above absolute zero (0 K, –273°C) (see Plate 1a). The total amount of energy an object emits is expressed by the *Stefan–Boltzman law* which states that

$$M = \sigma T^4$$

where M is the total exitance (emitted radiant flux per unit area) from the surface of the material (W m^{-2}), σ is the Stefan–Boltzman constant (5.6×10^{-8} W m^{-2} K^{-4}) and T is the absolute temperature of the emitting material (K). The total amount of energy emitted by an object therefore increases rapidly with temperature. This phenomenon is demonstrated by the larger area under the curve representing the electromagnetic spectrum (EMS) emitted by the Sun (at approximately 6000 K) than the corresponding area under the curve representing the EMS of the Earth (at approximately 300 K) (see Plate 1b). The figure also demonstrates the *Wien's displacement law*, or the shift towards emission of shorter wavelengths by an object at higher temperature (Colwell *et al.*, 1963; Monteith and Unsworth, 1990).

1.3. Atmospheric Transmittance, Spectral Response and Radiometer Design

Satellite-borne radiometers (or sensors) are instruments for measuring the intensity of EMR within a narrow range of wavelengths (or waveband), the resulting electronic signal from which, when processed, is often referred to as a channel. The measured EMR must travel through the atmosphere, which both scatters and absorbs EMR. These interactions are most significant close to the Earth's surface ($\sim$1–5 km) in the atmospheric boundary layer (ABL). Due to such interactions, atmospheric transmission of EMR is wavelength-dependent (see Plate 1b, top graph). Consequently, radiometers are often designed to maximize the information content of the signal received by operating in 'atmospheric windows' of maximal EMR transmission, thus reducing the effect of atmospheric attenuation.

The interaction of the EMR with the Earth surface (reflected, absorbed or transmitted), and what we can infer about surface properties from this interaction, is the essence of remote sensing. The reflected portion varies with both the material observed and the wavelength of EMR with which the measurement is taken. The composite nature of the reflected components is often referred to as a spectral response pattern (or signature). For example, the spectral response pattern of the ink on this book is designed to absorb EMR across a range of wavelengths and the page itself to reflect so that our eyes perceive a large contrast between black and white respectively, which makes for easy reading. The ideal, that each material on Earth be characterized by a unique spectral signature, is rarely achieved. Multi-temporal information and further processing of data into geophysical information can often assist in this discrimination and is discussed further (see also Curran *et al.*, this volume; Rogers, this volume; Goetz *et al.*, this volume).

2. ACTIVE REMOTE SENSING

Radar (radio detection and ranging) RS is the sub-set of active RS that is potentially useful in public health. Radar RS operates in the microwave proportion of the electromagnetic spectrum, generally considered to be at wavelengths of 1 mm to 1 m. Most modern radar sensors incorporate software routines on the sensor to improve spatial resolution mathematically and to cope with multiple pictures of the same object, and are hence referred to as Synthetic Aperture Radars (SARs) (Brown and Porcello, 1969; Sarder, 1997). SAR sensors are unique in that they can determine the polarization (horizontal or vertical) of electromagnetic energy they emit and receive, allowing increased information on retrieval surface properties. The atmosphere has almost complete transmittance at microwave wavelengths (see Plate 1b, top graph), and radar wavelengths are so resistant to atmospheric attenuation that images can be generated even through cloud. Furthermore, radars generate their own energy so they can operate day and night independent of solar insolation. These characteristics would seem to provide ideal data for a range of RS applications, but several factors have contributed to SAR image interpretation remaining a specialist discipline (Waring *et al.*, 1995; Kasischke *et al.*, 1997). These are (i) the difficulty of interpreting the information content of SAR imagery (see, for example, Oliver, 1991), (ii) historically the relative lack of calibrated SAR data over areas of interest, (iii) the lack of software to automate handling of the data generated, and (iv) the unique problems involving topography and image speckle. The specific capabilities of various sensors and polarizations have been considered by Schmullius and Evans (1997) and those of interest to public health include accurate classification and detection of change in land-cover, excellent discrimination of flooding extent and water bodies, and also soil moisture status (Waring *et al.*, 1995; Kasischke *et al.*, 1997). The routine use of SAR data in public health is not imminent and hence not evaluated in detail, but for the adventurous a daunting yet comprehensive starting point is Henderson and Lewis (1998).

3. PASSIVE REMOTE SENSING

3.1. Airborne Sensor Systems

The principals of RS from airborne and satellite platforms are very similar. To avoid duplication within this review I shall focus specifically on satellite sensor systems. Moreover, some aspects of airborne RS are detailed more fully by Curran *et al.* (this volume).

3.2. Satellite Sensor Systems

Radiometers can be carried on two broad categories of satellite, geostationary and polar-orbiting. Geostationary satellites are put into a high altitude orbit (~23 000–40 000 km) at the equator, with a speed equal to that of the Earth's rotation, so that they remain fixed above a particular point on Earth. Polar-orbiting satellites circle the globe repeatedly at much lower altitude orbits (~600–900 km) roughly perpendicular to the equator. Successive orbits therefore pass over a different section of the Earth as it rotates (Cracknell and Hayes, 1991).

Sensors receive EMR from the cone within which energy is focused on the detector (Figure 1). The relationship between the angle of this cone at the sensor (in radians), also called the instantaneous field of view (IFOV), β, the height of the sensor above the Earth, H, and the resulting diameter of the viewing area, D, often referred to as the spatial resolution, is given by:

$$D = H\beta$$

For example, the spatial resolution of the National Oceanic and Atmospheric Administration—Advanced Very High Resolution Radiometer (NOAA-AVHRR) with an average IFOV of 1.4 milliradians and an orbiting altitude of approximately 833 km can be found from the above equation as $D = 833\ 000 \times (1.4 \times 10^{-3}) = 1166$ m; close to the 1.1 km often quoted (Kidwell, 1998).

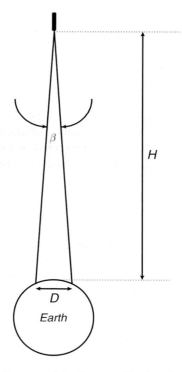

Figure 1 A schematic diagram of the factors affecting the spatial resolution of a radiometer. The instantaneous field of view (IFOV), β, is measured in radians; the height of the sensor above the Earth, H, and the resulting diameter of the viewing of the Earth, D, are measured in metres.

The data from satellite sensors are stored and transmitted as digital numbers, with each value referring to the smallest area for which the satellite sensor can record data. When viewed on a computer monitor these areas are called picture elements or 'pixels'. The assumption that for each pixel the digital number represents the mean spectral signal from all objects within the IFOV is generally correct for those sensors that try and match the pixel to the IFOV (Price, 1982) but this is a more complicated issue than it might first appear (Cracknell, 1998).

In polar-orbiting satellites the sensor scans across the track of the satellite as the orbit progresses to generate a series of contiguous scan lines which, when combined, form a two dimensional image (or scene). In geostationary satellites the radiometer itself must move perpendicularly to the plane of the scan line at regular intervals to generate the image. The swath width (breadth of the area over which data are recorded by a sensor) is determined by both the satellite altitude and sensor characteristics. For example the NOAA-AVHRR scans to ±55.4° from the point of Earth directly under the satellite (its nadir) which, at an altitude of approximately 833 km, results in a swath width of approximately 2700 km. Due to the curvature of the Earth, the effective distance between the Earth and the sensor increases with the scan angle so that the spatial resolution decreases to approximately 4×4 km toward the edge of the swath.

The repeat time, or the time taken between viewing the same part of the Earth's surface, varies between satellite systems and is determined by a combination of the swath width and orbital characteristics. Furthermore, data volume of satellite sensors is constrained by on-board storage media (especially in older satellites) and the limited opportunity for telemetry (transmission of data between satellite and receiving station) during the satellite overpass, so that RS images tend to have either a high-temporal resolution or a high-spatial resolution, but not both. Satellite sensor data therefore are limited in their spectral, spatial and temporal resolution by a variety of factors which reflect the compromise between the constraints of atmospheric effects, engineering limitations and the desired application. The compromises reached between the spectral, spatial and temporal resolutions by some of the satellite sensor systems currently used in public health applications are detailed in Tables 1 and 2, although advances in satellite sensor technology are constantly improving specifications and increasing the amount of data that can be generated, stored and transmitted.

3.2.1. *High Spatial Resolution Sensors*

In this review and throughout subsequent chapters, the ability to resolve areas smaller than 1×1 km is used as an arbitrary criterion for defining high spatial resolution sensors. The lower frequency of image capture for any point on the Earth for high spatial resolution satellite sensors means that such satellites sensors give few cloud-free images of the Earth's surface per year, especially over tropical regions. This, together with the cost of such imagery, has generally limited application of high-spatial resolution imagery to the production of habitat maps for relatively small areas and for a particular season of the year. The following examples do not form a comprehensive survey, particularly for very recent sensors which have not yet been adopted in public health application. Those sensors from which data are very difficult to obtain are also excluded. A more complete list of high spatial resolution satellite sensors is available (Morain and Budge, 1997).

(a) *The Landsat Series*. The launch of Landsat-1 in 1972 heralded a new era of high resolution RS, and changed the perception by experts and lay people alike of the possible ways in which to view the Earth (Lauer *et al.*, 1997; Lowman, 1999). The Landsat programme has generated a continuous supply of high resolution imagery, for the entire globe, from the first Multispectral

Table 1 The spectral, spatial and temporal resolution of high spatial resolution satellite sensors systems. See Morain and Budge (1997) for a more comprehensive list.

Satellite sensor system	Resolution		
	Spectral[a] (μm)	Spatial[b] (m)	Temporal (days)
Landsat-1, -2, -3 Return Beam Vidicon (RBV) camera	Ch 1–3 (0.475–0.830)[c]	79	18
Landsat-1, -2, -3, -4, -5 Multispectral Scanner (MSS)	Ch 4–7 (0.5–1.1)[d]	79/82[e]	16/18
Landsat-4, -5 Thematic Mapper (TM)	Ch 1–5 and 7 (0.45–2.35) Ch 6 (10.40–12.50)	30 120	16
Landsat-6[f], -7 Enhanced Thematic Mapper+ (ETM+)	Ch 1–5 and 7 (0.45–2.35) Ch 6 (10.40–12.50) Ch P (0.50–0.90)	30 60 15	16
Satellite pour l'Observation de la Terre-1, -2, -3 (SPOT) High Resolution Visible (HRV) Panchromatic Mode (HRV-PAN) Multispectral Mode (HRV-XS)	Ch 1 (0.51–0.73) Ch 2–4 (0.50–0.89)	10 20	26[g]

SPOT-4			
High Resolution Visible and Infrared (HRVIR)	Ch 1 (0.61–0.68)	10	
	Ch 2–5 (0.50–1.75)	20	26[g]
Terra			
Moderate Resolution Imaging Spectroradiometer (MODIS)	Ch 1–2 (0.620–0.876)	250	1–2
	Ch 3–7 (0.459–2.115)	500	
	Ch 8–36 (0.405–14.385)	1000	

[a] The spectral resolutions are the electromagnetic wavelength range in μm where 0.3 is at the visible and 14 at the thermal infrared end of the spectrum (Plate 1a).
[b] The spatial resolution is given as the diameter of the viewing area of the sensor, D, at nadir.
[c] Landsat-3 had a fourth panchromatic RBV channel (0.505–0.750) at a 40 m spatial resolution.
[d] Landsat-3 had an eighth MSS thermal channel (10.4–12.6) at 237 m spatial resolution.
[e] The spatial resolution is 79 m for Landsat-1 to -3 and 82 m for Landsat-4 and -5 since the satellite altitude changed from 920 km to 705 km. The temporal resolution also changed accordingly.
[f] Landsat-6 never achieved orbit and Landsat-7 is currently in operation.
[g] A pointing facility can increase the frequency of coverage.

Table 2 The spectral, spatial and temporal resolution of low spatial resolution satellite sensors systems. See Morain and Budge (1997) for a more comprehensive list.

Satellite sensor system	Resolution		
	Spectral[a] (μm)	Spatial[b] (km)	Temporal (h)
National Oceanic and Atmospheric Administration (NOAA) Advanced Very High Resolution Radiometer (AVHRR)	Ch 1–5 (0.58–11.50)	1.1	12
Meteosat-4, -5, -6 High Resolution Radiometer (HRR)	Ch 1 (0.40–1.10) Ch 2–3 (5.70–12.50)	2.5 5	0.5
Meteosat Second Generation-1 (MSG)[c] Spinning Enhanced Visible and Infrared Imager (SEVRI)	Ch 1 (0.6–0.9) Ch 2–12 (0.56–14.4)	1 3	0.25

[a] The spectral resolutions are the electromagnetic wavelength range in μm where 0.3 is at the visible and 14 at the thermal infrared end of the spectrum (Plate 1a).
[b] The spatial resolution is given as the diameter of the viewing area of the sensor, D, at nadir.
[c] Due to be launched at the end of 2000.

Scanner (MSS) aboard Landsat-1 to the latest Enhanced Thematic Mapper (ETM+) on board Landsat-7 (Mika, 1997) (Table 1). In so doing, it has offered unique insights into global terrestrial phenomena (Goward and Williams, 1997). During this time there has been a substantial evolution in the quality of the radiometers (Mika, 1997), their calibration (Thome *et al.*, 1997) and the development of multispectral data analysis techniques developed to process captured data (Landgrebe, 1997), all of which will continue (Ungar, 1997). Moreover, the novelty and conspicuous success of the Landsat programme forced issues regarding data distribution and cost (Draeger *et al.*, 1997) and the feasibility of commercial RS (Williamson, 1997) to be considered seriously for the first time. Many countries have understandably emulated and extended features of the Landsat programme, and other high resolution RS data sources are now increasingly available (Morain and Budge, 1997).

(b) *The SPOT Series.* The French Satellite pour l'Observation de la Terre (SPOT) programme began in 1986 with the launch of SPOT-1 with the High

Resolution Visible (HRV) sensor payload (Table 1). Data have been collected continuously since this time with SPOT-4, carrying the High Resolution Visible and Infrared (HRVIR) sensor that is now in operation. There are many similarities between these data and Landsat-TM imagery, but SPOT-HRV achieves a slightly higher spatial resolution with fewer spectral channels. Data accessibility, the degree of cloud contamination on scenes of interest, and cost, all help determine a researcher's choice between the different sorts of high resolution imagery available.

3.2.2. *Low Spatial Resolution Sensors*

In contrast to the sensors on board Landsat and SPOT, the sensors on board the NOAA series of polar-orbiting meteorological satellites and the Meteosat series of geostationary satellites have relatively high-temporal and low-spatial resolutions. The advantages of these features are detailed extensively in the following chapters and have led to these systems being very widely utilized by the RS community (Cracknell, 1999). Furthermore, these data are available free in the public domain to research institutes and will soon have increased spectral, temporal and spatial coverage (Hay *et al.*, 1996) (see Section 3.3).

(a) *The NOAA Satellite Series.* The NOAA series of polar-orbiting Television Infrared Observation Satellites (TIROSs) has been operational since 1978 (Hastings and Emery, 1992; Cracknell, 1997g). TIROS-N (later renamed NOAA-6) was the first satellite to carry the AVHRR and has been followed by seven satellites each achieving an operational lifetime of between 2 and 4 years. The 'very high resolution' refers to the 10 bit radiometric resolution of the sensor, which therefore has the ability to store data in the zero to 1023 range. The definitive description of the NOAA polar-orbiting satellites, their radiometer payloads and the data they generate is given in Kidwell (1998).

The NOAA satellites complete 14.1 near-polar, Sun-synchronous orbits per day at an altitude of 833–870 km. Since the number of orbits is not an integer, the orbital track over the Earth does not repeat on a daily basis. The even-numbered satellites have an ascending node with a north-bound equatorial crossing during the evening (19:30 Local Solar Time (LST)) and a descending node with a south-bound equatorial crossing in the morning (07:30 LST), whereas the odd-numbered satellites have an ascending node in the afternoon (14:30 LST) and a descending node at night (02:30 LST). At present the exception is the latest NOAA-14 satellite, launched in December 1994, which replaces NOAA-13 in functionality and thus has a daytime ascending node (Cracknell, 1997g).

As we have discussed, the NOAA-AVHRR can view a 2400 km swath of the Earth and, at this orbital frequency, daily data are recorded for the entire Earth surface. Radiation is measured in five distinct wavebands of the EMS so that five separate waveband images are recorded for each orbit (Table 2). The visible channel 1 and near infrared (NIR) channel 2 measure reflected solar radiation whereas the thermal channels 4 and 5 measure emitted thermal infrared (TIR). Channel 3, in the mid infrared (MIR), is a hybrid and sensitive to a combination of both reflected and emitted radiances.

The spatial resolution of the AVHRR is approximately 1.1 km beneath the track of the orbiting satellite (see Section 3.2). These nominal 1.1 km AVHRR data are continuously transmitted and may be received by stations along or near to the satellite's path, where they are referred to as high resolution picture transmission (HRPT) data (Cracknell, 1997d). On request to NOAA these data may also be recorded on an on-board tape storage system and later transmitted to Earth as the satellites pass over a network of receiving stations. The data are then referred to as Local Area Coverage (LAC) data. These 1.1 km data have found application in a very wide range of disciplines and are reviewed by Ehrlich *et al.* (1994) and Cracknell (1997b).

Two processing steps further reduce the spatial resolution of most of the AVHRR data available to the user community. The on-board tape system is incapable of holding global coverage data at 1.1×1.1 km spatial resolution. Instead, the information from each area of five (across-track) by three (along-track) pixels is stored as a single value, the average of the first four pixels only of the first row of the 5 by 3 block. The resulting imagery is referred to as Global Area Coverage or GAC data. GAC data, with a stated nominal spatial resolution of 4 x 4 km, are obviously far from ideal representations of the raw data (Justice *et al.*, 1989) and their method of sub-sampling has consequences for environmental modelling (Belward and Lambin, 1990; Belward, 1992). Nevertheless, GAC data are the form in which most of the AVHRR archive was collected and reasonable quality global data sets are available at a variety of spatial resolutions (4×4 km or coarser) from the early 1980s to date (Townshend and Tucker, 1984).

(b) *The Meteosat Satellite Series.* The European Organization for the Exploitation of Meteorological Satellites (EUMETSAT) geostationary Meteosat satellite series began with the launch of Meteosat-1 in 1977 (Morain and Budge, 1997). Experimental satellites were used until the launch of the first operational satellite, Meteosat-4, in June 1989. The spin-stabilized satellites are put into orbit at an altitude of 35 800 km over the Gulf of Guinea, at the crossing of the Equator and the Greenwich meridian (0°N, 0°E). In this position, images are captured for the full Earth's disc including Africa, Europe and the Middle East. A reserve satellite operates nearby in a standby condition (Anonymous, 1971).

The principal payload of the satellite is a High Resolution Radiometer (HRR). The radiometer operates in a broad visible waveband (channel 1), a thermal infrared waveband (channel 2) and a water vapour absorption infrared waveband (channel 3) (Table 2). The Meteosat satellites were designed for meteorological applications, so channel 3 is located in the thermal infrared area of maximal water vapour absorption and hence is ideal for monitoring clouds. At nadir the spatial resolution is 2.5×2.5 km for the visible images and 5×5 km for the thermal infrared and water vapour images. Further from the equator, the spatial resolution decreases so that over northern Europe it is 4×4 km in the visible wavebands and 8×8 km in the thermal infrared and water vapour wavebands. Each image is transmitted to the Earth in real time as each scan line is completed and new images are generated at 30-minute intervals (WMO, 1994).

3.3. Future Satellite Sensors

There are many planned enhancements to the satellite sensor systems described above (Beck *et al.*, 2000; Wood *et al.*, this volume). These improvements tend to be frequently modified and are therefore best reviewed at the relevant internet addresses for each satellite sensor series (Table 3). Of particular interest to those interested in RS applications in African public health will be the EUMETSAT Meteosat Second Generation (MSG) satellites that will carry the Spinning Enhanced Visible and Infrared Imager

Table 3 Useful universal resource locators (URLs) for common satellite sensor systems, image processing software and GPS manufacturers. The table is not comprehensive and does not endorse any specific company or product.

Remote sensing satellite systems

Landsat	http://landsat.gsfc.nasa.gov
SPOT	http://www.spotimage.fr
Terra	http://terra.nasa.gov
NOAA-AVHRR	http://daac.gsfc.nasa.gov/CAMPAIGN_DOCS/ LAND_BIO/GLBDST_main.html
Meteosat / MSG	http://www.esa.int/satellites

Image processing software

EASI/PACE	http://www.pcigeomatics.com
ENVI	http://www.envi-sw.com
ERDAS	http://www.erdas.com
ER Mapper	http://www.ermapper.com
TNTmips	http://www.microimages.com
WinDisp	http://www.fao.org/giews/english/windisp/ windisp.htm

Hand-held GPS manufacturers

Garmin	http://www.garmin.com
Magellan	http://www.magellangps.com
Trimble	http://www.trimble.com

(SEVRI) (Battrick, 1999). Located in the same geostationary orbit as the existing Meteosat satellites, the MSG-SEVRI will provide 1×1 to 3×3 km spatial resolution images at 15 minute intervals in 12 spectral wavebands, ranging from 0.56 μm in the visible to 14.4 μm in the infrared domain. This increases significantly the capabilities of this satellite sensor series, the main advantage being the frequency of data collection allowing less attenuated images to be rapidly composited on a daily or weekly basis (see Section 4.2).

In addition to improvements of existing satellite sensor series, new systems are also being continually developed. An example that will be widely adopted in public health is NASA's Terra, which will generate imagery with a range of on-board sensors, most noticeably the Moderate Resolution Imaging Spectroradiometer (MODIS) (Table 1). Benefits to RS applications will be threefold. First the range of data available will increase substantially with 36 spectral channels from which more accurate meteorological and other ecological variables may be derived. Moreover, the channels have been designed with smaller waveband ranges to exploit 'spectral windows' where atmospheric signal attenuation is minimal and will therefore have significantly improved signal-to-noise ratios. Secondly, MODIS will have a high temporal resolution (two-day repeat time) at significantly higher spatial resolution (250×250 to 1000×1000 m depending on the channel) than existing sensors, in effect resulting in hybrid data with characteristics of both the NOAA-AVHRR and Landsat-TM. Thirdly, the stated aim is to ingest, process and disseminate data within three days of acquisition, including many potentially useful products, such as improved spectral vegetation indices, land surface temperature and evapotranspiration estimates. This means that much of the routine data processing will be performed at source giving unparalleled, rapid access to contemporary data on large-area ecosystem processes.

This section is not intended to be comprehensive, as future advances in remote sensing pertinent to epidemiology are dealt with in more detail by Goetz *et al.* (this volume) and Myers *et al.* (this volume).

4. TURNING SATELLITE SENSOR DATA INTO GEOPHYSICAL DATA

This section details some of the fundamental problems experienced by an orbiting satellite sensor in the measurement of reflected and radiated EMR from a curved surface, through an atmosphere of spatially heterogeneous composition. It then deals with how such data are converted into calibrated geophysical variables at known geographical locations. At this stage we introduce the caveat that the information presented in Sections 4 and 5 is orientated toward the NOAA-AVHRR sensor. This is because the author's experience of remote sensing stems primarily from using such data and many of the future applications of RS in public health will utilize these and related satellite data. Thus, in many cases, while the exact details may not be accurate for a different satellite sensor system, the process of obtaining useful information from satellite-sensor-derived digital data will be informative. It is acknowledged that many applications in public health have been concerned with the classification of land cover and its extent, using single scenes. These aspects of remote sensing are considered in more depth by Curran *et al.* (this volume).

4.1. Image Registration

Raw digital data derived from satellite sensors need to be pre-processed geometrically to rectify (or register) them, usually to a base map at a particular scale and in a particular map projection (Snyder, 1987), or to other images in a series for monitoring change (also see Section 6). Registration is generally an automated process that uses an ephemeris model of the orbital parameters of the satellite and a time signal sent down with the satellite sensor imagery to predict the satellite's position relative to the Earth at the time of image capture (Brush, 1988; Emery *et al.*, 1989; Baldwin and Emery, 1995). Algorithms for positional calculations originally assumed that the satellites were in their correct attitude and were following precisely their intended orbits. Unfortunately, this is not the case, since satellites vary considerably in both their orbit and attitude (McGregor and Gorman, 1994). An extreme example is that the equator crossing time for NOAA-11, which was 14:20 LST when the satellite was launched in February 1985, had drifted to 16:07 LST by November 1988 (Kidwell, 1998). These deviations from design values, and variations in these deviations, mean that some of the resulting imagery is not accurately registered to the appropriate base maps.

Satellite sensor images also suffer from geometric distortions due to other factors which include: panoramic distortion, Earth curvature, atmospheric refraction, relief displacement and non-linearities in the sensor's field of view (Lillesand and Kiefer, 1994). These errors may be systematic or random. Geometric correction of systematic errors is usually done by modelling the sources of errors mathematically and applying the resulting corrective formulae (Wu and Liu, 1997).

Random distortions are overcome by measuring the shift of ground control points (GCP), distinctive geographical features of known location on the image, and resampling or reforming the original image to a new one accordingly. The functional relationship (f) between the X and Y file co-ordinates of the original satellite sensor image and the known latitude (x) and longitude (y) are determined by a least squares regression to determine the coefficients for two co-ordinate transform equations:

$$X = f_1(X, Y)$$
$$Y = f_2(X, Y)$$

After a geometrically correct geographical grid is defined in terms of the longitude and latitude, each cell in this grid is given values of x and y according to the co-ordinate transform equations above. The computer then maps the digital number from the pixel closest to this address in the raw image to the geometrically correct geographical grid. This last step can be done in a simple way, such as to the nearest neighbouring pixel, or by using more sophisticated and computer intensive methods, such as bilinear and cubic spline interpolations (Khan *et al.*, 1992, 1995). These spatial interpolation processes, however, may substantially alter the radiometric fidelity of the data so must be considered carefully (Goward *et al.*, 1991). The most recent systems developed for NOAA-AVHRR image navigation use an ephemeris model and GCP-based rectification simultaneously (Marçal, 1999).

The on-board sensors of geostationary satellites view the Earth as a disc, so that apart from spherical distortion there are few, if any, problems of geo-registering such images. The raw data from polar-orbiting satellite sensors, however, are a series of strips that must be co-registered and geometrically corrected before successive images can be joined together.

Image registration to a map can involve a loss of spatial resolution, the extent of which is usually increased to allow for the coarsest spatial resolution of the data rather than over sampling at the nadir pixel size. Final registration to a base map frequently has to be performed by visual inspection of the image with a map overlay (Krasnopolsky, 1994). The effects of these various stages of image resampling on NOAA-AVHRR data are considered in Khan *et al.* (1995) and reviewed in detail in Cracknell (1997f).

4.2. Reducing Cloud Contamination

Each of the high-temporal resolution images from the NOAA (and Meteosat) satellite sensors is as affected by cloud contamination as any single Landsat or SPOT sensor image, but their much higher frequency means that data quality can be improved by combining images over a relatively short period of time by compositing. The aim of compositing is to choose the most cloud free and/or least atmospherically contaminated radiance value within the compositing period.

Most compositing algorithms rely on the fact that one common image product from the NOAA-AVHRR, the Normalized Difference Vegetation Index (NDVI), produced from channels 1 and 2 (see Section 5.1 for details), has values that are generally reduced by cloud and other atmospheric contamination (Holben, 1986; Kaufman and Tanré, 1992). The highest

NDVI values recorded during any relatively short time period are therefore thought to occur when cloud cover is least, and such values are taken to represent the least attenuated pixel value for the period. This method of image production is called maximum value compositing (MVC), and is usually carried out over a ten-day (decadal) compositing period. It has the important consequence that maximum image values for adjacent pixels in a single image may have been collected on different days during the compositing period. MVC methods tend to degrade still further the spatial resolution of the final image product, primarily for geo-referencing reasons (Meyer, 1996; Robinson, 1996) so that the recorded value for any nominal 8×8 km pixel may in fact have been drawn from an area as large as 20×20 km. Further problems associated with MVC and alternatives are outlined in Stoms *et al.* (1997), but have not been widely acknowledged.

Selection of the least cloud-contaminated images in the other AVHRR channels usually depends upon the selection of the NDVI date chosen by MVC. The same image that is used to generate the NDVI for any pixel is also taken as the source of information for the other AVHRR channels for that pixel and period. Increasingly in land applications over the tropics, however, AVHRR channels 4 and 5 are composited separately, i.e. without reference to the NDVI, since the overlying clouds are generally colder than the land so that the highest thermal value in the series will probably be the least cloud contaminated (Lambin and Ehrlich, 1995, 1996).

4.3. Reducing Other Atmospheric Effects

During image registration to a base map, other corrections for atmospheric effects such as Rayleigh scattering caused by aerosols and absorption by water vapour, carbon dioxide and ozone, may also be applied, using ancillary information in the data stream from the satellite sensor (Vermote *et al.*, 1990; Tanré *et al.*, 1992). This is very important because atmospheric aerosols, which are highly spatially and temporally variable in the atmosphere (Holben *et al.*, 1991), scatter light particularly at short visible wavelengths (i.e. NOAA-AVHRR channel 1), whilst atmospheric water vapour absorbs particularly in the near infrared (i.e. NOAA-AVHRR channel 2) (Kaufman and Tanré, 1992). If corrections for atmospheric effects are not made at the time of image registration they generally cannot be made as accurately at a later stage. Instead, corrections are based on average values for a 'standard' atmosphere within a region (Hanan *et al.*, 1993; Goetz *et al.* 1995). Despite attempts to remove the effects of clouds and other contaminants by MVC of AVHRR imagery, continuous total or sub-pixel cloud cover and haze may still affect MVC images.

4.4. Satellite Sensor Drift

During their operational lifetime the AVHRR sensor characteristics change with use and as their components age (Gorman and McGregor, 1994) so that external or 'vicarious' calibration is required. The TIR channels are continuously calibrated against the 4 K space background temperature, measured by a thermistor on the baseplate of the satellite. Channels 1 and 2 are calibrated by making simultaneous measurements with aircraft under-flights (Smith *et al.*, 1988), or by examining the change in signal from relatively invariant reflectors such as deserts and high clouds, or from invariant reflective phenomena such as the molecular scattering of the visible signal over oceans and areas of sun glint (Che and Price, 1992; Kaufman and Holben, 1993). Correction factors can also be added to spectral vegetation indices (SVIs) without recourse to the visible channel data by assuming a linear degradation in sensor response, examples of which are given by Los (1993) for the NDVI. It should suffice to note that calibration is currently a contested term in RS and for those interested in the debate, guidelines can be found at http://www.tandf.co.uk/journals.

4.5. Satellite Orbit Drift

Even when corrections for atmospheric effects and instrumental drift have been made, the resulting imagery may still show periodic changes in the signal due to a precession of the satellite's orbit known as phasing. This results in cyclical variation of over-pass times, which, in the case of NOAA-AVHRR imagery, have a 17-day cycle (McGregor and Gorman, 1994). Signal variation can therefore be due to the changing angles between the Sun, the Earth and satellite sensors. Moreover, afternoon equitorial overpass times become progressively later after launch, causing artefactual 'cooling' trends in the brightness temperature time series (Price, 1991) because measurements occur later in the afternoon. Recent work has suggested elegant solutions to these problems for long-term archived RS data sets (Gutman, 1999a,b; Gleason *et al.* 2000).

5. TURNING GEOPHYSICAL DATA INTO INFORMATION FOR PUBLIC HEALTH

The previous section highlighted the problems of obtaining geographically registered satellite sensor data and the sources of error involved in the processes. This section describes how such information can be converted to vegetation, land surface temperature, atmospheric moisture and rainfall indices. The accuracy with which meteorological variables can be described is also discussed. The most recent advances and a more in-depth consideration of some of these issues are provided by Goetz *et al*. (this volume).

5.1. Spectral Vegetation Indices

Most SVIs (reviewed by Huh, 1991; Myneni *et al*., 1995b; Cracknell, 1997h; Lyon *et al*., 1998) exploit the fact that chlorophyll and carotenoid pigments in plant tissues absorb light in the visible red wavelengths (which corresponds to AVHRR channel 1), whereas mesophyll tissue reflects light in the near infrared wavelengths (which corresponds to AVHRR channel 2) (Sellers, 1985; Tucker and Sellers, 1986). A healthy and actively photosynthesizing plant will therefore look darker in the visible, and brighter in the near-infrared region, than an unhealthy or senescing plant. Furthermore, as vegetation coverage increases, there is more absorption of red radiation due to the increasing amount of pigmentation and more reflectance of near-infrared radiation due to increases in internal leaf scattering of mesophyll (Curran and Williamson, 1986). The reflectance from dry soil is less complex than that of vegetation, showing a general increase in reflectance with wavelength which is dependant on soil texture, structure, and water, organic carbon and iron oxide content (Huete and Escadafal, 1991). Since soils and vegetation exhibit very different spectral properties these features are used to differentiate between the two types of surfaces. SVIs are simply designed to maximize the contrast in reflectance and thereby identify the presence of vegetation in RS images.

The most simple SVI is the ratio of AVHRR channel 2 (Ch_2) over channel 1 (Ch_1) reflectances, called the ratio vegetation index (RVI) or simple ratio index (SRI). Other SVIs attempted to overcome the problem of reflectance from the (usually dark or reddish) soil backgrounds by dividing the difference between these two channels by their sum, to give the NDVI (Tucker, 1979), defined as follows:

$$\text{NDVI} = \frac{(Ch_2 - Ch_1)}{(Ch_2 + Ch_1)}$$

The values of the NDVI can theoretically range from −1 to +1, but in practice usually fall within 0.0–0.8 limits (Colwell, 1974; Tucker, 1979; Tucker *et al*., 1991). The NDVI, in common with all red/near-infrared indices, is a specific measure of chlorophyll abundance and light absorption (Myneni *et al*., 1995a), but its use has been extended to quantify herbaceous vegetation biomass (Tucker *et al*., 1983, 1985b), vegetation primary productivity (Prince and Goward, 1995; Goetz *et al*., 1999), vegetation coverage (Tucker *et al.*, 1985a; Goetz, 1997) and phenology (Justice *et al.*, 1985) in a range of ecosystems. NDVI measurements are particularly useful in areas of sparse vegetation coverage, where they have a larger dynamic range than the simpler SVIs such as RVI. The NDVI does, however, saturate in areas of full coverage such as forests (Huh, 1991). It is also less than ideal because of continuing problems with background soils (which are, for example, darkened by rainfall (Huete *et al.*, 1985)) and differential atmospheric effects on channel 1 and 2 radiances.

Alternative indices have been suggested to overcome some of these problems (Jackson and Huete, 1991; Leprieur *et al.*, 1996) such as the soil adjusted vegetation index (SAVI) (Huete, 1988) and the global environment monitoring index (GEMI) (Pinty and Verstraete, 1992). They have been much less widely applied to ecological and epidemiological problems, however, and are not considered further.

5.2. Land Surface Temperature Indices

The theoretical concept of a black body is used to describe any material that absorbs and emits radiation perfectly at all wavelengths. Such a hypothetical material is described as having a spectral emissivity of 1, i.e. the ratio of emission at temperature (T) versus emission at the standard temperature ($T_s = 273$ K) (Monteith and Unsworth, 1990). In ideal conditions therefore, the temperature of a black body can be determined by detecting the energy it emits at a particular wavelength. Natural surfaces do not behave as black bodies, however, and have emissivity values less than 1; usually 0.99 for water, 0.96–0.99 for vegetation and lower for soils (Salisbury and Daria, 1992). Furthermore, the radiometric brightness temperature (Becker and Li, 1990) measured by the satellite sensor is also affected by absorption characteristics of atmospheric constituents (particularly water vapour but also ozone, carbon dioxide and aerosols), as well as emission of radiation by the atmosphere itself. Attempts to estimate accurate surface temperatures from satellite sensor derived brightness temperature must therefore correct for atmospheric attenuation and the spatially heterogeneous nature of land surface emissivity. These are major areas of past and current RS research (Norman *et al.*, 1995; Goetz *et al.*, 1995).

Channels 4 and 5 of the AVHRR radiometer have long been used to measure water vapour attenuation in the 10–12 μm spectral window to increase the accuracy of sea-surface temperature determination (Prabhakara *et al.*, 1974). The attenuation is greater in channel 5 than in channel 4 so that the difference between the signal of these two channels can be used to estimate the amount of atmospheric water vapour attenuation and is used to reduce such effects. This simultaneous use of information from both channels to estimate surface brightness temperatures is described as a 'split-window' technique, because it is performed within the same radiance window of the atmosphere. Surface emissivity is more variable on land than over the relatively uniform sea surface and so allowance needs to be made for emissivity when comparing surface brightness temperatures of different land-surface types. Many split-window techniques have been developed, which largely rely on ancillary data to quantify atmospheric water content and surface emissivity (Prabhakara et al., 1974; Becker and Li, 1995; Norman et al., 1995).

Of the many split-window algorithms available (Becker and Li, 1990; Cracknell, 1997e; Qin and Karnieli, 1999), the only land surface temperature index (LSTI) that has been used in public health applications to date, and hence is explained here, requires only raw channel AVHRR data. Price (1984) derived a simple algorithm from radiative transfer theory to estimate land surface temperature, T (K), from the AVHRR channel 4, Ch_4 (K), and the AVHRR channel 5, Ch_5 (K) brightness temperatures that accounted for the emissivity of the land surface:

$$T = Ch_4 + A(Ch_5 - Ch_4)$$

where A is a constant determined by Price to be 3.33 for channels 4 and 5 of the NOAA-7 AVHRR, when channel 4 and 5 emissivities were assumed equal. This equation was stated to provide land surface temperature estimates accurate to ± 2–3 K after modelling potential error sources. This algorithm was later found to be accurate to ± 3 K using LAC data for a uniform tall grass prairie habitat in Kansas when a constant emissivity was assumed (Cooper and Asrar, 1989) and subsequently to be accurate to ± 4.5 K using LAC data from a similar habitat (Sugita and Brutasaert, 1993; Goetz, 1997). Continental scale application of this algorithm for monthly maximum temperatures shows LST determination accuracy equivalent to that of spatial interpolation of meteorological data, ± 4 K for both tropical Africa (Hay and Lennon, 1999) and temperate Europe (Green and Hay, 2000).

5.3. Atmospheric Moisture Indices

The total precipitable water content of the atmospheric column has been estimated according to a method proposed by Dalu (1986). Similar to the split-window algorithm, this method exploits the difference in atmospheric attenuation due to atmospheric water vapour between channels 4 and 5 of the NOAA-AVHRR. The algorithm was derived from atmospheric radiative transfer models over the ocean, where a surface relative humidity of 80% was assumed due to the natural equilibrium between evaporation and diffusion, and tested against measurements taken from ships. Based on a derived correction factor, a, and taking into account the changing atmospheric path length as a function of scan angle, θ, the total precipitable water content of the atmospheric column, U (kg m^{-2}), can be estimated as follows:

$$U = a \times (Ch_4 - Ch_5) \times \cos \theta$$

The estimates were stated to have an accuracy of ± 5 kg m^{-2} over the ocean. The accuracy of these estimates over the land surface will be influenced by varying emissivity, as well as by deviation from the assumption of 80% relative humidity at the surface. Justice *et al.* (1991), however, have noted agreement between values for atmospheric water content estimated using the above equation and those measured by photometers at several sites in the Sahel. Furthermore, the difference in the AVHRR brightness temperatures (channel 4 – channel 5) has been shown to have a linear relationship with total precipitable water in the atmospheric column (Eck and Holben, 1994) using balloon radiosonde and sun photometer data from three meteorological stations in Mali. The standard error of the total precipitable water estimate was between 0.31 and 0.48 kg m^{-2} and was found to increase for these sites when the above equation was applied to the same data. Total precipitable water was also calculated using the equation from Eck and Holben (1994) where

$$U = A + B(Ch_4 - Ch_5)$$

and, A and B are constants, 1.337 and 0.837 respectively. These were determined by a linear regression of (channel 4 – channel 5) against estimated precipitable water content of the atmospheric column using radiosonde data from the Gao meteorological station in Mali. The coefficient of determination for the relationship was 0.96. Note that the total perceptible water content of the atmospheric column (U) is often expressed as kg m^{-2}. These are units of pressure (i.e. mass per unit area) and are converted to the amount of water that would be precipitated from the atmospheric column in centimetres by dividing by 10, since the density of water is 1 g cm^{-3}.

The estimated precipitable water content, U (cm), is then converted to a near surface dew point temperature, T_d (°F), or the temperature to which a sample of air must be cooled for it to become saturated and condense using the following relationship (Smith, 1966):

$$T_d = \frac{\ln U - (0.113 - \ln(\lambda + 1))}{0.0393}$$

where λ is a variable that is a function of the latitude and the time of the year.

The dew point values can then be converted into Kelvins and used with the Price (1984) estimate of land surface temperature, T_p (K), to calculate the vapour pressure deficit, Vpd (kPa), using the equation provided in Prince and Goward (1995):

$$Vpd = 0.6111\left[\exp\left(17.27\times\frac{T_p-273}{T_p-36}\right)-\exp\left(17.27\times\frac{T_d-273}{T_d-36}\right)\right]$$

Applying this algorithm to approximately 200 meteorological stations across Africa in each month of 1990 showed that RS determination of Vpd could be measured to ±6 mb, which was about equivalent to that of spatial interpolation of the meteorological data (Hay and Lennon, 1999). This was also the case for a similar 3 year study expanded to temperate Europe (Green and Hay, 2000).

5.4. Rainfall Indices

In tropical latitudes where diurnal heating provides large reservoirs of potential energy, weather systems are dominated by atmospheric convection processes (Martyn, 1992; Emanuel, 1994). The most vigorous convection currents provide the strongest updrafts which result in clouds with higher water contents that are more likely to be rain-bearing (Byers and Barnham, 1948; Ba and Nicholson, 1998). These convection currents form deep clouds with high and cold tops which emit very low radiance values in the thermal infrared. These cloud-top temperatures can be recorded by channel 2 of the Meteosat satellite (Table 2). The relationship between cloud temperature and the probability of rainfall has been well established (Burt et al., 1995). The particular threshold temperature associated with rain-bearing clouds and the quantity of rain they deposit varies temporally and spatially, however, and must be established empirically (Milford and Dugdale, 1990; Laurent et al., 1998; Grimes et al., 1999). The pixels in a cold cloud duration (CCD) image therefore represent the time that that location was covered by rain-bearing clouds during the compositing period. Comparing CCD retrievals with spatially interpolated rainfall for each month of Africa showed RS to be much more accurate, with a root mean square error (r.m.s.e.) of ±38 mm (Hay and Lennon, 1999).

More sophisticated rainfall estimation techniques that relate cloud top reflectances and the growth and decay of cloud systems to rainfall amounts have been reviewed by Petty (1995). Significant advances are also being made using a combination of high spatial resolution radar, passive microwave and visible and infrared radiometer measurements by the tropical rainfall measuring mission (TRMM) (Theon, 1993; Kummerow et al., 1998). These data have not yet been adopted widely in public health and are not discussed in depth here.

5.5. Middle Infrared Radiation

Land surface applications of MIR have focused mainly on the detection of hot regions associated with forest, peat and straw fires and burn scars (Giglio et al., 1999). It has often been used in conjunction with visible radiation for both surface-temperature mapping and for land-cover discrimination (Kerber and Schutt, 1986), where it enhances the spectral separability of land-cover classes.

Despite these studies, the level of understanding and documentation regarding the interaction of MIR radiation with targets, relative to the visible and near infrared wavelengths, is limited. Moreover, the use of MIR wavelengths in land-cover mapping is at an early stage of development (Ehrlich et al., 1994). This is due to the hybrid nature of this spectral region (sensitive to both reflected and emitted radiation—Kidwell, 1998), which makes the interpretation of the signal returning from the target more difficult, and to historical difficulties of data access and instrument noise that can seriously contaminate the MIR signal in AVHRR sensors (Dudhia, 1989).

There are, however, good reasons for using MIR radiation for land-cover discrimination in the tropics. Primarily MIR suffers less attenuation in the atmosphere (Bernstein, 1982; Wooster et al., 1994) and can penetrate to a greater depth through smoke than the visible or NIR wavelengths (Kaufman and Remer, 1994). The MIR region also suffers little attenuation due to atmospheric water (Kerber and Schutt, 1986), making it particularly suitable for applications in the tropics. These factors, coupled with the known interactions of MIR with vegetation, help to justify the incorporation of such data in land-cover type discrimination.

Boyd and Curran (1998) have proposed an explanation for the interaction of MIR and the biophysical properties of vegetation canopies with particular reference to tropical forests. The primary factors include the water content, surface temperature and the structure and roughness of the vegetation target. Increases in each of these factors with increasing vegetation coverage are postulated to cause a decrease in the MIR signal. First, an increase in the amount of vegetation corresponds to an increase in liquid water that can absorb MIR (Kaufman and Remer, 1994), hence reducing the signal. Secondly, there is the effect of thermal emission which dominates the response in the MIR region. The decrease in MIR emitted with increasing amounts of vegetation occurs due to a decrease in the surface resistance to evapotranspiration (i.e. greater transpiration), and because canopy foliage temperatures are significantly lower than background soil surface temperatures due to their relative specific heat capacities (Lambin and Ehrlich, 1996). Thirdly, an increasingly complex canopy structure, dependent upon canopy depth, leaf orientation and distribution, has an effect on incoming MIR radiation by trapping photons and producing shadows which decrease the intensity of reflected MIR radiation (Dadhwal et *al.*, 1996). An excellent review of MIR with particular reference to the NOAA-AVHRR is provided in Cracknell (1997c).

6. GEODESY

Geodesy is the discipline concerned with the measurement of the size and shape of the Earth and positions on it. It is important to remote sensing and geographical information systems (GIS) as it underpins the essential process of georeferencing images and associated vector coverages to Earth surface locations. Geodesy starts simply and rapidly gets more complicated as the mathematical model used for the Earth become more realistic and compromises relating to the area, height and shape of objects become more specific. It is not appropriate to review the subject in detail and the reader is referred to some standard texts where the concepts of geoids, reference ellipsoids, datums, projections and grid referencing systems are outlined (Burkard, 1964; Snyder, 1987; Smith 1988). In this section I shall elaborate on the global positioning system (GPS), however, as the first stage of any analysis aiming to utilize RS or GIS is to determine the exact location of the phenomena of interest.

6.1. The Global Positioning System

In 1973 the US Department of Defence conceived a space-based navigation system that would enable US military forces continuously and accurately to determine their position, velocity and time in a common reference system anywhere on Earth. The present Navigation System with Timing and Ranging (NAVSTAR) GPS is the result of this initiative. The current constellation of 21 evenly-spaced satellites in circular 12 hour orbits, inclined at 55° to the equatorial plane, was found to be the most economic way to satisfy the condition that four satellites could be seen at any one time from any position on Earth (Herring, 1996).

The determination of a location on Earth by ranging from this satellite constellation is a simple concept (Hofmann-Wellenhof *et al.*, 1997b). If, for example, a satellite is exactly 20 000 km distant from an unknown point, this point must be somewhere on the edge of an imaginary 'sphere of position' (of 40 000 km diameter) surrounding that satellite. Because such satellites travel in relatively stable and predictable orbits, the location of the satellite and its hypothetical sphere are known precisely. If at the same time the distance to a second and third satellite (and thus their spheres of position) can be determined, it is theoretically possible to be in only one of two locations. Since one of these is often deep inside the Earth or far into space, a precise geo-location can be determined from three satellites. The problem then becomes how to determine the satellite distance or range accurately. This was achieved by making the satellites transmit signals at exact times so that the interval between transmission and reception of a signal could be used to determine distance, which in turn requires that the clock of the satellites and the receivers be exactly synchronized. The satellites contain highly accurate 'atomic clocks' and the ground receivers less expensive and accurate electronic clocks. Synchronization is achieved by using the signal from a fourth satellite. The receiver's clock is assumed to be approximately correct so that 'pseudo-ranges' to the four satellites can be calculated. The spheres of position calculated for the satellites will be slightly too large if the receiver's clock is slow and slightly too small if the receiver's clock is fast. There is one value that can be calculated, however, for amount of clock error that will make the spheres intersect exactly, which is therefore used to synchronize the receiver.

6.2. Selective Availability

The US government was anxious to protect the massive investment and perceived tactical advantage conferred by the NAVSTAR-GPS system and achieved this by altering the satellites' atomic clocks, known as 'dithering', according to a specific code known only to the military (Hofmann-Wellenhof *et al.*, 1997c). This 'selective availability' meant that civilian users and the enemy could navigate only to an accuracy of ±100 m horizontally and ±156 m vertically. Relatively quickly, however, systems were developed to overcome selective availability. The simplest system takes many readings at one location and averages them over an extended period of time (Arnaud and Fiori, 1998). Many different types of differential ranging use a signal from a stationary beacon of known location to send a signal that is used to correct the satellite times. This can be done in real-time in many parts of the world by subscribing to commercial providers, or retrospectively in more remote areas with a specialized system of two GPS receivers. One is kept stationary and used to correct for the apparent 'drift' in position of the roving unit by comparing the two signals on return (differential or DGPS). Such techniques can readily provide accuracy to ±1 m. Further technical complexity can provide accuracy to ±5 mm, but are of little relevance to public health and are not discussed further. The universal resource locators (URLs) for some common GPS manufacturers are given in Table 3.

These techniques have led to some considerable debate in the US as to why selective availability is maintained in times of peace. Discussion was especially vigorous when it was realized that, due to a shortage of military grade GPS receivers during the Persian Gulf War and the Haiti occupation, civilian devices were distributed to US Armed service personnel, and the selective availability turned off. A nearly identical Russian Global Navigation Satellite System (GLONASS) without any signal degradation is now available and is beginning to be incorporated by some commercial GPS systems (Hofmann-Wellenhof *et al.*, 1997a), which may also hasten the abandonment of NAVSTAR selective availability.[*]

7. CONCLUSIONS

The techniques of RS and geodesy have been reviewed in order to provide sufficient background for the research outlined in later chapters to be readily interpreted. Although it is impossible to address every question in such a limited number of pages, this chapter should at least illustrate where existing answers can be found. This is inevitably a biased account, towards my own perceptions of what aspects need explaining and highlighting for those working in public health. Finally, in addition to the specifications of the satellite systems mentioned in this chapter (Table 1 and 2), some key internet addresses are provided to assist in searches for the most recent information (Table 3).

ACKNOWLEDGEMENTS

I am grateful to the editors, Paul Curran, Scott Goetz, Byron Woods and Louisa Beck for comments on earlier drafts of this manuscript. Bernhard Bakker also helped locate the relevant literature for the discussion of active remote sensing. SIH is an Advanced Training Fellow funded by the Wellcome Trust (No. 056642).

REFERENCES

Anonymous. (1971). *Guide to Meteorological Instruments and Observing Practices*. Geneva: World Meteorological Organization.
Anonymous. (2000). Statement by the President regarding the United States decision to turn stop degrading global positioning system accuracy. Washington, DC: Office of the Press Secretary, The White House.

---

[*] Since going to press, selective availability has been turned off (1 May 2000) (Anonymous, 2000). This means standard GPS receivers should have a horizontal accuracy of ± 20 m. There is no corresponding improvement in DGPS.

Arnaud, M. and Fiori, A. (1998). Bias and precision of different sampling methods for GPS position. *Photogrammetric Engineering & Remote Sensing* **64**, 597–600.

Asrar, G. (1989). *Theory and Applications of Optical Remote Sensing.* New York: Wiley & Sons.

Ba, M.B. and Nicholson, S.E. (1998). Analysis of convective activity and its relationship to the rainfall over the Rift Valley lakes of East Africa during 1983–90 using the Meteosat infrared channel. *Journal of Applied Meteorology* **37**, 1250–1264.

Baldwin, D. and Emery, W.J. (1995). Spacecraft attitude variations of NOAA-11 inferred from AVHRR imagery. *International Journal of Remote Sensing* **16**, 531–548.

Barrett, E.C. and Curtis, L.F. (1999). *Introduction to Environmental Remote Sensing,* 4th edn. Cheltenham: Stanley Thornes.

Battrick, B. (1999). *Meteosat Second Generation: The Satellite Development.* Noordwijk: European Space Agency.

Beck, L.R., Lobitz, B.M. and Wood, B.L. (2000). Remote sensing and human health: new sensors and new opportunities. *Emerging Infectious Diseases* **6**, 217–226.

Becker, F. and Li, Z.-L. (1990). Towards a local split window method over land surfaces. *International Journal of Remote Sensing* **11**, 369–393.

Becker, F. and Li, Z.-L. (1995). Surface temperature and emissivity at various scales: definition, measurement and related problems. *Remote Sensing Reviews* **12**, 225–253.

Belward, A.S. (1992). Spatial attributes of AVHRR imagery for environmental monitoring. *International Journal of Remote Sensing* **13**, 193–208.

Belward, A.S. and Lambin, E. (1990). Limitations to the identification of spatial structures from AVHRR data. *International Journal of Remote Sensing* **11**, 921–927.

Bernstein, R.L. (1982). Sea-surface temperature estimation using the NOAA-6 satellite Advanced Very High-Resolution Radiometer. *Journal of Geophysical Research—Oceans and Atmospheres* **87**, 9455–9465.

Boyd, D.S. and Curran, P.J. (1998). Using remote sensing to reduce uncertainties in the global carbon budget: the potential of radiation acquired in the middle infrared wavelengths. *Remote Sensing Reviews* **16**, 293–327.

Brown, W.M. and Porcello, L. (1969). An introduction to Synthetic Aperture Radar. *IEEE Spectrum* **6**, 52–66.

Brush, R.J.H. (1988). The navigation of AVHRR imagery. *International Journal of Remote Sensing* **9**, 1491–1502.

Burkard, R.K. (1964). *Geodesy for the Layman.* St Louis, MO:USAF Aeronautical Chart and Information Center.

Burt, P.J.A., Colvin, J. and Smith, S.M. (1995). Remote sensing of rainfall by satellite as an aid to *Oedaleus senegalensis* (Orthoptera, Acrididae) control in the Sahel. *Bulletin of Entomological Research* **85**, 455–462.

Byers, H.R. and Barnham, R.R.J. (1948). Thunderstorm structure and circulation. *Journal of Meteorology* **5**, 71–86.

Che, N. and Price, J.C. (1992). Survey of radiometric calibration results and methods for visible and near-infrared channels of NOAA-7, NOAA-9, and NOAA-11 AVHRRs. *Remote Sensing of Environment* **41**, 19–27.

Colwell, J.E. (1974). Vegetation canopy reflectance. *Remote Sensing of Environment* **3**, 175–183.

Colwell, R.N. (1983a). *Manual of Remote Sensing. I. Theory, Instruments and Techniques,* 2nd edn. Falls Church, VA: American Society of Photogrammetry.

Colwell, R.N. (1983b). *Manual of Remote Sensing. II. Interpretation and Applications.* Falls Church, VA: American Society of Photogrammetry.

Colwell, R.N., Brewer, W., Landis, G. *et al.* (1963). Basic matter and energy relationships involved in remote reconnaissance. *Photogrammetric Engineering* **29**, 761–799.

Cooper, D.I. and Asrar, G. (1989). Evaluating atmospheric correction models for retrieving surface temperatures from the AVHRR over a tallgrass prairie. *Remote Sensing of Environment* **27**, 93–102.

Cracknell, A.P. (1997a). *The Advanced Very High Resolution radiometer (AVHRR).* London: Taylor & Francis.

Cracknell, A.P. (1997b). Applications. In: *The Advanced Very High Resolution Radiometer*, pp. 343–463. London: Taylor & Francis.

Cracknell, A.P. (1997c). Channel 3, the neglected channel. In: *The Advanced Very High Resolution Radiometer*, pp. 321–342. London: Taylor & Francis.

Cracknell, A.P. (1997d). The data. In: *The Advanced Very High Resolution Radiometer*, pp. 45–132. London: Taylor & Francis.

Cracknell, A.P. (1997e). Earth surface temperatures. In: *The Advanced Very High Resolution Radiometer*, pp. 181–231. London: Taylor & Francis.

Cracknell, A.P. (1997f). Pre-processing. In: *The Advanced Very High Resolution Radiometer*, pp. 133–180. London: Taylor & Francis.

Cracknell, A.P. (1997g). The spacecraft and instruments. In: *The Advanced Very High Resolution Radiometer*, pp. 1–43. London: Taylor & Francis.

Cracknell, A.P. (1997h). Vegetation. In: *The Advanced Very High Resolution Radiometer*, pp. 233–320. London: Taylor & Francis.

Cracknell, A.P. (1998). Synergy in remote sensing—what's in a pixel? *International Journal of Remote Sensing* **19**, 2025–2047.

Cracknell, A.P. (1999). Twenty years of publication of the *International Journal of Remote Sensing. International Journal of Remote Sensing* **20**, 3469–3484.

Cracknell, A.P. and Hayes, L.W.B. (1991). *Introduction to Remote Sensing.* London: Taylor and Francis.

Curran, P.J. (1985). *Principles of Remote Sensing.* London: Longman.

Curran, P.J. and Williamson, H.D. (1986). Sample size for ground and remotely sensed data. *Remote Sensing of Environment* **20**, 31–41.

Dadhwal, V.K., Parihar, J.S., Medhavy, T.T., Ruhal, D.S., Jarwal, S.D. and Khera, A.P. (1996). Comparative performance of thematic mapper middle-infrared bands in crop discrimination. *International Journal of Remote Sensing* **17**, 1727–1734.

Dalu, G. (1986). Satellite remote sensing of atmospheric water vapor. *International Journal of Remote Sensing* **7**, 1089–1097.

Draeger, W.C., Holm, T.M., Lauer, D.T. and Thompson, R.J. (1997). The availability of Landsat data: past, present, and future. *Photogrammetric Engineering & Remote Sensing* **63**, 869–875.

Dudhia, A. (1989). Noise characteristics of the AVHRR infrared channels. *International Journal of Remote Sensing* **10**, 637–644.

Eastman, J.R. (1999a). *Idrisi32. Guide to GIS and Image Processing*, vol. 1. Worcester, MA: Clark University.

Eastman, J.R. (1999b). *Idrisi32. Guide to GIS and Image Processing*, vol. 2. Worcester, MA: Clark University.

Eck, T.F. and Holben, B.N. (1994). AVHRR split window temperature differences and total precipitable water over land surfaces. *International Journal of Remote Sensing* **15**, 567–582.

Ehrlich, D., Estes, J.E. and Singh, A. (1994). Applications of NOAA-AVHRR 1 km data for environmental monitoring. *International Journal of Remote Sensing* **15**, 145–161.

Emanuel, K.A. (1994). Observed characteristics of precipitating convection. In: *Atmospheric Convection*, pp. 230–279. Oxford: Oxford University Press.

Emery, W.J., Brown, J. and Nowak, Z.P. (1989). AVHRR image navigation: summary and review. *Photogrammetric Engineering & Remote Sensing* **55**, 1175–1183.

Gibson, P. and Power, C. (2000a). *Introducing Remote Sensing: Principles and Concepts*. Andover: Routledge.

Gibson, P. and Power, C. (2000b). *Introducing Remote Sensing: Digital Image Processing and Applications*. Andover: Routledge.

Giglio, L., Kendall, J.D. and Justice, C.O. (1999). Evaluation of global fire detection algorithms using simulated AVHRR infrared data. *International Journal of Remote Sensing* **20**, 1947–1985.

Gleason, A.C.R., Prince, S.D. and Goetz, S.J. (2000). Effects of orbital drift on observations of land surface temperature recovered from the AVHRR sensors. *Remote Sensing of Environment*, in press.

Goetz, S.J. (1997). Multi-sensor analysis of NDVI, surface temperature, and biophysical variables at a mixed grassland site. *International Journal of Remote Sensing* **18**, 71–94.

Goetz, S.J., Halthore, R., Hall, F.G. and Markham, B.L. (1995). Surface temperature retrieval in a temperate grassland with multi-resolution sensors. *Journal of Geophysical Research* **100**, 25397–25410.

Goetz, S.J., Prince, S.D., Goward, S.N., Thawley, M.M., Small, J. and Johnston, A. (1999). Mapping net primary production and related biophysical variables with remote sensing: application to the BOREAS region. *Journal of Geophysical Research* **104**, 27719–27733.

Gorman, A.J. and McGregor, J. (1994). Some considerations for using AVHRR data in climatological studies: II. Instrument performance. *International Journal of Remote Sensing* **15**, 549–565.

Goward, S.N. and Williams, D.L. (1997). Landsat and Earth systems science: development of terrestrial monitoring. *Photogrammetric Engineering & Remote Sensing* **63**, 887–900.

Goward, S.N., Markham, B., Dye, D.G., Dulaney, W. and Yang, J. (1991). Normalized difference vegetation index measurements from the Advanced Very High Resolution Radiometer. *Remote Sensing of Environment* **35**, 257–277.

Green, R.M. and Hay, S.I. (2000). Mapping of climate variables across tropical Africa and temperate Europe using meteorological satellite sensor data. *Remote Sensing of Environment*, in press.

Grimes, D.I.F., Pardo-Iguzquiza, E. and Bonifacio, R. (1999). Optimal areal rainfall estimation using raingauges and satellite data. *Journal of Hydrology* **222**, 93–108.

Gutman, G.G. (1999a). On the monitoring of land surface temperatures with the NOAA/AVHRR: removing the effect of satellite orbit drift. *International Journal of Remote Sensing* **20**, 3407–3413.

Gutman, G.G. (1999b). On the use of long-term global data on land surface reflectances and vegetation indices derived from the Advanced Very High Resolution Radiometer. *Journal of Geophysical Research* **104**, 6241–6255.

Hanan, N.P., Prince, S.D. and Franklin, J. (1993). Reflectance properties of West African savanna trees from ground radiometer measurements. II. Classification of components. *International Journal of Remote Sensing* **14**, 1081–1097.

Hastings, D.A. and Emery, W.J. (1992). The advanced very high resolution radiometer (AVHRR): a brief reference guide. *Photogrammetric Engineering & Remote Sensing.* **58**, 1183–1188.

Hay, S.I. and Lennon, J.J. (1999). Deriving meteorological variables across Africa for the study and control of vector-borne disease: a comparison of remote sensing and spatial interpolation of climate. *Tropical Medicine & International Health* **4**, 58–71.

Hay, S.I., Tucker, C.J., Rogers, D.J. and Packer, M.J. (1996). Remotely sensed surrogates of meteorological data for the study of the distribution and abundance of arthropod vectors of disease. *Annals of Tropical Medicine and Parasitology* **90**, 1–19.

Henderson, F.M. and Lewis, A.J. (1998). *Principles and Applications of Imaging Radar.* New York: Wiley & Sons.

Herring, T.A. (1996). The global positioning system. *Scientific American* **274**, 44–50.

Hofmann-Wellenhof, B., Lichtenegger, H. and Collins, J. (1997a). Future of GPS. In: *Global Positioning System: Theory and Practice*, 4 edn., pp. 345–352. Wien: Springer Verlag.

Hofmann-Wellenhof, B., Lichtenegger, H. and Collins, J. (1997b). Introduction. In: *Global Positioning System: Theory and Practice*, 4 edn., pp. 1–10. Wien: Springer Verlag.

Hofmann-Wellenhof, B., Lichtenegger, H. and Collins, J. (1997c). Overview of GPS. In: *Global Positioning System: Theory and Practice*, 4 edn., pp. 11–26. Wien: Springer Verlag.

Holben, B.N. (1986). Characteristics of maximum-value composite images from temporal AVHRR data. *International Journal of Remote Sensing* **7**, 1417–1434.

Holben, B.N., Eck, T.F. and Fraser, R.S. (1991). Temporal and spatial variability of aerosol optical depth in the Sahel region in relation to vegetation remote sensing. *International Journal of Remote Sensing* **12**, 1147–1163.

Huete, A.R. (1988). A soil-adjusted vegetation index (SAVI). *Remote Sensing of Environment* **25**, 295–309.

Huete, A.R. and Escadafal, R. (1991). Assessment of biophysical soil properties through spectral decomposition techniques. *Remote Sensing of Environment* **35**, 149–159.

Huete, A.R., Jackson, R.D. and Post, D.F. (1985). Spectral response of a plant canopy with different soil backgrounds. *Remote Sensing of Environment* **17**, 37–53.

Huh, O.K. (1991). Limitations and capabilities of the NOAA satellite advanced very high resolution radiometer (AVHRR) for remote sensing of the Earth's surface. *Preventive Veterinary Medicine* **11**, 167–183.

Jackson, R.D. and Huete, A.R. (1991). Interpreting vegetation indices. *Preventive Veterinary Medicine* **11**, 185–200.

Justice, C.O., Townshend, J.R.G., Holben, B.N. and Tucker, C.J. (1985). Analysis of the phenology of global vegetation using meteorological satellite data. *International Journal of Remote Sensing* **6**, 1271–1318.

Justice, C.O., Markham, B.L., Townshend, J.R.G. and Kennard, R.L. (1989). Spatial degradation of satellite data. *International Journal of Remote Sensing* **10**, 1539–1561.

Justice, C.O., Eck, T.F., Tanre, D. and Holben, B.N. (1991). The effect of water vapour on the normalized difference vegetation index derived for the Sahelian region from NOAA AVHRR data. *International Journal of Remote Sensing* **12**, 1165–1187.

Kasischke, E.S., Melack, J.M. and Dobson, M.C. (1997). The use of imaging radars for ecological applications: a review. *Remote Sensing of Environment* **59**, 141–156.

Kaufman, Y.J. and Holben, B.N. (1993). Calibration of the AVHRR visible and near-IR bands by atmospheric scattering, ocean glint and desert reflection. *International Journal of Remote Sensing* **14**, 21–52.

Kaufman, Y.J. and Remer, L.A. (1994). Detection of forests using mid-IR reflectance: an application for aerosol studies. *IEEE Transactions on Geosciences and Remote Sensing* **32**, 672–683.

Kaufman, Y.L. and Tanré, D. (1992). Atmospherically resistant vegetation index (ARVI) for EOS-MODIS. *IEEE Transactions on Geoscience and Remote Sensing* **30**, 261–270.

Kerber, A.G. and Schutt, J.B. (1986). Utility of AVHRR channel 3 and 4 in land-cover mapping. *Photogrammetric Engineering & Remote Sensing* **52**, 1877–1883.

Khan, B., Hayes, L. and Cracknell, A.P. (1992). The optimisation of higher-order resampling methods in the multiprocessor environment. *Parallel Computing* **18**, 1335–1347.

Khan, B., Hayes, L.W.B. and Cracknell, A.P. (1995). The effects of higher-order resampling on AVHRR data. *International Journal of Remote Sensing* **16**, 147–163.

Kidwell, K.B. (1998). *NOAA Polar Orbiter Data User's Guide (TIROS-N, NOAA-6, NOAA-7, NOAA-8, NOAA-9, NOAA-10, NOAA-11, NOAA-12, NOAA-13 and NOAA-14)*. Suitland, MD: National Oceanic and Atmospheric Administration.

Krasnopolsky, V.M. (1994). The problem of AVHRR image navigation revisited. *International Journal of Remote Sensing* **15**, 979–1008.

Kummerow, C., Barnes, W., Kozu, T., Shiue, J. and Simpson, J. (1998). The Tropical Rainfall Measuring Mission (TRMM) sensor package. *Journal of Atmospheric and Oceanic Technology* **15**, 809–817.

Lambin, E.F. and Ehrlich, D. (1995). Combining vegetation indexes and surface temperature for land-cover mapping at broad spatial scales. *International Journal of Remote Sensing* **16**, 573–579.

Lambin, E.F. and Ehrlich, D. (1996). The surface temperature-vegetation index space for land cover and land-cover changes analysis. *International Journal of Remote Sensing* **17**, 463–487.

Landgrebe, D. (1997). The evolution of Landsat data analysis. *Photogrammetric Engineering & Remote Sensing* **63**, 859–867.

Lauer, D.T., Morain, S.A. and Salomonson, V.V. (1997). The Landsat program: its origins, evolution, and impacts. *Photogrammetric Engineering & Remote Sensing* **63**, 831–838.

Laurent, H., Jobard, I. and Toma, A. (1998). Validation of satellite and ground-based estimates of precipitation over the Sahel. *Atmospheric Research* **48**, 651–670.

Leprieur, C., Kerr, Y.H. and Pichon, J.M. (1996). Critical assessment of vegetation indexes from AVHRR in a semiarid environment. *International Journal of Remote Sensing* **17**, 2549–2563.

Lillesand, T.M. and Kiefer, R.W. (2000). *Remote Sensing and Image Interpretation*, 4th edn. New York: Wiley & Sons.

Los, S.O. (1993). Calibration adjustment of the NOAA AVHRR normalized difference vegetation index without recourse to component channel 1 and 2 data. *International Journal of Remote Sensing* **14**, 1907–1917.

Lowman, P.D. (1999). Landsat and Apollo: the forgotten legacy. *Photogrammetric Engineering & Remote Sensing* **65**, 1143–1147.

Lyon, J.G., Yuan, D., Lunetta, R.S. and Elvidge, C.D. (1998). A change detection experiment using vegetation indices. *Photogrammetric Engineering & Remote Sensing* **64**, 143–150.

Marçal, A.R.S. (1999). A new method for high accuracy navigation of NOAA AVHRR imagery. *International Journal of Remote Sensing* **20**, 3273–3280.

Martyn, D. (1992). The climates of Africa. In: *Developments in Atmospheric Sciences, 18. Climates of the World*, pp. 119–261. Amsterdam: Elsevier.

Mather, P.M. (1999). *Computer Processing of Remotely-Sensed Images: An Introduction*, 2nd edn. Chichester: Wiley & Sons.

McGregor, J. and Gorman, A.J. (1994). Some considerations for using AVHRR data in climatological studies: I. Orbital characteristics of NOAA satellites. *International Journal of Remote Sensing* **15**, 537–548.

Meyer, D.J. (1996). Estimating the effective spatial resolution of an AVHRR time series. *International Journal of Remote Sensing* **17**, 2971–2980.

Mika, A.M. (1997). Three decades of Landsat instruments. *Photogrammetric Engineering & Remote Sensing* **63**, 839–852.

Milford, J.R. and Dugdale, G. (1990). Monitoring of rainfall in relation to the control of migrant pests. *Philosophical Transaction of the Royal Society London B* **328**, 689–704.

Monteith, J.L. and Unsworth, M.H. (1990). *Principles of Environmental Physics*. London: Edward Arnold.

Morain, S.A. and Budge, A.M. (1997). *Earth Observing Platforms and Sensors*. Bethesda, MD: American Society of Photogrammetry and Remote Sensing.

Myneni, R.B., Hall, F.G., Sellers, P.J. and Marshak, A.L. (1995a). The interpretation of spectral vegetation indices. *IEEE Transactions on Geosciences and Remote Sensing* **33**, 481–486.

Myneni, R.B., Maggion, S., Iaquinto, *et al.* (1995b). Optical remote sensing of vegetation: modelling, caveats, and algorithms. *Remote Sensing of Environment* **51**, 169–188.

Norman, J.M., Divakarla, M. and Goel, N.S. (1995). Algorithms for extracting information from remote thermal-IR observations of the Earth's surface. *Remote Sensing of Environment* **51**, 157–168.

Oliver, C.J. (1991). Information from SAR images. *Journal of Physics D—Applied Physics* **24**, 1493–1514.

Petty, G.W. (1995). The status of satellite-based rainfall estimation over land. *Remote Sensing of Environment* **51**, 125–137.

Pinty, B. and Verstraete, M.M. (1992). GEMI: a non-linear index to monitor global vegetation from satellites. *Vegetation* **101**, 15–20.

Prabhakara, C., Dalu, G. and Kunde, V.G. (1974). Estimation of sea surface temperature from remote sensing in the 11- to 13- μm window region. *Journal of Geophysical Research* **79**, 5039–5044.

Price, J.C. (1982). On the use of satellite data to infer surface fluxes at meteorological scales. *Journal of Applied Meteorology* **21**, 1111–1122.

Price, J.C. (1984). Land surface temperature measurements from the split window channels of the NOAA 7 advanced very high resolution radiometer. *Journal of Geophysical Research* **89**, 7231–7237.

Price, J.C. (1991). Timing od NOAA afternoon passes. *International Journal of Remote Sensing* **12**, 193–198.

Prince, S.D. and Goward, S.N. (1995). Global primary production: a remote sensing approach. *Journal of Biogeography* **22**, 815–835.

Qin, Z. and Karnieli, A. (1999). Progress in the remote sensing of land surface temperature and ground emissivity using NOAA-AVHRR data. *International Journal of Remote Sensing* **20**, 2367–2393.

Quattrochi, D.A. and Goodchild, M.F. (1997). *Scale in remote sensing and GIS*. Boca Raton, FL: CRS.

Rencz, A.N. (1999). *Remote Sensing for the Earth Sciences*. New York: Wiley & Sons.

Richards, J.A. and Jia, X. (1999). *Remote Sensing and Digital Image Analysis: An Introduction*. Berlin: Springer-Verlag.

Robinson, T.P. (1996). Spatial and temporal accuracy of coarse resolution products of NOAA-AVHRR NDVI data. *International Journal of Remote Sensing* **17**, 2303–2321.

Sabins, F.F. (1997). *Remote Sensing: Principles and Interpretation*, 3rd edn. New York: W.H. Freeman.

Salisbury, J.W. and Daria, D.M. (1992). Emissivity of terrestrial materials in the 8–14 μm atmospheric window. *Remote Sensing of Environment* **42**, 83–106.

Sarder, A.M. (1997). The evolution of space-borne imaging radar systems: a chronological history. *Canadian Journal of Remote Sensing* **23**, 276–280.

Schmullius, C.C. and Evans, D.L. (1997). Synthetic Aperture Radar (SAR) frequency and polarization requirements for applications in ecology, geology, hydrology, and oceanography: a tabular status quo after SIR-C/X-SAR. *International Journal of Remote Sensing* **18**, 2713–2722.

Sellers, P.J. (1985). Canopy reflectance, photosynthesis and transpiration. *International Journal of Remote Sensing* **6**, 1335–1372.

Smith, G.R., Levin, R.H., Abel, P. and Jacobowitz, H. (1988). Calibration of the solar channels of NOAA-9 AVHRR using high altitude aircraft measurements. *Journal of Atmospheric and Oceanic Technology* **5**, 631–639.

Smith, J.R. (1988). *Basic Geodesy: An Introduction to the History and Concepts of Modern Geodesy without Mathematics*. Rancho Cordova, CA: Landmark Enterprises.

Smith, W.L. (1966). Note on the relationship between total precipitable water and surface dewpoint. *Journal of Applied Meteorology* **5**, 726–727.

Snyder, J.P. (1987). *Map projections—a working manual. U.S. Geological Survey professional paper 1395*. Washington DC: United States Government Printing Office.

Stoms, D.M., Bueno, M.J. and Davis, F.W. (1997). Viewing geometry of AVHRR image composites derived using multiple criteria. *Photogrammetric Engineering & Remote Sensing* **63**, 681–689.

Sugita, M. and Brutasaert, W. (1993). Comparison of land surface temperatures derived from satellite observations with ground truth during FIFE. *International Journal of Remote Sensing* **14**, 1659–1676.

Swain, P.H. and Davis, S.M. (1978). *Remote Sensing: The Quantitative Approach*. New York: McGraw-Hill.

Tanré, D., Holben, B.N. and Kaufman, Y. (1992). Atmospheric correction algorithm for NOAA-AVHRR products: theory and application. *IEEE Transactions on Geoscience and Remote Sensing* **30**, 231–248.

Theon, J.S. (1993). The Tropical Rainfall Monitoring Mission (TRMM). *Advances in Space Research* **14**, 159–165.

Thome, K., Markham, B., Barker, J., Slater, P. and Biggar, S. (1997). Radiometric calibration of Landsat. *Photogrammetric Engineering & Remote Sensing* **63**, 853–858.

Townshend, J.R.G. and Tucker, C.J. (1984). Objective assessment of Advanced Very High Resolution Radiometer data for land cover mapping. *International Journal of Remote Sensing* **5**, 497–504.

Tucker, C.J. (1979). Red and photographic infrared linear combinations for monitoring vegetation. *Remote Sensing of Environment* **8**, 127–150.

Tucker, C.J., Vanpraet, C., Boerwinkel, E. and Gaston, A. (1983). Satellite remote sensing of total dry matter production in the Senegalese Sahel. *Remote Sensing of Environment* **13**, 461–474.

Tucker, C.J., Townshend, J.R.G. and Goff, T.E. (1985a). African land-cover classification using satellite data. *Science* **227**, 369–375.

Tucker, C.J., Vanpraet, C.L., Sharman, M.J. and Van Ittersum, G. (1985b). Satellite remote sensing of total herbaceous biomass production in the Senegalese Sahel: 1980–1984. *Remote Sensing of Environment* **17**, 233–249.

Tucker, C.J. and Sellers, P.J. (1986). Satellite remote sensing of primary production. *International Journal of Remote Sensing* **7**, 1395–1416.

Tucker, C.J., Newcomb, W.W., Los, S.O. and Prince, S.D. (1991). Mean and inter-year variation of growing-season normalized difference vegetation index for the Sahel 1981–1989. *Remote Sensing of Environment* **8**, 127–150.

Ungar, S.G. (1997). Technologies for future Landsat missions. *Photogrammetric Engineering & Remote Sensing* **63**, 901–905.

Vermote, E., Tanre, D. and Herman, M. (1990). Atmospheric effects on satellite imagery, correction algorithms for ocean color or vegetation monitoring. *International Society for Photogrammetry and Remote Sensing* **28**, 46–55.

Waring, R.H., Way, J.B., Hunt, E.R., *et al.* (1995). Biologists' toolbox—Imaging radar for ecosystem studies. *Bioscience* **45**, 715–723.

Williamson, R.A. (1997). The Landsat legacy: remote sensing policy and the development of commercial remote sensing. *Photogrammetric Engineering & Remote Sensing* **63**, 877–885.

WMO. (1994). *Information on Meteorological and other Environmental Satellites.* Geneva: World Meteorological Organization.

Wooster, M.J., Sear, C.B., Patterson, G. and Haigh, J. (1994). Tropical lake surface temperatures from locally received NOAA-11 AVHRR data—comparison with *in-situ* measurements. *International Journal of Remote Sensing* **15**, 183–189.

Wu, B.F. and Liu, H.Y. (1997). A simplified method of accurate geometric correction for NOAA AVHRR 1B data. *International Journal of Remote Sensing* **18**, 1795–1808.

Linking Remote Sensing, Land Cover and Disease

P.J. Curran, P.M. Atkinson, G.M. Foody and E. J. Milton

*Department of Geography, University of Southampton, Highfield,
Southampton SO17 1BJ, UK*

ABSTRACT

Land cover is a critical variable in epidemiology and can be characterized
remotely. A framework is used to describe both the links between land cover

ADVANCES IN PARASITOLOGY VOL 47
0065-308-X $30.00

and radiation recorded in a remotely sensed image, and the links between land cover and the disease carried by vectors. The framework is then used to explore the issues involved when moving from remotely sensed imagery to land cover and then to vector density/disease risk. This exploration highlights the role of land cover; the need to develop a sound knowledge of each link in the predictive sequence; the problematic mismatch between the spatial units of the remotely sensed and epidemiological data and the challenges and opportunities posed by adding a temporal mismatch between the remotely sensed and epidemiological data. The paper concludes with a call for both greater understanding of the physical components of the proposed framework and the utilization of optimized statistical tools as prerequisites to progress in this field.

1. INTRODUCTION

The application of remote sensing to epidemiology is based on the development of a logical sequence that links measures of radiation made by a sensor on board an aircraft, or more usually a satellite, to measures of a disease and its vector (e.g. Crombie *et al.*, 1999). At its most general, a sequence could be: (i) remotely sensed data can be used to provide information on land cover (e.g. different vegetation types or classes of vegetation amount) and thereby habitat (Innes and Koch, 1998), (ii) the spatial distribution of a vector-borne disease is related to the habitat of that vector (Pavlovsky, 1966) and (iii) therefore, remotely sensed data can be used to provide information on the spatial distribution of the vector-borne disease (Hay *et al.*, 1997).

In certain circumstances the first two propositions can be well-founded. For example, where remotely sensed data and disease data are both related to climate (Hugh-Jones, 1991a; Thomson *et al.*, 1996; Hay *et al.*, 1996, 1998a). When this is the case researchers have sought to use remote sensing as a direct and instrumental tool (Curran, 1987) to predict and map the location of some of the major diseases affecting human health (Bailey and Linthicum, 1989; Hay *et al.*, 1997, 1998b; Malone *et al.*, 1997; Connor, 1999).

> Satellite (*sensor*) images can pinpoint the breeding grounds of the mosquitoes that cause malaria, pick out the tsetse fly's favourite haunts and perhaps even identify places where there is a risk of cholera.
>
> (Kleiner, 1995, p. 9)

In support of such predictions, research has been undertaken in different environments, for different diseases and vectors and with various combinations of imagery and ancillary data (Hugh-Jones and O'Neil, 1986;

Hay *et al.*, 1997; Estrada-Peña, 1998). Some 30 years of experience (Cline, 1970) have shown that while the remote sensing of disease is certainly viable (Linthicum *et al.*, 1987; Hugh-Jones, 1991a; Rogers and Williams, 1993) it will not be a robust and reliable epidemiological technique until we have developed a sound understanding of *all* of the links between remotely sensed data and variables of epidemiological significance. This paper will attempt to provide a framework for this understanding that is focused on the role of land cover.

2. A FRAMEWORK FOR THE REMOTE SENSING OF DISEASE

The logical sequence linking remotely sensed data to diseases and their vectors (Section 1) can be expanded into a framework (Figure 1). The framework highlights the links that span the gap between image(s) and

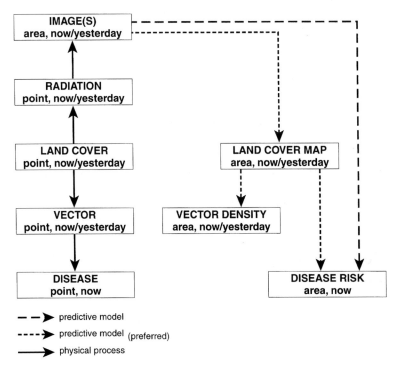

Figure 1 A framework linking remotely sensed images with disease. Note that land cover refers to both type (e.g. forest, grass) and amount (e.g. high biomass, low biomass).

Table 1 Four illustrative examples of the framework in Figure 1

	Example 1	Example 2	Example 3	Example 4
Image(s)	Landsat TM coverage of a region for a given day, last month	NOAA AVHRR coverage of a country, maximum NDVI values from last month	JERS-1 SAR coverage of a country for summer and winter last year	Landsat MSS coverage of an agricultural region, 10 years ago
Radiation	Six wavebands (visible to middle infrared) for 30 × 30 m area for a given day, last month	Two wavebands (visible and near infrared) for 1.1 × 1.1 km area for various times last month	Backscattered microwaves for 18 × 18 m areas for two dates, last year	Four wavebands (visible to near infrared) for 80 × 80 m areas, ten years ago
Land cover	Mix of water, pasture and villages for a region, last month	Amount of green biomass for a country, last month	Deciduous forest for a country, last year	Proportion of land cover that is open scrub within 5 km of any village, 10 years ago
Vector	Point data on mosquitoes, now	Point data on tsetse flies, last month	Point data on ticks in spring of this year	Point observations of vole presence, 10 years ago
Disease	Point data on malaria, next month	Point data on sleeping sickness, today	Point data on Lyme disease in summer of this year	Point data on liver disease, today

disease; the role of land cover; the variation in space and time between image and disease data and the simplifying assumptions that need to be made if remotely sensed data are to be used to predict the density of disease vectors or risk of disease, either directly or via the mapping of land cover (Curran *et al.*, 1998). The permutations of imagery (Rees, 1999; Hay, this volume), land cover, disease vectors and types of disease in both space and time are enormous (Washino and Wood, 1994). However, to illustrate the preferred predicted model in the framework (Figure 1) only four permutations have been chosen (Table 1). In *example one* radiation in visible to middle infrared wavelengths was recorded last month on several Landsat Thematic Mapper (TM) images for a region in South East Asia. The radiation measurements were pre-processed and then classified to produce a land cover map comprising water (a possible home to mosquito larvae), pasture (location of cattle) and villages (location of people). The mix of land cover was, assuming little change, related to the number of mosquitoes at points now and thereby the number of people in the villages who could fall victim to malaria within the next month or so. In *example two* radiation in red (R) and near infrared (NIR) wavelengths was recorded last month on several National Oceanic and Atmospheric Administration (NOAA) Advanced Very High Resolution Radiometer (AVHRR) images for a country in Africa. Following image pre-processing the monthly maximum normalized difference vegetation index (NDVI = (NIR – R)/ (NIR + R)) was used to estimate and map the amount of green biomass for last month. The amount of green biomass would have been responding to the same climatic triggers as tsetse flies. Therefore, the amount of green biomass would have been related to the number of tsetse flies at points last month and thereby the number of people who would fall victim to sleeping sickness today. In *example three* backscattered microwave radiation was recorded during the winter and summer of last year on two Japanese Earth Resources Satellite (JERS-1) Synthetic Aperture Radar (SAR) images for a country in northern Europe. The images were pre-processed, classified and then used to produce a map of last year's deciduous forest cover. The area of deciduous forest cover would have been related to the number of tick hosts (deer, pheasants) and the total number of ticks on such hosts is known to peak in the spring. Therefore, the area of deciduous forest cover would be related to this year's point data on ticks in springtime and thereby will be related to the number of people who will fall ill with Lyme disease during this year's summer. In *example four* radiation in visible to near infrared wavelengths was recorded ten years ago on several Landsat Multispectral Scanning System (MSS) images for a region in central Asia. The images were pre-processed and then used to produce a land cover map comprising open shrub (wild vole habitat) and villages (people). The

Table 2 A selection of papers from 1988 to 1999 to illustrate the range of remotely sensed imagery, characterization of land cover and spatial units involved in the remote sensing of disease. Definitions of the remote sensing abbreviations are provided in the text

Reference	Location(s)	Remotely sensed data to study land	Remotely sensed spatial sampling unit	Land cover	Disease/vector	Disease/vector spatial sampling unit
Hugh-Jones et al. (1988)	Guadeloupe	Landsat TM	30 m	Land cover	African bont tick	967 cattle
Welch et al. (1989a)	Texas USA	Colour infrared aerial photography	Various	Land and water cover	Mosquito	286×10^3 acres
Welch et al. (1989b)	Texas USA	Colour infrared aerial photography	Various	Land and water cover	Mosquito	232 cm^2
Wood et al. (1991)	California USA	TM Simulator/ Landsat TM	?/ 30 m	Water and vegetation cover	Mosquito larvae	104×2000 m^2 fields
Rogers and Randolph (1991)	West Africa (Côte d'Ivoire, Burkina Faso)	AVHRR NDVI	1.1 km	Vegetation amount	Trypanosomiasis/ tsetse	700 km long transect
Pope et al. (1992)	Kenya	Airborne SAR/ Landsat TM (classify)	2.4 m/30 m	Topography	Rift valley fever virus/mosquito	Dambos in river valleys
Wood et al. (1992)	California USA	TM Simulator (mean NDVI)	?	Water and vegetation cover	Mosquito larvae	104/2000 m^2 fields

Study	Location	Sensor	Resolution	Variable	Disease/vector	Notes
Rogers and Randolph (1993)	Zimbabwe Kenya Tanzania	AVHRR NDVI	1.1 km	Vegetation amount	Trypanosomiasis/ tsetse, ticks	?
Beck et al. (1994)	Mexico	Landsat TM	30 m	Land cover	Malaria/mosquito	Landscape around 40 villages
Ahearn and DeRooy (1996)	Kwara State Nigeria	Landsat TM	30 m	Vegetation amount	Guinea worm/ water flea	?
Roberts et al. (1996)	Belize	SPOT HRV	20 m	Vegetation amount	Malaria/mosquito	Four legs at 20 sample points along highway
Rogers et al. (1996)	Côte d'Ivoire and Burkina Faso	AVHRR NDVI, temp	7.6 km	Vegetation amount	Trypanosomiasis/ tsetse	0.250°
Robinson et al. (1997)	Malawi Mozambique Zambia Zimbabwe	AVHRR NDVI	1.1 km	Vegetation amount	Trypanosomiasis/ tsetse	?
Hay et al. (1998a)	Kenya	AVHRR NDVI, temp, MIR	8 km	Vegetation amount	Malaria/mosquito	Catchment of hospitals in three villages
Thomson et al. (1999)	Gambia	AVHRR NDVI	1.1 km	Vegetation amount	Malaria/mosquito	2039 children from 65 villages

presence of open scrub cover within around 5 km of a village was likely to provide a suitable hunting ground for domesticated dogs that ingest the wild voles and their tapeworms. The tapeworms are excreted, some carry liver disease and the incubation time is around ten years. Therefore, the proportion of open shrub cover ten years ago would be related to the number of people in the villages who are now suffering with liver disease.

As these four simplified examples of prediction illustrate, there is no one methodology for the remote sensing of disease. The four examples have used data from different satellite sensors, recorded for various areas, spatial resolutions and times in the past both to identify land cover (e.g. forests, pasture, cereal) and to quantify land cover (e.g. high or low biomass) that is linked, ultimately, to data on disease at the specific location recorded at some moment in time. The remainder of this text will restrict itself to data collected in optical (visible to thermal) wavelengths and although discussion will cover both the identification and quantification of land cover, emphasis will be on identification in which land cover is treated as a categorical rather than a continuous variable.

To emphasize the diversity that is encompassed by the framework (Figure 1), Table 2 lists 15 contemporary studies, all of which have had to contend with a major mismatch in the spatial sampling units of remotely sensed and epidemiological data. These studies have a complete spatial coverage of remotely sensed data broken down into spatial sampling units of various size. The data for these units are then used to produce, or more usually infer, a contiguous but simple land cover representation that is, in turn, related to vector density or disease incidence data that tends to be spatially and numerically incomplete and misplaced spatially from the location of disease contraction.

The links that are being employed in the framework (Figure 1), exemplars and examples (Tables 1 and 2) vary from the physical and well-understood link between radiation and an image to the empirical link between a vector in a field and, say, a patient in a hospital. As Table 2 indicates, this latter link is weakest (Roberts *et al.*, 1991) particularly in the developing world, if only because 'the most difficult data to obtain are reliable field data, especially of disease' (Hugh-Jones, 1991b, pp. 202–203).

For the next three stages of the paper we will explore, in detail, some of the more important facets of the framework (Figure 1). We will start with a remotely sensed image, or images, of landscapes (Hay, this volume); and move to a possibly (multi-temporal) spectral mosaic (Section 3) and thence to a map of land cover (Section 4). This will serve to illustrate the physical processes that link remotely sensed data to the landscape (Figure 1). Section 2.5 will move to the prediction side of the framework (Figure 1) and focus on the use of remotely sensed data to estimate vector density and disease risk.

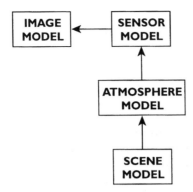

Figure 2 The remote sensing information system (discussion in text).

3. FROM REMOTELY SENSED IMAGES OF LANDSCAPE TO A MULTI-TEMPORAL SPECTRAL MOSAIC

The preceding section established that land cover is a central issue in the use of remote sensing in epidemiology, whether explicitly, through the relationship between vectors and their habitat, or implicitly, through the use of remote sensing to provide proxy ecological indicators such as the NDVI (Table 1) (Crombie *et al.*, 1999). Irrespective of location, the focus of epidemiological interest usually concerns landscapes in which green vegetation, bare soil and water are present. The interaction between these generalized land covers, either through time or across space is central to understanding ecological and therefore disease patterns using remote sensing.

The source of remotely sensed data for epidemiological research is typically one or more digital images of part of the Earth's surface. The aim of this section is to show how reliable data on the reflectance properties of the Earth's surface may be derived from such images. We begin by breaking down what can be termed the 'remote sensing information system' into four linked models (Figure 2): scene model, sensor model, atmosphere model and image model. The scene model is a representation of the Earth's surface that acts as the gateway between the information domain (i.e. the biophysical properties of the real world that we are interested in) and the data domain (i.e. spatial patterns and temporal fluxes of electromagnetic radiation). The sensor model and atmosphere model are certainly relevant to the epidemiologist, but are less accessible or open to modification because they are physically-based or determined by engineering decisions made by data providers. The image model is the output of the system and makes the necessary translation back from the data domain to the information domain. There are as many different information domains as there are users, so there are many possible image models. Some properties of the image model may be fixed by system parameters (e.g. images from a given sensor have a given pixel size), but others may be under the control of the user (e.g. the bands may be programmable or only a few are selected).

The aim of the remote sensing information system shown in Figure 2 is to address questions such as 'What is the land cover at point *a*?', 'Has it changed from last month?', and 'How much of material *b* (e.g. green biomass) is there at point *c*?'. To understand how to optimize the remote sensing information system to answer such questions accurately and to appreciate its limitations as well as its possibilities, it is necessary to consider each of the four models in turn (scene, atmosphere, sensor, image) and we will do this by following the flow of data through the system from the scene comprising a landscape of different land covers to the image(s) used by the epidemiologist (Figure 1).

3.1. The Scene Model

The interaction of electromagnetic radiation with various components of the landscape is complex (Curran *et al.*, 1998). Spectral measurements of individual leaves made using laboratory spectrophotometers have proved to be useful aids for understanding interactions between electromagnetic radiation (EMR) and plant materials, but they are insufficient in themselves to explain the spectral reflectance of a plant canopy comprising leaves, branches, shadow and soil background (Colwell, 1974). Vegetation canopies comprise many individual scene components (e.g. leaves, stems, flowers) arranged throughout a three-dimensional volume which is difficult to simulate in the laboratory. Furthermore, in the 'real world', a vegetation canopy is situated within a particular radiation environment comprising direct illumination from the Sun and diffuse illumination from skylight. Accurate simulation of the illumination geometry and spectral distribution of sunlight is exceptionally difficult to achieve in the laboratory. The same limitations affect our understanding of the reflectance of bare soil surfaces. Whilst it is relatively easy to measure the spectral reflectance of crushed or powdered samples in the laboratory there is ample evidence to confirm that one of the primary influences on soil reflectance is surface roughness, through its influence upon shadowing (Irons *et al.*, 1992). These problems may be addressed either by measuring the spectral reflectance of plant canopies and soils in the 'real' radiation environment (i.e. the field) (Milton, 1987) or by using computer modelling to simulate the reflectance properties of appropriate scene components within a virtual radiation environment (Goel, 1988).

Passive remote sensing systems use the Sun as an energy source, thus the position of the Sun in the sky at the time of sensing controls which scene components are illuminated and the location and extent of shadows. The geometric variables associated with remote sensing measurements are defined in Figure 3. The three-dimensional nature of vegetation canopies means that the contribution of different canopy components to spectral reflectance varies depending upon the angle at which the canopy is viewed and illuminated. For a forest or shrub canopy, for example, the proportion of green leaves presented to a remote sensing system increases as the view angle from nadir (θ_r) increases, whereas the proportion of shadow decreases. The combined dependence of the reflected flux upon the zenith and azimuth angles of the sensor (θ_r, Φ_r) and those of the Sun (θ_i, Φ_i) is recognized by the use of the prefix 'bidirectional' to describe spectral reflectance measurements made in the field environment (Milton, 1987). Bidirectional reflectance is an intrinsic property of the surface and is defined as the ratio of energy reflected from the surface when illuminated by direct light from the Sun as a proportion of that which would be reflected from a perfectly diffuse, 100% reflecting, surface at the same location. A single measurement of bidirectional reflectance is actually the integration over the solid angle subtended by the sensor of a more fundamental property: the 'bidirectional reflectance distribution function' (BRDF) (Nicodemus et al., 1977).

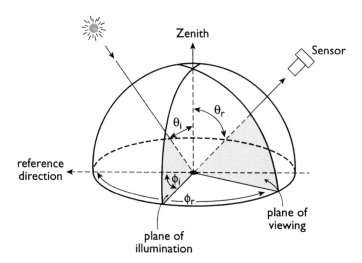

Figure 3 The geometric variables associated with remote sensing measurements (adapted from Kriebel, 1976).

3.2. The Atmosphere Model

The interaction between electromagnetic radiation and the atmosphere is also complex. Several different strategies for reducing or 'correcting' for the influence of the atmosphere upon remotely sensed data have been developed, however (Teillet *et al*., 1990). Different methods are appropriate depending upon the region of the electromagnetic spectrum being sensed, the altitude and orbit of platform being used, the availability of ancillary information and the accuracy required. For some epidemiological applications, the main concern is monitoring or detection of land cover change, in which case absolute accuracy (e.g. unbiased and precise measurements of ground-leaving radiation in physical units of spectral radiance) may be less important than the repeatability of the measurements.

Atmospheric correction procedures range from methods based solely on image data (*e.g.* Chavez, 1996), to vicarious methods that require measurements from the ground surface at the time of sensor overpass (Slater, 1984), to methods that use the geometry of viewing to infer the effect of the atmosphere (e.g. Mackay *et al*., 1998), to physically-based radiative transfer models (e.g. Tanré *et al*., 1990).

3.3. The Sensor Model

Most electro-optical sensors are designed to have a wavelength-specific linear relationship between the physical variable being measured (e.g. radiance or temperature) and the digital number (DN) value. Satellite sensors are calibrated in the laboratory before launch but are prone to drift over time following the trauma of the launch and as the exposed optical surfaces degrade in the space environment. Measurements from spectrally-stable desert environments have been used to derive a correction factor for the NOAA AVHRR sensor based on the number of days since launch and other schemes proposed for the Landsat sensors include using the Moon as a stable reference target (Kieffer and Wildey, 1985). Depending upon how the data are to be used, it may not be necessary, or indeed desirable, to convert DN to values of radiance. The results from even simple analytical techniques, however, such as the use of multispectral ratios (e.g. NDVI) may *require* that values of relative DN are converted to absolute radiance and, in addition, correction is made for the effect of the atmosphere (Crippen, 1988).

3.4. The Image Model

The three-dimensional (*x*, *y*, *z*) spatial assemblage of scene components constitutes the scene model from which the remote sensor draws its measurements. The resulting image is a representation of the scene at a particular moment in time and the degree to which the complexity of the scene is captured in the image depends upon the size of the scene components in relation to the instantaneous field-of-view (IFOV) of the sensor and their spectral contrast with each other (e.g. between canopy and background). At one extreme are scenes composed of spectrally-distinct, temporally-variable objects larger than the IFOV and at the other are scenes composed of spectrally-indistinct, temporally invariant objects smaller than the IFOV. Strahler *et al.* (1986) described these extremes and showed how the image model chosen influences the subsequent analysis and the conclusions derived.

There are many image models that are potentially useful for epidemiological remote sensing, but before giving some examples of these we should note the trade-off which occurs in satellite remote sensing between the revisit interval of the satellite and the IFOV of the sensor. For example, a single NOAA satellite has a revisit interval of 12 h and so provides two AVHRR images per day of a site but these have a fairly coarse spatial resolution (nominally 1.1 km). By contrast a Landsat satellite has a revisit interval of 16 days but carries the TM which provides data with a fairly fine spatial resolution (nominally 30 m in all except thermal wavelengths). Two other features of satellite sensor data are relevant in an operational context: the size of area covered by a single image and the total length of the observation record.

Most epidemiological research that has used remote sensing has been based on a very restricted set of wavelengths; often just two bands, one in the red part of the spectrum and one in the near infrared (e.g. example two in Table 1). These are well-established regions for the study of vegetation, although the physical interpretation of the commonly-used NDVI is still problematic (e.g. Sellers, 1987; Verstraete and Pinty, 1996). For example, some researchers would prefer to include green wavelengths instead of red to avoid problems of radiation saturation when recording large amounts of biomass (e.g. Gitelson *et al.*, 1996) or middle infrared wavelengths instead of red to increase sensitivity to canopy water content (Boyd and Curran, 1998).

The power of spectroscopy as a tool to investigate and classify materials is well-established in the laboratory and this is now being extended to sensors on board aircraft and satellites (Curran *et al.*, 1997; Verstraete *et al.*, 1999). Airborne imaging spectrometers have been developed which can provide images of the Earth's surface in which each pixel contains a complete reflectance spectrum over a broad spectral region in hundreds of contiguous spectral bands (Curran, 1994). Spaceborne spectrometers have been demonstrated (Goetz *et al.*, 1982) and imaging spectrometers will soon be operational on satellites to be launched by the National Aeronautics and Space Administration (NASA) and the European Space Agency (ESA).

Eventually, it may be possible to use data from such systems in physically-based plant growth models, since imaging spectrometry offers the possibility of measuring directly the biochemical (e.g. lignin and cellulose) composition of plant canopies (Curran, 1989).

The advent of spaceborne imaging spectrometers raises a key question of how many bands are required in order to capture the land cover information of relevance to epidemiology? The strong correlation between spectral bands means that the intrinsic dimensionality (Curran *et al.*, 1998) of such data is much less than the number of bands might suggest. Of 224 bands acquired by the airborne visible/infrared imaging spectrometer (AVIRIS) from a vegetated landscape in California, Boardman (1994) found the data could sometimes be represented within as few as four independent dimensions. This result accords with those from other studies, including work by Price (1990) which approached the same problem using spectra measured in the field, and it seems likely that most spectral information from vegetation and soils can be captured by relatively few spectral features (Curran *et al.*, 1998). These features may be discrete spectral bands or they may be combinations of bands, such as the NDVI.

Although the spectral requirements of epidemiological remote sensing are quite modest, the spatial and temporal requirements are much more demanding, particularly if progress is to be made towards linking remotely sensed data to the rates rather than the outcomes of ecological processes. Traditional land cover surveys using remote sensing provide a static view of the Earth's surface. Some may include information on the seasonal variability of classes, however, by combining data collected at different times of the year, but they do not, in general, capture the dynamic processes which drive the biological system and therefore are important in epidemiology, such as plant phenology, temperature change and rainfall input (Roughgarden *et al.*, 1991). To study plant canopy and landscape dynamics over timescales that are relevant to epidemiology it is necessary to acquire remotely sensed data at frequent time intervals. In theory, this is possible with sensors carried on Earth resources satellites such as Landsat or SPOT (Rees, 1999). In practice, however, at least for the humid tropics, the high probability of cloud cover and the 16-day or longer revisit period conspire to make this unlikely (Foody and Curran, 1994). The pointing capability of the SPOT High Resolution Visible (HRV) sensor and the future availability of many more Earth resources satellites will increase data availability, though probably at the expense of increased problems of data inter-calibration.

For many years researchers have used low cost NOAA AVHRR data to study vegetation phenology at regional to global scales (Justice *et al.*, 1985). The few spectral bands available from this sensor are suited to monitoring vegetation and the twice daily revisit interval greatly increases the chance of obtaining cloud-free data at the time of the year when they are required. The finest spatial resolution image model possible using such data is 1.1 km, but this need not be the ultimate limit to the information content of such data. An image model developed at any spatial resolution always has the potential to be used to estimate the within-pixel proportions of objects smaller than a pixel in size, using techniques such as the spectral unmixing described by Adams *et al.* (1993). Pixel unmixing is used widely in remote sensing for mineral exploration and regional and global scale mapping of fractional vegetation cover (Adams *et al.*, 1995) but although it has much to offer (Crombie *et al.*, 1999), it has been little used in epidemiology (see discussion in Section 4).

This section has discussed the links between land cover and an image or images and in doing so reference has been made to models of the scene, atmosphere, sensor and image. To make predictions using remotely sensed data we need to reverse this sequence. The first step is to remove as many extraneous factors as possible from the remotely sensed measures of radiation by means of radiometric, atmospheric, geometric and perhaps even topographic correction (Richards, 1993; Cracknell, 1997; Schowengerdt, 1997; Mather, 1999a; Barrett and Curtis, 1999). This will enable the production of a spectral mosaic of the landscape. The next step is to turn this single date or multi-temporal spectral mosaic into map(s) of land cover properties.

4. FROM A MULTI-TEMPORAL SPECTRAL MOSAIC TO LAND COVER

The production of a map of land cover type is based on the existence of a specific spectral response for each land cover class. To enhance the ability to classify land cover accurately it is, therefore, important that, as far as possible, only the land cover and not some characteristic of the sensor or atmosphere, influence the remotely sensed response derived from the imagery (see Figure 1). That is, the DN or tone of a pixel in a particular waveband should be related directly to the magnitude of the radiation measured from the area represented by the pixel in that waveband. Furthermore, since the end product is a thematic map depicting land cover type (e.g. different vegetation types or classes of vegetation amount) it is often desirable to alter the image geometrically to a suitable map projection. Since the temporal dimension of remote sensing can be beneficial, if not essential, to some epidemiological studies (Hugh-Jones, 1989; Riley, 1989; Hay *et al.*, 1997), especially in relation to prediction and control activities (Linthicum *et al.*, 1999; Myers *et al.*, this volume), the need to correct the imagery such that one image can be compared with another should be seen as important and fundamental to the use of remotely sensed data. Therefore, as was noted at the end of Section 3, the basis for epidemiological research should not be an image of a landscape but a carefully corrected spectral mosaic of that landscape at one or more points in time. This spectral mosaic may then be used to derive maps of land cover. In epidemiological studies the desire can be to map either land cover type (e.g. forest, pasture, cereal, urban) or some quantitative series of land cover classes (e.g. high, medium and low biomass) (Section 3, Tables 1 and 2). This section outlines the identification of *land cover type* and quantification of *land cover amount* from remotely sensed imagery, reviewing conventional and widely used approaches, as well as illustrating some recent developments in the field. The emphasis in the sections that follow, however, is on the identification of land cover type in which land cover is treated as a categorical rather than a continuous variable.

4.1. Quantifying Land Cover (e.g. Vegetation Amount)

The amount of vegetation, expressed typically in terms of biomass or leaf area index (LAI), has a strong relation to variables of epidemiological significance (Table 1). Consequently, maps of vegetation amount may be used in epidemiological studies as a surrogate variable for key environmental conditions or variables associated with diseases (Figure 1). For example, many studies have utilized vegetation indices, such as the NDVI, that are both physically related and correlated strongly with vegetation amount (Cross *et al.*, 1996; Hay *et al.*, 1997; Linthicum *et al.*, 1999; Thomson *et al.*, 1999). For some quantitative analyses and comparison between studies or over time it is essential that vegetation indices be calculated from appropriately pre-processed imagery (Price, 1987; Guyot and Gu, 1994; Mather, 1999a). The vegetation index image derived from the remotely sensed data may be interpreted as a map depicting vegetation amount as a continuous variable. Alternatively, a set of vegetation index thresholds may be defined to map discrete classes of vegetation amount (Curran, 1983). Although simple, such approaches can yield information of epidemiological interest. For example, vegetation amount at a local scale is often related to climate, whereas at the regional scale it is often related to broad classes of land cover (Table 1). The accuracy of vegetation amount maps may be evaluated if ground data on vegetation amount are available (Curran and Williamson, 1985). The mapping of vegetation amount, or more usually a surrogate, is well developed in regional scale epidemiology (Hay *et al.*, 1997). To understand the processes that underpin the pattern of diseases and their vectors, however, it may be necessary to map the land cover type and this necessitates classification. The remainder of this paper will therefore concentrate on the identification and mapping of land cover type.

4.2. Conventional Classification Approaches for Mapping Land Cover Type

Classification techniques are used typically in the mapping of land cover type from remotely sensed imagery. This applies to mapping based on both visual and digital analysis of the imagery. Visual analysis makes use of the ability of the human eye–brain system to interpret imagery rapidly and accurately. The identification of land cover type is based generally on the use of tonal, textural and contextual information often with the aid of a classification key that gives illustrative examples of the appearance of the land cover classes (Barrett and Curtis, 1999; Kelly *et al.*, 1999; Lillesand and Kiefer, 1999). This approach has been used frequently in epidemiological studies. Visual interpretation of remotely sensed imagery, including aerial photographs and satellite sensor images, has, for example, been used in the surveillance and control of arthropod vectors (Riley, 1989; Washino and Wood, 1994). Although visual interpretation is still used widely and can be accurate, it is not without problems. Key concerns are subjectivity in the interpretation; difficulty of handling the large multispectral data sets acquired by many remote sensors and the problem of maintaining a standardized classification over time and space (Philipson, 1997). Computer-based mapping techniques can make fuller use of the data set and be undertaken in a more objective manner. In addition, digital analysis eases integration of the imagery and derived map data with other digital spatial data sets within a geographical information system (GIS) aiding the development of relations between the environment and variables associated with diseases (Rejmankova *et al.*, 1995). Digital image classification is one of the most commonly performed analyses of remotely sensed imagery (Jensen, 1996). As with visual interpretation it aims to convert the image into a thematic map and to this end two broad approaches are used; unsupervised and supervised.

4.2.1. Unsupervised Classification

Unsupervised classifiers are effectively clustering algorithms that seek spectral clusters in the imagery. The classification algorithm groups together pixels with similar spectral properties and gives them a class label. Since class-specific spectral responses are commonly observed and are fundamental to the use of classification as a tool for mapping, an unsupervised classifier may be expected to identify land cover classes. This approach to classification has been used widely in the identification of land covers and habitats associated with intermediate hosts and disease vectors (Hugh-Jones, 1991a; Beck *et al.*, 1994; Pope *et al.*, 1994; Rejmankova *et al.*, 1995; Thomson *et al.*, 1999). A major problem with unsupervised classification, however, is that the analyst must relate the clusters defined by the classification to the land cover type after the classification and there is no guarantee that the clusters will correspond to land cover classes of epidemiological significance. Since the classes of interest are often well-known (e.g. those that represent habitats associated with particular hosts or disease vectors) an unsupervised classification is often inappropriate and a supervised classification should be used.

4.2.2. *Supervised Classification*

A supervised classification uses examples of the land cover classes, derived typically from within the imagery, to direct the classification algorithm (Campbell, 1996; Lillesand and Kiefer, 1999; Mather, 1999a). This is useful as the land cover classes and/or habitats of disease vectors are often known (Hugh-Jones, 1989; Riley, 1989; Hay *et al.*, 1997) and consequently this approach has been used in epidemiological studies (e.g. Hayes *et al.*, 1985). The classification comprises three key stages. First comes the training stage, in which sites of known land cover class are identified in the image and characterized spectrally. The end product of this stage is a set of training statistics that describe the spectral properties of the classes to be mapped. The second stage of the classification involves the use of training statistics together with a classification algorithm to allocate each pixel in the image to a land cover class. This typically involves a comparison of the pixel's spectral properties with those of the classes derived in the training stage and allocation of the pixel to the class with which it has greatest similarity. In this way, the remotely sensed image is converted into a thematic map depicting the spatial distribution of the land cover classes of interest. The final stage of the classification is the testing stage in which the accuracy of the classification is assessed. This aims to derive a quantitative measure of the accuracy with which land cover has been mapped. Classification accuracy is evaluated generally from a cross-tabulation of the actual and predicted class membership of a set of pixels that were not used in training the classification (Figure 4). The end product of the supervised classification is, therefore, a map of known accuracy that depicts the spatial distribution of land cover classes of interest.

Although a popular means of deriving maps of land cover type, many factors affect the accuracy of a supervised classification and hence its utility for epidemiological studies. A series of such factors may be associated with each stage of the classification (Campbell, 1996; Arora and Foody, 1997; Foody and Arora, 1997). Much attention has focused on issues concerned with the classification algorithm used. To date the majority of digital image classifications have used a conventional statistical classification algorithm. Probabilistic techniques such as the maximum likelihood classifier and discriminant analysis have been particularly popular. These approaches have firm statistical foundations and allocate each case (e.g. pixel) to the class with which it has the highest probability of membership (Peddle, 1993; Mather 1999a). Although this is an intuitively appealing approach and can be accurate, the correct application of such classifications requires the satisfaction of several assumptions that are not always tenable (Section 4.3) and it is sometimes difficult to integrate ancillary data into the analysis. Moreover, it has often proved difficult to produce reliable maps of land cover type with an accuracy that is acceptable operationally (Townshend, 1992). Consequently, considerable attention has been directed at the development and evaluation of alternative classification approaches. This has included the use of a range of non-parametric classifiers, with approaches based on evidential reasoning or neural networks proving popular (Benediktsson *et al.*, 1990; Srinivasan and Richards, 1990; Fischer *et al.*, 1997). Comparative studies using a suite of classification methods have shown repeatedly that for the majority of imagery and landscapes, neural networks can provide the most accurate classification of land cover (Benediktsson et *al.*, 1990; Peddle *et al.*, 1994; Paola and Schowengerdt, 1995). From the range of network types, feedforward networks such as the multi-layer perceptron are now the most commonly used methods in mapping land cover type from remotely sensed data (Figure 5). This network uses a learning algorithm (e.g. backpropagation) and the training set to adjust its internal properties until it can identify accurately the land cover classes of interest from the remotely sensed imagery (and any available ancillary data) presented to it.

4.3. Problems with Conventional Classification Approaches for Mapping Land Cover Type

Despite the considerable advances made in the development of classification algorithms, including neural networks, the accuracy of land cover type maps derived from remotely sensed imagery is still often insufficient for operational application (Wilkinson, 1996). Further increases in accuracy may be made by refinements to the training and testing stages of the classification (e.g. training set refinement to remove or down-weight ambiguous training samples, acquisition of larger training and testing sets). A major limit to the accuracy of digital image classifications has, however, been the tendency to use only spectral information in the classification

Predicted Class

Actual Class		Pasture	< 2	2 - 3	3 - 6	6 - 14	> 14	Σ	Producer's accuracy (%)
	Pasture	192	0	0	0	0	0	192	100.00
	< 2 years	0	75	49	13	37	0	174	43.10
	2 - 3 years	1	0	31	0	0	0	32	96.88
	3 - 6 years	0	1	0	55	74	0	130	42.31
	6 - 14 years	0	0	0	12	180	0	192	93.75
	> 14 years	0	0	0	0	0	192	192	100.00
	Σ	193	76	80	80	291	192	912	
	User's accuracy (%)	99.48	98.68	38.75	68.75	61.86	100.00		

Figure 4 Classification accuracy assessment with a confusion or error matrix. The matrix is a cross-tabulation of the class label derived from the classification and the corresponding actual class label for a set of testing pixels. The ground data providing the actual class labels may, like any data set, contain uncertainties and error. Indeed many factors influence the quality of the ground data used in remote sensing investigations and the data are unlikely to be error-free (Curran and Williamson, 1985; Steven, 1987; Foody, 1991; Bauer *et al.*, 1994). The accuracy assessment is, therefore, in reality a measure of the degree of agreement or correspondence between the labels derived from the classification and those depicted in the ground data. The quality of the ground data set is, therefore, critical and must be considered when deriving a classification accuracy statement. Once formed, however, a range of measures to indicate the accuracy of the classification may be derived from the confusion matrix. The matrix shown is taken from a study that classified tropical vegetation, including forests, into a variety of (non-overlapping) age classes and is reported in detail in Foody *et al.* (1996). The main diagonal of the matrix represents the instances where the class labels derived from the classification corresponded to those in the ground data. The sum of these correctly allocated cases may be used, relative to the grand total of cases in the testing set, to derive the percentage of cases correctly allocated. In this example the overall percentage of correctly allocated pixels was 79.49%. This gives a simple measure of the overall accuracy of the classification. If the main focus of attention is on the accuracy with which a particular class has been classified, rather than the overall accuracy, this may be estimated on the basis of the ratio of correctly allocated cases to the relevant column or row total in the matrix depending on the analyst's perspective (Story and Congalton, 1986). These accuracies may vary markedly for the same class. For example, the accuracy with which the forests aged 2–3 years class was classified from the producer's and

user's perspectives was 96.88% and 38.75% respectively. Although simple, such measures are not without problems. The sample design used to derive the testing set, for example, can influence significantly the accuracy statement (Campbell, 1996; Stehman and Czaplewski, 1998). A major concern, however, is that the simple measures of accuracy focus only on the main diagonal of the matrix (highlighted in grey) and make no compensation for chance agreement. To reduce these problems, the kappa coefficient of agreement (Cohen, 1960) has been widely used in remote sensing (Rosenfield and Fitzpatrick-Lins, 1986; Congalton, 1991). For the matrix illustrated, the kappa coefficient is 0.7476. This measure may also be derived for individual classes of interest. The calculation of the kappa coefficient of agreement may also be weighted to accommodate variations in the magnitude of error (Cohen, 1968), which is valuable as the confusion between some classes may be more damaging than others in remote sensing studies. With a weight associated with all possible allocations that indicates the relative severity of the misallocations, a weighted kappa coefficient may be derived. In the example (Foody *et al.*, 1996), this gave a weighted kappa coefficient of 0.8569. The kappa coefficient is also not always ideal (Foody, 1992). In particular it is inappropriate for the evaluation of fuzzy classifications where a pixel may have membership of more than one class. As classification accuracy assessment is still a field of active research, care should be taken to provide a thorough description of the data sets and methods used in evaluating a classification to help others interpret the quality of the classification and its appropriateness for the application in-hand.

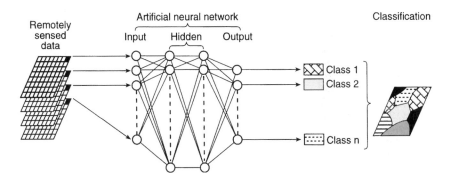

Figure 5 A typical neural network used for image classification. The neural network acts to convert the remotely-sensed imagery into a thematic classification. The network itself consists of a large number of simple processing units arranged in a layered architecture with each unit in a layer linked to every unit in adjacent layers by a weighted connection. There is an input unit associated with each discriminating variable (e.g. spectral waveband) used, one or more hidden layers of processing units and an output layer with one unit associated with each class. The training data and a learning algorithm (e.g. backpropagation) are used to help the network adjust the value of the weighted connections between units until the network can correctly identify class membership from the input data presented to it. Discussion of neural network processing may be found in Schalkoff (1992) and Bishop (1995) and examples of their application in mapping land cover from remote sensing in Benediktsson *et al.* (1990) and Foody (1996).

(Curran *et al.*, 1998). The classifications have, therefore, tended to use only the amount of radiation recorded for each pixel in each spectral waveband to discriminate between land cover types. Thus, while human interpretation uses tone, texture and context, most digital image classifications have used only tonal information which may be relatively uninformative. Image texture, which describes simply the local variability of image tone, can be quantified with relative ease from remotely sensed imagery (Mather, 1999a). Popular measures of image texture range from simple measures of tonal variability within a local window, particularly those based on grey level co-occurrence matrices (Haralick and Shanmugam, 1974; Holmes *et al.*, 1984) through to geostatistical descriptors of the local variability of image tone (Miranda and Carr, 1994; Berberoglu *et al.*, 2000; Lloyd *et al.*, 2001). Context describes a range of features such as the size, shape and location of objects to be classified. Although it is difficult to quantify context from remotely sensed imagery it may be used to increase significantly the accuracy with which land cover type is mapped (Gurney and Townshend, 1983; Harris, 1985; Groom *et al.*, 1996). In addition to image-based information there may be various ancillary data that can enhance class separability. For example, ancillary data on altitude or soil type may help discriminate vegetation communities that have a similar appearance in the imagery but are known to be located in different environments. The integration of ancillary information into the analysis has often been found to increase the accuracy of a variety of classifications (Strahler, 1980; Hutchinson, 1982; Peddle *et al.*, 1994). Despite these refinements to the techniques available no classification is ideal, this is because there remain several fundamental problems with classification as a tool for land cover mapping (Foody, 1999; Mather, 1999b).

Classification, by whatever method, makes several basic assumptions. Two critical assumptions, of relevance here, are that each case to be classified is pure (i.e. represents an area on the ground covered by a single class) and that the classes to be mapped are discrete and exclusive mutually. Both of these basic assumptions are often unsatisfied when mapping land cover type from remotely sensed imagery. Often, for example, pixels of mixed land cover class composition may be abundant in an image. Thus, for instance, vegetation classes may be continuous and inter-grade gradually with many areas of mixed class composition, particularly near imprecise or 'fuzzy' class boundaries. Alternatively, as a pixel is an arbitrary spatial unit, the size, shape and location of which is dependent more on the properties of the sensor than the land surface, it may represent an area on the ground that comprises more than one discrete land cover class. This often occurs when the area represented by the pixel straddles the boundaries of two or more classes and is common in landscapes composed of small spatial units and/or when using coarse spatial resolution imagery (Crapper, 1984; Campbell, 1996). Irrespective of their origin, mixed pixels are a major problem because a conventional supervised classification algorithm will force the allocation of a mixed pixel to one class, which need not even be one of the component classes (Campbell, 1996). Since the conventional classification output is 'hard', comprising only the code of the allocated class, such techniques cannot be used appropriately to represent the land cover type of the area represented by a mixed pixel. A large proportion of mixed pixels in an image can be a major problem in the production of accurate maps of land cover type from remotely sensed data. Errors in the representation of land cover type may also propagate into analyses based on the map. This includes the evaluation of change in land cover type through post-classification comparison methods and analyses using co-registered data sets within a GIS. Problems associated with mixed pixels are usually apparent when using coarse spatial resolution data sets. These data sets, however, are often used for mapping land cover type at regional to global scales because (i) remote sensing is the only feasible source of data on land cover type at such scales, or (ii) there is a need to monitor change in land cover type at a high temporal frequency. Some epidemiological applications (Table 2) have focused on relatively small areas and used fine spatial resolution imagery (e.g. from the SPOT HRV and Landsat TM) and so problems associated with the presence of mixed pixels would have been reduced but not removed. When mixed pixels are common, therefore, specific methods are required to represent land cover accurately. Two commonly used methods are spectral unmixing and soft or fuzzy classification.

4.4. Spectral Mixture Modelling

Spectral unmixing is likely to be of direct relevance to the use of remotely sensed data in epidemiology. This is because contemporary sensors onboard the NASA's Terra or the ESA's Envisat have a ground resolution that represents areas in the order of tens of hectares and yet epidemiologists may wish to use such data to derive the land cover of fields only a few hectares in size. The principle of spectral unmixing may be illustrated easily by reference to the early technique of Adams et al. (1993). This technique makes certain simplifying assumptions about the scene. First, that the scene is composed of a geometric arrangement of identifiable components, such as green leaves, bare soil and shadow; second, that energy only interacts with one of these components on its passage from the Sun to the sensor; and third, that the reflectance properties of the different components are known, either from data acquired in the field or laboratory, or from measurements made from the scene itself. We can, for example, use these data on the spectral properties of the scene components to determine the optimum 'feature space' (Curran et

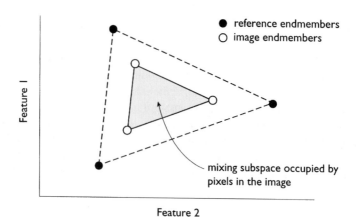

Figure 6 A hypothetical mixing space between three spectral endmembers in a two-dimensional feature space.

al., 1998) to unmix the data, which is the one in which the components are most separated from each other, as is shown diagrammatically in Figure 6. In practice, the optimum feature space would most likely be composed of combinations or mathematical transformations of radiation in wavebands measured by the sensor. Within the optimum feature space the data from the scene mix together within a subspace, the vertices of which are known as 'endmembers'. The relative proportion of endmembers within a pixel may then be calculated using its location within the mixing subspace (Figure 6). Endmembers may be of two types, depending upon whether they are defined from pure components of the image (image endmembers, such as a particular set of radiation values) or pure components of the scene (reference endmembers, such as trees, water, grass) (Milton, 1999). This process of spectral unmixing may be described mathematically as:

$$r_i = \sum_{j=1}^{n} a_{ij} f_j + e_i \tag{1}$$

in which r_i is the reflectance of the pixel in band i of m. The total number of components within the pixel is n, f_j is the proportion of endmember j in the pixel, and a_{ij} is the reflectance of endmember j in spectral band i. The term e is an error term which expresses the difference between the observed reflectance and the reflectance computed from the model (Mather, 1999a).

To solve the pixel unmixing equation given above, the total number of components within the pixel must be no more than $n + 1$, where n is the number of spectral features (i.e. dimensionality of the mixing space). Whilst this limitation can be a problem in some applications (e.g. mineral exploration), for epidemiological research this would still allow the unmixing of green vegetation, soil background and water (or shadow) from a feature space which contains only two dimensions. Because of the strongly contrasting spectral properties of vegetation, water, and most types of soil it is reasonable to expect any satellite sensor with bands in R and NIR wavelengths to be able to generate a suitable 2-D feature space.

Roberts *et al.* (1998) have modified the basic unmixing model described above by allowing the set of endmembers against which a given mixed pixel is labelled to vary across the image. This has the effect of permitting many more than the theoretical limit of $n + 1$ scene components to be estimated, making it feasible to map the proportions of different types of vegetation, or live versus dead vegetation even though the data are distributed in relatively few dimensions. Spectral unmixing, however, is not the only way to deal with the problems of mixed pixels.

4.5. Soft or Fuzzy Classification

Since the conventional (hard) classification output cannot represent the continuous graduation of some land cover types or mixing of classes within the area represented by a pixel, alternative classification-based approaches have been sought (Wang, 1990). Much attention has focused on the potential of soft or fuzzy classifications that allow a pixel to have partial and multiple class membership (Wang, 1990; Foody, 1996; Bastin, 1997). There are a variety of methods for deriving a fuzzy classification. The most obvious is a range of fuzzy classifiers that can be used to derive estimates of sub-pixel scale class membership (Cannon *et al.*, 1986). Alternatively, a fuzzy classification may be achieved by 'softening' the output of a 'hard' classification. For instance, measures of the strength of class membership, rather than just the code of the most likely class of membership, may be output. Thus, for example, with a maximum likelihood classification, a probability vector containing the probability of membership of a pixel to each defined class could be the output. In this probability distribution, the partitioning of the class membership probabilities between the classes would, ideally, reflect the land cover composition of a mixed pixel (Foody *et al.*, 1992). This type of output makes fuller use of the information on class membership generated in the classification and may be considered to be fuzzy, as an imprecise allocation may be made and a pixel can display membership to all classes. The remotely sensed data must still, however, satisfy the assumptions and requirements of the classification technique used, which is often unlikely with the widely used probability-based classifiers. The lack of distribution assumptions and ability to integrate ancillary data acquired at a low level of measurement precision are major attractions of alternative classifiers such as neural networks. Although used generally to produce a hard classification, the output of an artificial neural network may be softened to provide measures of the strength of class membership which may sometimes better represent land cover than a traditional 'hard' classification (Foody, 1996).

Although fuzzy classifications have been used to provide an appropriate representation of land cover type that may be considered to be fuzzy, they do not fully resolve the mixed pixel problem (Foody, 1999). A fuzzy classification provides only a means of appropriately representing land cover type that may be considered fuzzy at the scale of the pixel. Thus, while the class allocation made by a fuzzy classification may accommodate appropriately mixed pixels, it is only one of the three stages of the supervised classification process, with the classification still requiring training and testing. Relatively little attention has focused on the accommodation of mixed pixels in the training and testing stages of a supervised classification. In both of these stages of the classification, the ground data on class membership are generally related to the remotely sensed data at the scale of the pixel and so may be fuzzy. Since a large proportion of image pixels may be mixed it is important that this be recognized and accommodated in the classification. In testing the classification, for example, pure pixels are used generally, as the conventional measures of accuracy assessment were designed for application to 'hard' classifications. As the majority of pixels may be mixed, failure to include them in the accuracy assessment may result in an inappropriate and inaccurate estimation of classification accuracy. A variety of methods may be used to accommodate for fuzziness in the classification output (Gopal and Woodcock, 1994; Maselli *et al.*, 1994) and, under certain circumstances, in the ground data as well (Foody, 1996). Indeed methods exist to accommodate mixed pixels in each stage of the classification (Foody and Arora, 1996; Foody, 1997). The degree to which fuzziness is accommodated in each stage may also vary (e.g. the classification output could be the degree of membership to each class or a crisper product comprising the most likely class of membership and a secondary class label). Consequently, a classification may be undertaken along a continuum of classification fuzziness (Foody, 1999) ranging from the conventional 'hard' classification, in which no action is made to accommodate mixed pixels, to fully-fuzzy classifications, in which mixed pixels are accommodated in each stage. The design of a classification should, therefore, be based on the nature of the available data sets as well as on the desired end product. The derived map of land cover type may then be integrated with data, typically point based, on disease and disease vectors. Relationships between land cover and disease variables may then be established from these data. These relationships may then be extrapolated spatially (and, if a time series of imagery is available, temporally) using maps of land cover type derived from remotely sensed imagery.

These two sets of links from land cover to image and land cover to disease (see Figure 1) are fundamental to our understanding of how remote sensing can be used in epidemiology. In many cases, however, these links become secondary to the pragmatic task of using remotely sensed imagery to estimate the density of disease vectors and the degree of disease risk. The remainder of this paper will therefore build on this background (Sections 3 and 4) and focus on the task of prediction. In doing so, several recommendations will be made for best practice in the remote sensing of vector density and disease risk.

5. PREDICTING VECTOR DENSITY AND DISEASE RISK FROM LAND COVER DATA

As outlined in the introduction, and summarized in Figure 1, a complex set of inter-relationships exist between remotely sensed imagery of the land surface and disease risk distributed spatially on that land surface. We have seen in previous sections that as the amount of radiation reflected (at a given wavelength) from a point on the Earth's surface depends on the nature of the material (land cover type) at that point, it is possible to predict land cover spatially from remotely sensed imagery. This prediction usually depends on knowledge (data) of the *in situ* land cover via a model of the relation between the imagery and the land cover. This spatial prediction of land cover is based on explicit physical processes (see Figure 1). Similarly, the links between other environmental variables that have been employed in the remote sensing of disease risk (e.g. cold cloud duration, CCD, Hay *et al.*, 1996) and remotely sensed imagery are based on physical processes.

The link between land cover and vector density (or disease risk) is based on processes that are less easily described in terms of the physics of the processes involved, but which may be described well by stochastic models. Clearly, there are many different types of disease and vector. In recent years, much research has been undertaken on the use of remotely sensed data to predict malaria and trypanosomiasis risk (see Table 2) and for ease of presentation the following section will focus mainly on the malaria-mosquito disease-vector combination (Horsfall, 1955). As suggested by the dashed predictive lines in Figure 1, it is useful to describe, first, the link between land cover and vector density and, second, the link between vector density and disease risk.

Unfortunately, the relations between land cover and vector density are not well understood. Most studies describe the relations between habitat (which may or may not be synonymous with land cover) and vector (Beck *et al.*, 1994). It is generally accepted, however, that for common vectors (such as various species of mosquito) proximity to water is important, particularly in the breeding phase. Habitats (represented as land covers or combinations of land covers) that are favourable to mosquitoes tend to be aquatic or at least close to water bodies. However, the physical processes underlying the link between land cover and vector density will, in practice, depend on (i) internal processes relating to the population, (ii) interrelations between the vector and vertebrate populations and (iii) many environmental influences, particularly microclimatic variables that may be difficult to measure.

The relation between vector density and disease risk is yet more complex. For example, mosquitoes depend on vertebrates for blood which may, or may not, carry a pathogen or parasite (say, malaria). The vectors take up the pathogen/parasite and may, or may not, pass it on to the next vertebrate on which it feeds. In these circumstances, we can define three populations, that of the mosquitoes, that of the vertebrates and that of the disease. The inter-dependencies between the three populations are complex. These can again be split into (i) those internal to each population (e.g. density of vertebrates per unit area), (ii) those describing the relations between populations (e.g. proximity of vertebrates to vectors) and (iii) those external to the populations (e.g. proximity to water).

The effect of meteorological variables on vector populations and disease risk are less complex only in the sense that meteorological variables tend to vary spatially over larger distances than does land cover. Thus, for local studies, spatial variation in properties such as CCD may be negligible. Rather, meteorological data are generally used to predict future vector densities and disease risk. For example, in a study by Hay *et al.* (1998a), monthly mean percentage of annual malaria admissions in The Gambia was found to be highly correlated with CCD lagged by two months. The implication is that if CCD is monitored it can be used to predict malaria risk two months in advance (Myers *et al.*, this volume). Thus, whereas land cover data provide a wealth of *spatial* information, meteorological variables are more likely to provide *temporal* information on vector density or disease risk.

5.1. Predicting Vector Density or Disease Risk?

The researcher will need to be clear as to what primary variables are to be estimated and how these variables are to be organized.

5.1.1. *Choice of Variables and Predictive Route*

Several researchers have sought to predict vector density or some other vector-related variable based on remotely sensed imagery. Others have sought to predict disease risk or some other disease-related variable directly from imagery. Clearly, if disease risk is predicted directly, the simple correlation obtained between the secondary (explanatory) variable(s) and the risk will depend on the vectors for which no information is available (see Figure 1).

Some studies have involved prediction of vector density or disease risk based on the direct use of remotely sensed reflectance in several wavebands. Others have involved the processing of remotely sensed imagery into vegetation indices (such as the NDVI) or the classification of land cover type based on the imagery prior to prediction.

In this paper, we have so far implied and will now argue, that the most natural model is one in which, first, some land cover variable is predicted from remotely sensed imagery and second, this information on land cover is used to predict vector density or disease risk.

5.1.2. Organizing the Primary Variable

Data on the primary phenomenon of interest, whether it be the vector or the disease, will generally be required to build a statistical model of its relation to the secondary variable (say, land cover type). These data may be acquired in a variety of formats. If the interest is in the vector, then the variable should be defined as a density, that is, number of vectors per unit area. The support (the size, geometry and orientation of the space on which each observation is defined) (Curran and Atkinson, 1999) should be constant across the region of interest. Thus, depending on the instrumentation and technique used to acquire data, the measured variable may need to be transformed prior to modelling. Where the secondary (explanatory) variable(s) are provided by remotely sensed imagery it may be advantageous to transform the primary vector variable to a grid with the same spatial resolution as the imagery. This step will facilitate the later co-location of the data within a GIS (Longley et al., 1999).

If the interest is disease risk, the primary data will often be in the form of number of patients admitted to a local clinic or hospital and diagnosed or treated for the disease. Ideally, data on location should be obtained for each individual so that the local spatial distribution of the disease can be mapped for each local medical centre. Then, the number of patients per unit area should be divided by the background population (ideally, the number of susceptibles) per unit area to produce an attack rate (or disease risk). As for the vector density, the disease risk can be produced on cells that match the pixels of remotely sensed imagery on which the predictive model will be constructed.

5.2. Sampling Issues

5.2.1. *Choosing a Spatial Resolution*

A choice that the investigator will face early in any study involving remotely sensed imagery is that relating to spatial resolution. The many studies involving remote prediction of vectors or disease reviewed by Hay *et al*. (1997, and this volume) can be divided into those involving fine spatial resolution and coarse spatial resolution imagery. Fine spatial resolution is used to refer to imagery such as that provided by SPOT HRV (10 m for panchromatic imagery), Landsat TM (30 m, in non-thermal wavelengths) and Landsat MSS (80 m). Coarse spatial resolution imagery is used to refer to imagery such as provided by the NOAA AVHRR (1.1 km and 8 km). Clearly, there is a large difference (two to three orders of magnitude) between the two groups. The choice (in the first instance between the two groups) will depend on several factors. By far the most important criterion, however, is the spatial detail that needs to be resolved. Since the primary variable of interest can be defined as a density on any desired cell size (e.g. through interpolation) the choice of spatial resolution of imagery is a difficult one.

Important variables influencing the choice of spatial resolution include the spatial extent of the region of interest, the quality of data that will be generated, the nature of the strategy intended to counter the disease (e.g. environmental measures) and the ability to (and utility of) targeting medical care spatially. In addition, the number of vertebrates carrying the disease and the number of susceptibles will decrease as the spatial resolution increases such that estimates of risk may become unreliable for small cell sizes. That is, very small cell sizes may be precluded on statistical grounds. We would suggest, however, that the most important criterion is the frequency of the spatial variation in the vector density (or disease risk). The spatial resolution should be chosen in relation to the scale(s) of spatial variation implicit in the variable of interest (Curran and Atkinson, 1999). The suggestion is that such scale(s) of variation should be quantified using a function such as the covariance function or variogram to aid in the choice of spatial resolution (Woodcock and Strahler, 1987; Atkinson and Curran, 1997).

5.2.2. *Spatial Sampling of the Primary Variable*

Having chosen a spatial resolution (which can often mean a particular sensor), the next task will be to select a source of data on the primary variable of interest. Often the investigations will be constrained by sample size and typically have to use archive sources. Nevertheless, it is worth considering here the effect of the sampling strategy on the resulting data. All data are a function of (i) the underlying property of interest and (ii) the sampling strategy. Therefore, the actual observed spatial variation will depend on the sampling strategy. Generally, systematic sampling is preferred over random sampling because it is more efficient where data are spatially dependent (proximate data pairs tend to be more similar than more distant data pairs) and this is usually the case (Atkinson, 1997). Whatever the circumstances, the investigator should strive to provide an even spatial coverage, at least of at-risk areas.

A second consideration relates to the support of the primary variable. This is likely to be approximated as a point in space and is unlikely to be the same as that of the remotely sensed imagery (land cover data). In practice, a variety of problems arise. Take the example of Hay *et al.* (1998a, b) in which five local communities in Kenya and the Gambia were monitored for malaria infection. The data were referenced geographically as five *points* in space. Each datum, however, actually has a support that is much larger than a point. Further, the support is likely to have a bell-shaped centre weighting since more cases are likely to be reported by people living close to a medical centre than by people living further away. Ideally, the locations of individuals should be recorded allowing the estimation of cell-based averages as described above, although this is a very difficult undertaking.

If the data supports can be treated as points then the investigator has a choice. First, the point data could be used in modelling directly. The only consequence of this is that the model and the predictions will be less accurate than had the support been equal to the pixel. Second, the investigator could choose to interpolate to the support of the imagery (image pixels), for example, using block kriging (see Section 5.3.2) (Atkinson, 1997). Any reorganisation of the data based on interpolation algorithms, however, will introduce the serious problem known as 'smoothing'. All linear weighted estimators involve smoothing. Smoothing means that the extent of the support of the estimates is larger than the desired support. In practice, smoothing alters the cumulative distribution function (c.d.f.) of the estimated data set: specifically, the c.d.f. of the estimated data has a smaller variance than that of the original data. This has serious implications where the objective is to build a model of the relation between the interpolated primary variable (i.e. vector density, disease risk) and the remotely sensed land cover and apply that model to predict the primary variable at other places or times.

A third consideration relates to the number of observations required. Where the statistical model of interest is regression (see below), this number can be small (say 30 observations that cover the region such that the full range of possible values is included). Where the model is classification-based (see below), this number can be uncomfortably large (say 30 observations per class, perhaps with more than ten classes).

5.3. Models for Predicting Vector Density and Disease Risk

Many models are possible for predicting vector density and disease risk. Here we will comment on five such models.

5.3.1. *Temporal Prediction: Simple Correlation on Monthly Averages*

As noted earlier in Section 5, Hay *et al.* (1998a) conducted research in which CCD lagged by two months was related to monthly percentage total of annual malaria admissions ($r^2 = 0.86$). The CCD data were obtained from remotely sensed data of the Atlantic coast of The Gambia provided by NOAA AVHRR and Meteosat High Resolution Radiometer (HRR) imagery. As part of the same study, the images were processed to estimate the NDVI, and the relation to monthly percentage total of annual malaria admissions was found to be strong ($r^2 = 0.66$). Since the remotely sensed NDVI provided a spatial distribution on a monthly basis, it was possible to extrapolate the prediction spatially to map areas at risk.

The fundamental problem with such a statistical approach based on temporal data is that relationships observed are not necessarily based on a physical link as discussed by Hay *et al.* (1998a). Multiple-year data are preferred to characterize the relation between the date and magnitude of peak NDVI and the date and magnitude of peak disease risk.

5.3.2. *Spatial Prediction: Hard Land Cover Classes*

Where the remotely sensed imagery has been used to estimate hard land cover classes (e.g. using a supervised maximum likelihood classification, Section 4.2.2) the investigator needs to know how to use such information on land cover type to predict vector density or disease risk. A variety of approaches could be employed depending on several factors including how the primary variable is organized. For example, if the primary variable is treated as a binary indicator of presence or absence of the disease (e.g. below or above some critical threshold) then logistic regression could be used to predict from hard land cover classes (Section 4.2).

Treating the primary variable as continuous, the simplest method is to obtain data on risk for each land cover class and assign the means of the class-specific distributions to the classes. The distributions can be used to estimate the uncertainty associated with each mean and also, via a classical analysis of variance, to determine whether the differences between the means are significant (Webster and Oliver, 1990). Further, the distributions can be used to simulate different spatial realizations of vector density or disease risk.

An optimal, albeit more sophisticated, alternative would be to use the geostatistical technique known as kriging (Matheron, 1965, 1971). The term 'kriging' refers to a set of generalized least-squares regression algorithms. All kriging algorithms are variants of the least squares regression estimator (Goovaerts, 1997):

$$\hat{Z}(\mathbf{V}) - m(\mathbf{V}) = \sum_{\alpha=1}^{n(\mathbf{V})} \lambda_\alpha(\mathbf{V})[Z(\mathbf{x}_\alpha) - m(\mathbf{x}_\alpha)] \tag{2}$$

where λ_α are the weights applied to data $z(\mathbf{x}_\alpha)$ interpreted as realizations of the regionalized variable (RV) $Z(\mathbf{x}_\alpha)$ and $m(\mathbf{V})$ and $m(\mathbf{x}_\alpha)$ are the means of the RVs $Z(\mathbf{V})$ and $Z(\mathbf{x}_\alpha)$. In practice, prediction is achieved using *only* the $n(\mathbf{V})$ point or quasi-point data $z(\mathbf{x}_\alpha)$ at locations $\mathbf{x}_\alpha$ within *a local neighbourhood* $W(\mathbf{V})$.

The objective of all kriging estimators is to minimize the estimation variance under the constraint of unbiasedness. That is,

$$\sigma_E^2(\mathbf{V}) = \text{var}\{\hat{Z}(\mathbf{V}) - Z(\mathbf{V})\} \tag{3}$$

is minimized under the constraint that

$$E\{\hat{Z}(\mathbf{V}) - Z(\mathbf{V})\} = 0 \tag{4}$$

The actual type of kriging estimator adopted varies depending on the model adopted for the random function (RF) and, in particular, the model adopted for the mean. In general, the RF can be decomposed into two components as follows:

$$Z(\mathbf{V}) = R(\mathbf{V}) + m(\mathbf{V}) \tag{5}$$

where $R(\mathbf{V})$ is modelled as having zero mean or expectation and its variation is modelled using the variogram. The component $m(\mathbf{V})$ is the mean of the RF $Z(\mathbf{V})$ and this can be modelled in various ways depending on the type of kriging estimator adopted.

Where the secondary data are in the form of what is termed a categorical variable (that is, a 'hard' land cover), they can be used directly as a model for the mean component. Kriging can then be applied to the residuals from that mean. This procedure is referred to as kriging with an external drift (Deutsch and Journel, 1992). Most often, the external drift is a two-dimensional polynomial (that is, trend surface), but it can equally be a categorical variable (Goovaerts, 1997; Pebesma and Wesseling, 1998). The main requirement for kriging in this form is sufficient data on the primary variable distributed spatially within the region of interest, but this requirement can be difficult to meet in epidemiological studies (Section 2).

By classifying first the land cover and, secondly, the vector density (or disease risk), the model follows more closely the physical processes occurring in reality (see Figure 1). Thus, such a two-stage approach is recommended over a direct estimation of disease risk from remotely sensed imagery.

5.3.3. *Spatial Prediction: Soft Land Cover Classes*

Traditional hard supervised classification (Section 4.2.2), while still popular in remote sensing, is being replaced, albeit gradually, by the more informative soft or fuzzy classification. As described in Section 4.4, soft or fuzzy classification can be used to estimate the proportion of a given pixel belonging to each of the available classes. In these circumstances, the secondary data (of land cover proportion) are a set of variables constrained in the interval 0–1. A range of approaches could be adopted to model the relation between these data and the vector density or disease risk (Webster and Oliver, 1990). None stand out clearly as the most appropriate, however.

One alternative would be to model the relation between vector density (or disease risk) and a hard class, and then estimate the mean per cell based on the fuzzy memberships (that is, estimate the mean as an average of each class, weighted by the proportion of the cell covered by each class). Kriging with an external drift could then be applied based on this spatially-varying estimate of the mean.

Another alternative would be to treat the secondary data as a set of continuous variables. The disadvantage of doing this is that the distribution of the data may need to be transformed prior to further analysis (e.g. using a logistic or arcsin transform) and that the fuzzy proportions will clearly be correlated between classes. If these problems are not serious, then cokriging (Myers, 1982; Atkinson *et al.*, 1992, 1994) could be used to estimate the primary variable and successful application would once again (Section 5.3.2) depend on the availability of primary (e.g. vector or disease) data.

5.3.4. *Spatial Prediction: Vegetation Amount*

In some cases it may be preferable to replace a land cover classification with a map of some quantitative land cover variable such as vegetation amount. In remote sensing, the estimation of vegetation-related variables has traditionally involved the use of multiple wavebands to calculate a vegetation index. The use of a vegetation index (e.g. NDVI) as a secondary variable has the advantage that multiple waveband information is compressed into a single variable. Further, indices such as the NDVI standardize for the amount of incoming radiation and information is provided as a continuous variable. Vegetation indices such as the NDVI, are highly correlated with properties such as leaf area index (LAI), and vector density or disease risk may be correlated with LAI, conditional upon specific climatic events (e.g.

rains following drought). In these circumstances, an indirect link is established between the vegetation index and vector density or disease risk, such that the vegetation index may be used as a secondary variable for prediction. Since both the primary and secondary variables are continuous, the natural choice of statistical model for prediction is simple linear regression. Alternatively, where there are sufficient spatial data on the primary variable, kriging with an external drift or cokriging could be applied to make full use of the available spatial information.

Despite the apparent simplicity of this predictive route there are potential problems. First, the correlation will depend on climate, as mentioned above. Second, the correlation may vary geographically (e.g. mosquitoes may favour certain plants and proximity to water bodies). Third, and most importantly, there is no reason to expect a linear relation. Relations between the NDVI and vector density or disease risk, when they do occur, are likely to be non-linear, especially when one considers that the relation between the NDVI and LAI tends to be asymptotic. Such non-linear relations would need to be included in any model used for prediction.

5.3.5. Spatial Prediction: GIS Variables

Often, secondary information exists in a form where the primary variable is not a simple linear function of the secondary data expressed on a raster grid. For example, mosquito breeding depends on proximity to water bodies, proximity to vertebrates and so on. To obtain a simple linear relation, it is necessary to obtain a second-order variable 'proximity to water' as a function of the first-order variable 'water'. A GIS can be used to provide the new variable. For example, in the case of proximity to water, the new variable can be estimated by a GIS procedure known as 'buffering' (Longley et al., 1999) applied to water bodies (themselves estimated through remote sensing and land cover classification). Care should be taken when deriving variables of this kind to ensure that the relations between the second-order variable and the primary variable of interest are linear. Then, the techniques described above (e.g. geostatistics) can be applied, treating the second-order information together with, say, land cover type.

6. DISCUSSION AND CONCLUSION

The framework in Figure 1 has focused our attention on the links between remotely sensed imagery and disease. In doing so, it has highlighted the uncomfortable assumptions and gaps in understanding that have to be traversed if we are to use measures of radiation to predict vector density or, better still, disease risk. It is not unusual for scientific research to progress by *observing* and *using* relationships prior to *understanding* relationships (Curran, 1987). Such is the case in the use of remote sensing in epidemiology where, not surprisingly, early research concentrated on the development of correlations between images and disease and then the use of images to predict disease. It is clear, however, that progress will rest on the development of physical understanding (Sections 3 and 4) and the utilization of optimized procedures for prediction (Section 5). Such progress is vital because 'despite the growing enthusiasm, disease prevention by satellite has still to prove itself for actual disease control' (Baringa, 1993, p. 32)

Specifically, this review makes four observations that will, it is hoped, help make remote sensing a vital component of the epidemiologists' tool kit:

(1) Land cover is central to the links between remote sensing and epidemiology. The accurate identification and quantification of land cover could be a useful part of prediction methodologies in epidemiology.
(2) Remotely sensed data contain spectral, spatial and temporal information that can be used to characterize the Earth's surface. To exploit remotely sensed data to the full we must have a clear understanding of the scene, atmosphere, sensor and image and, perhaps more importantly, links between them.
(3) Many techniques are available for mapping land cover type from remotely sensed data. The most suitable techniques for epidemiological studies may be soft classifications in which the proportion of a class is attributed to each pixel. These will be derived from an approach to spectral unmixing or through the application of a soft or fuzzy image classifier.
(4) Predicting vector density and/or disease risk is hampered by a spatial and temporal mismatch between remotely sensed data and vector/disease data.

Optimal use of remote sensing will involve the optimization of both spatial sampling and the models used for prediction in the context of the environment, the available data and the epidemiological problem.

REFERENCES

Adams, J.B., Smith, M.O. and Gillespie, A.R., (1993). Imaging spectroscopy: interpretation based on spectral mixture analysis. In: *Remote Geochemical Analysis: Elemental and Mineralogical Composition* (C.M. Pieters and P.A.J. Englert, eds), pp. 145–166, Cambridge: Cambridge University Press.

Adams, J.B., Sabol, D.E., Kapos, V. *et al.* (1995). Classification of multispectral images based on fractions of endmembers: application to land-cover change in the Brazilian Amazon. *Remote Sensing of Environment* **52**, 137–154.

Ahearn, S.C. and De Rooy, C. (1996). Monitoring the effects of *Dracunculiasis* remediation on agricultural productivity using satellite data. *International Journal of Remote Sensing* **5**, 917–929.

Arora, M.K and Foody, G.M (1997). Log-linear modelling for the evaluation of the variables affecting the accuracy of probabilistic, fuzzy and neural network classifications. *International Journal of Remote Sensing* **18**, 785–798.

Atkinson, P.M. (1997). Scale and spatial dependence. In: *Scaling-up: From Cell to Landscape* (P.R. van Gardingen, G.M. Foody and P.J. Curran, eds), pp. 35–60. Cambridge: Cambridge University Press.

Atkinson, P.M. and Curran, P.J. (1997). Choosing an appropriate spatial resolution for remote sensing investigations. *Photogrammetric Engineering & Remote Sensing* **63**, 1345–1351.

Atkinson, P.M., Webster, R. and Curran, P.J. (1992). Cokriging with ground-based radiometry. *Remote Sensing of Environment* **41**, 45–60.

Atkinson, P.M., Webster, R. and Curran, P.J. (1994). Cokriging with airborne MSS imagery. *Remote Sensing of Environment* **50**, 335–345.

Bailey, C.L. and Linthicum, K.J. (1989). Satellite remote sensing: the newest technology for monitoring vector populations and predicting arbovirus outbreaks. *Proceedings, 5th Symposium, Arbovirus Research in Australia*, pp. 111–116. Brisbane, Qld: CSIRO Division of Tropical Animal Production.

Barinaga, M. (1993). Satellite data rocket disease control efforts into orbit. *Science* **261**, 31–32.

Barrett, E.C. and Curtis, L.F. (1999). *Introduction to Environmental Remote Sensing*, 4th ed. Cheltenham: Stanley Thornes.

Bastin, L. (1997). Comparison of fuzzy c-means classification, linear mixture modelling and MLC probabilities as tools for unmixing coarse pixels. *International Journal of Remote Sensing* **18**, 3629–3648.

Bauer, M.E., Burk, T.E., Ek, A.R. *et al.* (1994). Satellite inventory of Minnesota forest resources. *Photogrammetric Engineering & Remote Sensing* **60**, 287–298.

Beck, L.R., Rodriguez, M.H., Dister, S.W. *et al.* (1994). Remote sensing as a landscape epidemiologic tool to identify villages at high risk for malaria transmission. *American Journal of Tropical Medicine and Hygiene* **5**, 271–280.

Benediktsson, J.A., Swain, P.H. and Ersoy, O.K. (1990). Neural network approaches versus statistical methods in classification of multisource remote sensing data. *IEEE Transactions on Geoscience and Remote Sensing* **28**, 540–551.

Berberoglu, S., Lloyd, C.D., Atkinson, P.M. and Curran, P.J. (2000). The integration of spectral and textural information using neural networks for land cover mapping in the Mediterranean. *Computers and Geosciences* **26**, 385–396.

Bishop, C.M. (1995). *Neural Networks for Pattern Recognition*. Oxford: Clarendon Press.

Boardman, J.W. (1994). Automating linear mixture analysis of imaging spectrometry data. In: *Proceedings of the International Symposium on Spectral Sensing Research* (R.B. Gomez ed.), pp. 302–309. Washington DC: US Army Corps of Engineers.

Boyd, D.S. and Curran, P.J. (1998). Using remote sensing to reduce uncertainties in the global carbon budget: The potential of radiation acquired in midle infrared wavelengths. *Remote Sensing Reviews* **16**, 293–327.

Campbell, J.B. (1996). *Introduction to Remote Sensing*, 2nd edn. London: Taylor and Francis.

Cannon, R.L., Dave, J.V., Bezdek, J.C. and Trivedi, M.M. (1986). Segmentation of a thematic mapper image using the fuzzy *c*-means clustering algorithm. *IEEE Transactions on Geoscience and Remote Sensing* **24**, 400–408.

Chavez, P.S. (1996). Image-based atmospheric corrections—revisited and improved. *Photogrammetric Engineering & Remote Sensing* **62**, 1025–1036.

Cline, B.L. (1970). New eyes for epidemiologists: aerial photography and other remote sensing techniques. *American Journal of Epidemiology* **92**, 85–89.

Cohen, J. (1960). A coefficient of agreement for nominal scales. *Educational and Psychological Measurement* **20**, 37–46.

Cohen, J. (1968). Weighted kappa. *Psychological Bulletin* **70**, 213–220.

Colwell, J.E. (1974). Vegetation canopy reflectance. *Remote Sensing of Environment* **3**, 175–183.

Congalton, R.G. (1991). A review of assessing the accuracy of classifications of remotely sensed data. *Remote Sensing of Environment* **37**, 35–46.

Connor, S.J. (1999). Malaria in Africa: the view from space. *Biologist* **46**, 22–25.

Cracknell, A.P. (1997) *The Advanced Very High Resolution Radiometer (AVHRR)*. London: Taylor and Francis.

Crapper, P.F. (1984). An estimate of the number of boundary cells in a mapped landscape coded to grid cells. *Photogrammetric Engineering & Remote Sensing* **50**, 1497–1503.

Crippen, R.E. (1988). The dangers of underestimating the importance of data adjustments in band ratioing. *International Journal of Remote Sensing* **9**, 767–776.

Crombie, M.K., Gilles, R.R., Arvidson, R.E. *et al.* (1999). An application of remotely derived climatological fields for risk assessment of vector-borne disease: a spatial study of *Filariasis* prevalence in the Nile delta, Egypt. *Photogrammetric Engineering & Remote Sensing* **65**, 1401–1409.

Cross, E.R., Newcomb, W.W. and Tucker, C.J. (1996). Use of weather data and remote sensing to predict the geographic and seasonal distribution of *Phlebotomus papatasi* in southwest Asia. *American Journal of Tropical Medicine and Hygiene* **54**, 530–536.

Curran, P.J. (1983). Multispectral remote sensing for the estimation of green leaf area index. *Philosophical Transactions of the Royal Society, London, Series A* **309**, 257–270.

Curran, P.J. (1987). Remote sensing methodologies and geography. *International Journal of Remote Sensing* **8**, 1255–1275.

Curran, P.J. (1989). Remote sensing of foliar chemistry. *Remote Sensing of Environment* **30**, 271–277.

Curran, P.J. (1994). Imaging spectrometry. *Progress in Physical Geography* **18**, 247–266.

Curran, P.J. and Atkinson, P.M. (1999). Issues of scale and optimal pixel size. In: *Spatial Statistics in Remote Sensing* (A. Stein and F. van der Meer, eds), pp. 115–133. Dordrecht: Kluwer.

Curran, P.J. and Williamson, H.D. (1985). The accuracy of ground data used in remote sensing investigations. *International Journal of Remote Sensing* **6**, 1637–1651.

Curran, P.J., Kupiec, J.A. and Smith, G.M. (1997). Remote sensing the biochemical composition of a slash pine canopy. *IEEE Transactions on Geoscience and Remote Sensing* **35**, 415–420.

Curran, P.J., Milton, E.J., Atkinson, P.M. and Foody. G.M. (1998). Remote sensing from data to understanding. In: *Geocomputation: A Primer* (P. Longley, S. Brooks, W. Macmillan and R. McDonnell eds), pp. 33–59. Chichester: Wiley & Sons.

Danson, F.M., Craig, P.S. and Man, W. (2000). Land use changes and disease transmission in central China. In: *Tapeworm Zoonoses—An Emergent Global Problem*, NATO Science Series. Amsterdam: IOS Press, in press.

Deutsch, C.V. and Journel, A.G. (1992). *GSLIB Geostatistical Software Library*. Oxford: Oxford University Press.

Estrada-Peña, A. (1998). Geostatistics and remote sensing in predictive tools of tick distribution: a cokriging system to estimate *Ixodes scapularis* (Acari: Ixodidae) habitat suitability in the United States and Canada from advanced very high resolution radiometer data. *Journal of Entomology* **35**, 989–995.

Fischer, M.M., Gopal, S., Staufer, P. and Steinnocher, K. (1997). Evaluation of neural pattern classifiers for a remote sensing application. *Geographical Systems* **4**, 195–225.

Foody, G.M. (1991). Soil moisture content ground data for remote sensing investigations of agricultural regions. *International Journal of Remote Sensing* **12**, 1461–1469.

Foody, G.M. (1992). On the compensation for chance agreement in image classification accuracy assessment. *Photogrammetric Engineering & Remote Sensing* **58**, 1459–1460.

Foody, G.M. (1996). Approaches for the production and evaluation of fuzzy land cover classifications from remotely sensed data. *International Journal of Remote Sensing* **17**, 1317–1340.

Foody, G.M. (1997). Fully fuzzy supervised classification of land cover from remotely sensed imagery with an artificial neural network. *Neural Computing and Applications* **5**, 238–247.

Foody, G.M. (1999). The continuum of classification fuzziness in thematic mapping. *Photogrammetric Engineering & Remote Sensing* **65**, 443–451.

Foody, G.M. and Arora, M.K. (1996). Incorporating mixed pixels in the training, allocation and testing stages of supervised classifications. *Pattern Recognition Letters* **17**, 1389–1398.

Foody, G.M. and Arora, M.K. (1997). An evaluation of some factors affecting the accuracy of classification by an artificial neural network. *International Journal of Remote Sensing* **18**, 799–810.

Foody, G.M. and Curran, P.J. (1994). Estimation of tropical forest extent and regenerative stage using remotely sensed data. *Journal of Biogeography* **21**, 223–244.

Foody, G.M., Campbell, N.A. Trodd, N.M. and Wood, T.F. (1992). Derivation and applications of probabilistic measures of class membership from the maximum likelihood classification. *Photogrammetric Engineering & Remote Sensing* **58**, 1335–1341.

Foody, G.M., Palubinskas, G, Lucas, R.M., Curran, P.J. and Honzak, M. (1996). Identifying terrestrial carbon sinks: classification of successional stages in regenerating tropical forest from Landsat TM data. *Remote Sensing of Environment* **55**, 205–216.

Gitelson, A.A., Kaufman, Y.J. and Merzlyak, M.N. (1996). Use of a green channel in remote sensing of global vegetation from EOS-MODIS. *Remote Sensing of Environment* **58**, 289–298.

Goel, N.S. (1988). *Models of Vegetation Canopy Reflectance and Their Use in Estimation of Biophysical Parameters*. London: Harwood Academic Publishers.

Goetz, A.F.H., Rowan, L.C. and Kingston, M.J. (1982). Mineral identification from orbit: initial results from the Shuttle Multispectral Infrared Radiometer. *Science* **218**, 1020–1024.

Goovaerts, P. (1997). *Geostatistics for Natural Resources Evaluation*. Oxford: Oxford University Press.

Gopal, S. and Woodcock, C. (1994). Theory and methods for accuracy assessment of thematic maps using fuzzy sets. *Photogrammetric Engineering & Remote Sensing* **60**,181–188.

Groom, G.B., Fuller, R.M. and Jones, A.R. (1996). Contextual correction: techniques for improving land cover mapping from remotely sensed images. *International Journal of Remote Sensing* **17**, 69–89.

Gurney, C.M. and Townshend, J.R.G. (1983). The use of contextual information in the classification of remotely sensed data. *Photogrammetric Engineering & Remote Sensing* **49**, 55–64.

Guyot, G. and Gu, X.F. (1994). Effect of radiometric corrections on NDVI determined from SPOT HRV and Landsat TM. *Remote Sensing of Environment* **49**, 169–180.

Haralick, R.M. and Shanmugam, K.S. (1974). Combined spectral and spatial processing of ERTS imagery data. *Remote Sensing of Environment* **3**, 3–13.

Harris, R. (1985). Contextual classification processing of Landsat data using a probabilistic relaxation model. *International Journal of Remote Sensing* **6**, 847–866.

Hay, S.I., Packer, M.J. and Rogers, D.J. (1997). The impact of remote sensing on the study and control of invertebrate intermediate hosts and vectors for disease. *International Journal of Remote Sensing* **18**, 2899–2930.

Hay, S.I., Snow, R.W. and Rogers, D.J. (1998a). Predicting malaria seasons in Kenya using multitemporal meteorological satellite sensor data. *Transactions of the Royal Society of Tropical Medicine and Hygiene* **92**, 12–20.

Hay, S.I., Snow, R.W. and Rogers, D.J. (1998b). From predicting mosquito habitat to malaria seasons using remotely sensed data: practices, problems and perspectives. *Parasitology Today* **14**, 306–313.

Hay, S.I., Tucker, C.J., Rogers, D.J. and Packer, M.J. (1996). Remotely sensed surrogates of meteorological data for the study of the distribution and abundance of arthropod vectors of disease. *Annals of Tropical Medicine and Parasitology* **90**, 1–19.

Hayes, R.O., Maxwell, E.L., Mitchell, C.J. and Woodzick, T.L. (1985). Detection, identification, and classification of mosquito larval habitats using remote sensing scanners in earth-orbiting satellites. *Bulletin of the World Health Organization* **63**, 361–374.

Holmes, Q.A., Nuesch, D.R. and Shuchman, R.A. (1984). Textural analysis and real-time classification of sea-ice types using digital SAR data. *IEEE Transactions on Geoscience and Remote Sensing* **22**, 113–120.

Horsfall, W.R. (1955). *Mosquitos: Their Bionomics and Relation to Disease*. New York: Ronald Press.

Hugh-Jones, M. (1989). Applications of remote sensing to the identification of the habitats of parasites and disease vectors. *Parasitology Today* **5**, 244–251.

Hugh-Jones, M. (1991a). The remote recognition of tick habitats. *Journal of Agricultural Entomology* **8**, 309–315.

Hugh-Jones, M. (1991b). Satellite imaging as a technique for obtaining disease-related data. *Review Scientifique et Technique de l'Office International Epizooties* **10**, 197–204.

Hugh-Jones, M. and O'Neil, P. (1986). The epidemiological uses of remote sensing and satellites. *Proceedings of the Fourth International Symposium on Veterinary Epidemiology*, pp. 113–118. Singapore: Singapore Veterinary Association.

Hugh-Jones, M.E., Barre, N., Nelson, G. *et al.* (1988). Remote recognition of *Amblyomma variegatum* habitats in Guadeloupe using Landsat-TM imagery. *Acta Veteriinaria Scandinavia* **84**, 259–261.

Hutchinson, C.F. (1982).Techniques for combining Landsat and ancillary data for digital classification improvement. *Photogrammetric Engineering & Remote Sensing* **48**, 123–130.

Innes, J.L. and Koch, B. (1998). Forest biodiversity and its assessment by remote sensing. *Global Ecology and Biogeography Letters* **7**, 397–419.

Irons, J.R., Campbell, G.S., Norman, J.M., Graham, D.W. and Kovalick, W.M. (1992). Prediction and measurement of soil bidirectional reflectance. *IEEE Transactions on Geoscience and Remote Sensing* **30**, 249–260.

Jensen, J.R. (1996). *Introductory Digital Image Processing*, 2nd edn. New Jersey: Prentice Hall.

Justice, C.O., Townshend, J.R.G., Holben, B.N. and Tucker, C.J. (1985). Analysis of the phenology of global vegetation using meteorological satellite data. *International Journal of Remote Sensing* **6**, 1271–1318.

Kelly, M., Estes, J.E. and Knight, K.A. (1999). Image interpretation keys for validation of global land-cover data sets. *Photogrammetric Engineering & Remote Sensing* **65**, 1041–1049.

Kieffer, H.H. and Wildey, R.L. (1985). Absolute calibration of Landsat instruments using the moon. *Photogrammetric Engineering & Remote Sensing* **51**, 1391–1393.

Kleiner, K. (1995). Satellites wage war on disease. *New Scientist* **148**, 9.

Kriebel, K.T. (1976). On the variability of the reflected radiation field due to differing distributions of the irradiation. *Remote Sensing of Environment* **4**, 257–264.

Lillesand, T.M. and Kiefer, R.W. (1999). *Remote Sensing and Image Interpretation*, 4th edn. New York: Wiley & Sons.

Linthicum, K.J., Bailey, C.L., Glyn Davies, F. and Tucker, C.J. (1987). Detection of rift valley fever viral activity in Kenya by satellite remote sensing imagery. *Science* **235**, 1656–1659.

Linthicum, K.J., Anyamba, A., Tucker, C.J., Kelley, P.W., Myers, M.F. and Peters, C.J. (1999). Climate and satellite indicators to forecast rift valley fever epidemics in Kenya. *Science* **285**, 397–400.

Lloyd, C.D., Berberoglu S., Curran, P.J. and Atkinson, P.M. (2001). The use of geostatistical structure functions and the co-occurrence matrix for per-field land cover mapping in the Mediterranean. *Remote Sensing of Environment,* in press.

Longley, P., Goodchild, M., Maguire, D. and Rhind, D. (eds) (1999). *Geographical Information Systems: Principles, Techniques, Management and Applications*. New York: Wiley & Sons.

Mackay, G., Steven, M.D. and Clark, J.A. (1998). An atmospheric correction procedure for the ATSR-2 visible and near-infrared land surface data. *International Journal of Remote Sensing* **19**, 2949–2968.

Malone, J.B., Huh, O.K., Abdel-Rahman, M.S. *et al.* (1997). Geographic information systems and the distribution of *Shistosoma manosoni* in the Nile delta. *Parasitology Today* **13**, 112–119.

Maselli, F., Conese, C. and Petkov, L. (1994). Use of probability entropy for the estimation and graphical representation of the accuracy of maximum likelihood classifications. *ISPRS Journal of Photogrammetry and Remote Sensing* **49**, 13–20.

Mather, P.M. (1999a). *Computer Processing of Remotely-Sensed Images: An Introduction*, 2nd edn. Chichester: Wiley & Sons.

Mather, P.M. (1999b). Land cover classification revisited. In: *Advances in Remote Sensing and GIS Analysis*. (P.M. Atkinson and N.J. Tate, eds). pp. 7–16. Chichester: Wiley & Sons.

Matheron, G. (1965). *Les Variables Régionalisées et Leur Estimation*. Paris: Masson.

Matheron, G. (1971). *The Theory of Regionalized Variables and its Applications*. Fontainbleau: Centre de Morphologies Mathématique de Fontainbleau.

Milton, E.J. (1987). Principles of field spectroscopy. *International Journal of Remote Sensing* **8**, 1807–1827.

Milton, E.J. (1999). Image endmembers and the scene model. *Canadian Journal of Remote Sensing* **25**, 112–120.

Miranda, F.P. and Carr, J.R. (1994). Application of the semivariogram textural classifier (STC) for vegetation discrimination using SIR-B data on the Guiana Shield, northwestern Brazil. *Remote Sensing Reviews* **10**, 155–168.

Myers, D.E. (1982). Matrix formulation of cokriging. *Mathematical Geology* **14**, 249–250.

Nicodemus, F.F., Richmond, J.C., Hsia, J.J., Ginsberg, I.W. and Limperis, T.L. (1977). *Geometrical Considerations and Nomenclature for Reflectance*. National Bureau of Standards Monograph 160. Washington DC: US Government Printing Office.

Paola, J.D. and Schowengerdt, R.A. (1995). A detailed comparison of backpropagation neural network and maximum likelihood classification for urban land use classification. *IEEE Transactions on Geoscience and Remote Sensing* **33**, 981–996.

Pavlovsky, E. (1966). *Natural Nidality of Transmissible Diseases with Special Reference to the Landscape Epidemiology of Zooanthroponoses* (Translated by F. Plous Jr; N. Levine ed.). Urbana, IL: University of Illinois Press.

Pebesma, E.J. and Wesseling, C.G. (1998) GSTAT—a program for geostatistical modelling, prediction and simulation. *Computers and Geosciences* **24**, 17–31.

Peddle, D.R. (1993). An empirical comparison of evidential reasoning, linear discriminant analysis, and maximum likelihood algorithms for alpine land cover classification. *Canadian Journal of Remote Sensing* **19**, 31–44.

Peddle, D.R., Foody, G.M., Zhang, A., Franklin, S.E. and LeDrew, E.F. (1994). Multisource image classification II: an empirical comparison of evidential reasoning and neural network approaches. *Canadian Journal of Remote Sensing* **20**, 397–408.

Philipson, W.R. (ed.) (1997) *Manual of Photographic Interpretation*, 2nd edn. Bethesda, MD: American Society for Photogrammetry and Remote Sensing.

Pope, K.O., Sheffner, E.J., Linthicum, K.J. *et al.* (1992). Identification of central Kenyan rift valley fever virus vector habitats with Landsat TM and evaluation of their flooding status with airborne imaging radar. *Remote Sensing of Environment* **40**, 185–196.

Pope, K.O., Rejmankova, E., Savage, H.M., Arrendondo-Jimenez, J.I., Rodriguez, M.H. and Roberts, D.R. (1994). Remote sensing of tropical wetlands for malaria control in Chiapas, Mexico. *Ecological Applications* **4**, 81–90.

Price, J.C. (1987). Calibration of satellite radiometers and the comparison of vegetation indices. *Remote Sensing of Environment* **21**, 15–27.

Price, J.C. (1990). On the information content of soil reflectance spectra. *Remote Sensing of Environment* **33**, 113–121.

Rees, G. (1999). *The Remote Sensing Data Book*. Cambridge: Cambridge University Press.

Rejmankova, E., Roberts, D.R., Pawley, A., Manguin, S. and Polanco, J. (1995). Predictions of adult *Anopheles albimanus* densities in villages based on distances to

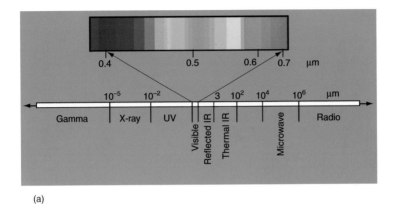

(a)

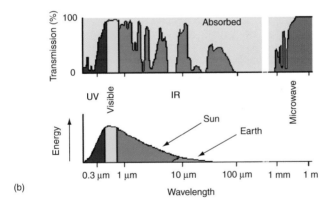

(b)

Plate 1 (a) The electromagnetic spectrum (image courtesy of Dr R. Douglas Ramsey, Utah State University). (b) The spectral characteristics of Sun and Earth electromagnetic radiation sources (bottom graph) and atmospheric sinks (top graph). Note that the wavelength scale is logarithmic. IR, infrared; UV, ultraviolet. See Hay (this volume).

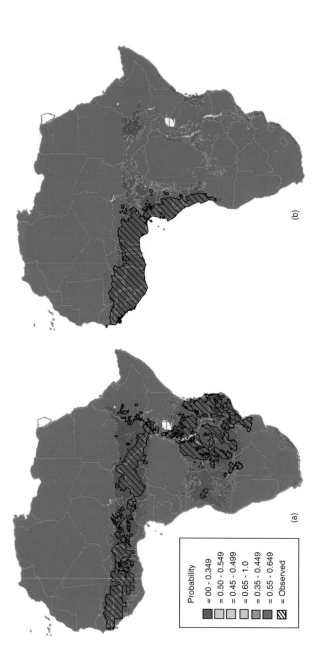

Plate 2 Examples of statistical modelling of the continental distributions of tsetse flies using AVHRR satellite sensor data and a digital elevation model. (a) *G. morsitans* (three subspecies, *G. m. submorsitans, G. m. centralis, G. m. morsitans*), three absence and three presence clusters. Predictions are 91% correct, with 7% false positives and 2% false negatives; sensitivity 0.950, specificity 0.857. (b) *G. palpalis* (two subspecies, (*G. p. palpalis* and *G. p. gambiensis*) also with three absence and three presence clusters. Predictions are 96% correct, with 3% false positives and 1% false negatives; sensitivity 0.97, specificity 0.94. The observed distributions are cross-hatched in black (from Ford and Katondo, 1977) and the predictions are on the colour-coded probability scales shown in (a). See Rogers (this volume).

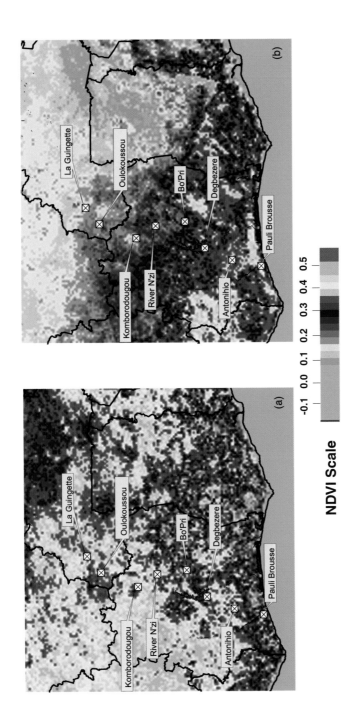

Plate 3 False-colour composite Normalized Difference Vegetation Index (NDVI) imagery for Côte d'Ivoire and Burkina Faso in the (a) wet and (b) dry seasons of 1984 showing locations of study sites where the tsetse *Glossina palpalis* were collected at ~100 km intervals. Satellite data are for the month preceding each transect sample. The false colour scaling of NDVI is shown beneath the plate. See Figure 4, Rogers (this volume).

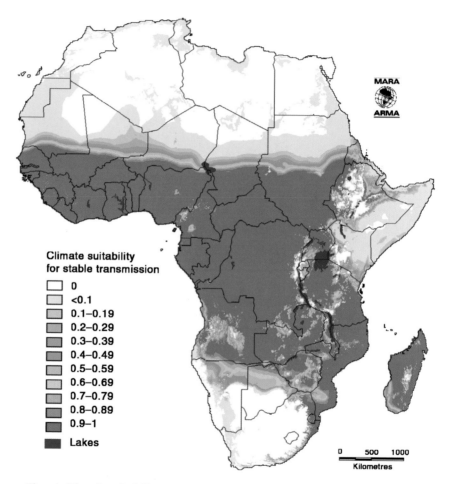

Plate 4 Climatic suitability for stable malaria. Fuzzy value: 0, malaria unstable or absent; 1, malaria probably endemic; between 0 and 1, increasing climatic suitability, increasing chance of stable malaria. See Hay *et al.* (this volume).

remotely sensed larval habitats. *American Journal of Tropical Medicine and Hygiene* **53**, 482–488.

Richards, J.A. (1993). *Remote Sensing Digital Image Analysis: An Introduction*, 2nd edn. Berlin: Springer Verlag.

Riley, J.R. (1989). Remote sensing in entomology. *Annual Review of Entomology* **34**, 247–271.

Roberts, D., Rodriguez, M., Rejmankova, E. *et al.* (1991). Overview of field studies for the application of remote sensing to the study of malaria transmission in Tapachula, Mexico. *Preventive Veterinary Medicine* **11**, 269–275.

Roberts, D.R., Paris, J.F., Manguin, S. *et al.* (1996). Predictions of malaria vector distribution in Belize based on multispectral satellite data. *American Journal of Tropical Medicine and Hygiene* **54**, 304–308.

Roberts, D.A., Gardner, M., Church, R., Ustin, S., Scheer, G. and Green, R.O. (1998). Mapping chaparal in the Santa Monica Mountains using multiple endmember spectral mixture models. *Remote Sensing of Environment* **65**, 267–279.

Robinson, T., Rogers, D. and Williams, B. (1997). Univariate analysis of tsetse habitat in the common fly belt of Southern Africa using climate and remotely sensed vegetation data. *Medical and Veterinary Entomology* **11**, 223–234.

Rogers, D.J. and Randolph, S.E. (1991). Mortality rates and population density of tsetse flies correlated with satellite imagery. *Nature* **351**, 739–741.

Rogers, D.J. and Randolph, S.E. (1993). Distribution of tsetse and ticks in Africa: past, present and future. *Parasitology Today* **9**, 266–271.

Rogers, D.J. and Williams, B.G. (1993). Monitoring trypanosomiasis in space and time. *Parasitology* **106**, 77–92.

Rogers, D.J., Hay, S.I. and Packer, M.J. (1996). Predicting the distribution of tsetse flies in West Africa using temporal Fourier processed meteorological satellite data. *Annals of Tropical Medicine and Parasitology* **90**, 225–241.

Rosenfield, G.H. and Fitzpatrick-Lins, K. (1986). A coefficient of agreement as a measure of thematic classification accuracy. *Photogrammetric Engineering & Remote Sensing* **52**, 223–227.

Roughgarden, J., Running, S.W. and Matson, P.A. (1991). What does remote sensing do for ecology? *Ecology* **72**, 1918–1922.

Schalkoff, R.J. (1992). *Pattern Recognition. Statistical Structural and Neural Approaches.* New York: Wiley & Sons.

Schowengerdt, R.A. (1997). *Remote Sensing Models and Methods for Image Processing*, 2nd edn. San Diego: Academic Press.

Sellers, P.J. (1987). Canopy reflectance, photosynthesis, and transpiration. II. The role of biophysics in the linearity of their interdependence. *Remote Sensing of Environment* **21**, 143–183.

Slater, P.N. (1984). The importance and attainment of accurate absolute radiometric calibration. *Proceedings of SPIE—The International Society for Optical Engineering*, SPIE, **475**, 34–40.

Srinivasan, A. and Richards, J.A. (1990). Knowledge-based techniques for multi-source classification. *International Journal of Remote Sensing* **11**, 505–525.

Stehman, S.V. and Czaplewski, R.L. (1998). Design and analysis for thematic map accuracy assessment: fundamental principles. *Remote Sensing of Environment* **64**, 331–344.

Steven, M.D. (1987). Ground truth – an underview. *International Journal of Remote Sensing* **8**, 1033–1038.

Story, M. and Congalton, R.G. (1986). Accuracy assessment: a user's perspective. *Photogrammetric Engineering & Remote Sensing* **52**, 397–399.

Strahler, A.H. (1980). The use of prior probabilities in maximum likelihood classification of remotely sensed data. *Remote Sensing of Environment* **10**, 135–163.

Strahler, A.H., Woodcock, C.E. and Smith, J.A. (1986). On the nature of models in remote sensing. *Remote Sensing of Environment* **20**, 121–139.

Tanré, D., Deroo, C., Duhaut, P., Perbos, J. and Deschamps, P.Y. (1990). Description of a computer code to simulate the satellite signal in the solar spectrum: the 5S code. *International Journal of Remote Sensing* **11**, 659–668.

Teillet, P.M., Slater, P.M., Ding, Y., Santor, R.P., Jackson, R.D. and Moran, M.S. (1990). Three methods for the absolute calibration of the NOAA AVHRR sensors in-flight. *Remote Sensing of Environment* **31**, 105–120.

Thomson, M.C., Connor, S.J., Milligan, P.J.M. and Flasse, S.P. (1996). The ecology of malaria—as seen from Earth-observation satellites. *Annals of Tropical Medicine and Parasitology* **90**, 243–264.

Thomson, M.C., Connor, S.J., D'Alessandro, U. *et al.* (1999). Predicting malaria infection in Gambian children from satellite data and bed net use surveys: the importance of spatial correlation in the interpretation of results. *American Journal of Tropical Medicine and Hygiene* **61**, 2–8.

Townshend, J.R. G. (1992). Land cover. *International Journal of Remote Sensing* **13**, 1319–1328.

Verstraete, M.M. and Pinty, B. (1996). Designing optimal spectral indexes for remote sensing applications. *IEEE Transactions on Geoscience and Remote Sensing* **34**, 1254–1265.

Verstraete, M.M., Pinty, B. and Curran, P.J. (1999). MERIS potential for land applications. *International Journal of Remote Sensing* **20**, 1747–1756.

Wang, F. (1990). Fuzzy supervised classification of remote sensing images. *IEEE Transactions on Geoscience and Remote Sensing* **28**, 194–201.

Washino, R.K. and Wood, B.L. (1994). Application of remote sensing to arthropod vector surveillance and control. *American Journal of Tropical Medicine and Hygeine* **50**, 134–144.

Welch, J.B., Olson, J.K., Hart, W.G., Ingle, S.G. and Davis, M.R. (1989a). Use of aerial color infrared photography as a survey technique for *Psorophora columbiae* oviposition habitats in Texas ricelands. *Journal of the American Mosquito Control Association* **5**, 147–160.

Welch, J.B., Olson, J.K., Yates, M.M., Benton Jr., A.R. and Baker, R.D. (1989b). Conceptual model for the use of aerial color infrared photography by mosquito control districts as a survey technique for *Psorophora columbiae* oviposition habitats in Texas ricelands. *Journal of the American Mosquito Control Association* **5**, 369–373.

Wilkinson, G.G. (1996). Classification algorithms—where next?. In: *Soft Computing in Remote Sensing Data Analysis* (E. Binaghi, P.A. Brivio and A. Rampini, eds), pp. 93–99, Singapore: World Scientific.

Wood, B.L., Washino, R., Beck, L. *et al.* (1991). Distinguishing high and low *anopheline*-producing rice fields using remote sensing and GIS technologies. *Preventive Veterinary Medicine* **11**, 277–288.

Wood, B.L., Beck, L.R., Washino, R.K., Hibbard, K.A. and Salute, J.S. (1992). Estimating high mosquito-producing rice fields using spectral and spatial data. *International Journal of Remote Sensing* **13**, 2813–2826.

Woodcock, C.E., and Strahler, A.H. (1987). The factor of scale in remote sensing. *Remote Sensing of Environment* **21**, 311–322.

Spatial Statistics and Geographical Information Systems in Epidemiology and Public Health

T.P. Robinson

International Livestock Research Institute, PO Box 30709, Nairobi, Kenya

ADVANCES IN PARASITOLOGY VOL 47
0065-308-X $30.00

ABSTRACT

This chapter surveys the principles behind spatial statistics and geographic information systems (GIS), and their application to epidemiology and public health. Like the other introductory chapters, it is aimed mainly to facilitate understanding in the chapters specific to certain diseases that follow, and to provide a short introduction to the field.

A brief overview of spatial statistics and GIS is provided in the introduction. The sections that follow explore the ways in which we can map the distribution of disease, ways in which we can look for spatial patterns in the distribution of disease, and ways in which we can apply spatial statistics and GIS to the problem of identifying the causal factors of observed patterns. In the last section I discuss some of the ways in which these techniques have been applied to assist decision making for disease intervention, and conclude by discussing future developments in the field, and some of the issues surrounding the integration of spatial statistics and GIS.

1. INTRODUCTION

1.1. Outline of the Chapter

Whilst the subject is of general scientific interest and the models developed can be applied to many other ecological systems, the principal reason for exploring the spatial characteristics of diseases and their causal agents is ultimately to assist decision making for disease intervention. By predicting the spatial and temporal distribution of disease risk we can target areas for intervention, and develop appropriate intervention strategies. To this end, through spatial analysis and geographical information systems (GIS), we can address the following sequence of questions: What is the spatial distribution of the disease? Can we detect patterns in this distribution? What are the causal factors of these patterns? And finally, how can we change these factors to improve the health of people and their livestock?

Following an overview of spatial statistics and GIS in the context of epidemiology, the chapter addresses the following issues: (i) disease mapping, (ii) disease pattern recognition and (iii) exploration of disease correlates. The second section, on disease mapping, reviews some of the issues of disease data collection, representation and visualization. The third section provides a review of the analytical methods that are available to explore spatial patterns in the distribution of disease. The fourth section discusses some of the methods that can be used to investigate relationships between spatial variables, in the search for causal factors behind the distribution of disease.

The penultimate section discusses the ways in which spatial analysis and GIS have been used to assist decision making in disease prevention and intervention. The concluding section draws these issues together with a discussion on the topics reviewed, how the use of spatial analysis and GIS in epidemiology and public health should progress, and some of the issues surrounding the integration of spatial analysis and GIS.

The review is structured by methodology; in each section a series of spatial methods is described and, following each one, selected examples from the literature are provided to illustrate their applications. The descriptions of methodologies are not detailed, the emphasis being on explaining the concepts in simple terms, with the aid of illustrations where appropriate. The formulae behind the methods have been avoided, but in all cases references to detailed descriptions are provided.

1.2. Spatial Data, Spatial Analysis and GIS

Spatial data comprise observations or measurements taken at specific locations, or within specific regions, and are characterized by including the relative positions of data points, usually associated with some geographical reference system, such as latitude and longitude, as well as the values of their attributes.

Analysis of spatial data differs from typical data analysis in that spatial information is included in models and predictions. Spatial structuring in data can arise from measurement error, spatial heterogeneity and from ecological and epidemiological processes that are space- and time-dependent. As a result, the values of a variable in one geographical position are correlated with those at neighbouring positions, and it is this that makes spatial analysis distinct and challenging. Reviews on the application of spatial analysis to epidemiology and public health include Gesler (1986), Haining (1990),

Hungerford (1991), Elliott *et al.* (1992), Liebhold *et al.* (1993) and Douven and Scholten (1995).

Geographic information systems have procedures to input, store, manipulate, analyse and output spatial data (Openshaw, 1996; Burrough and McDonnell, 1998). One of the main strengths of GIS is their ability to integrate different types of spatial data. A major failing with proprietary GIS programmes, however, is their currently limited, but improving, analytical capability. Factors that have contributed to the rapid growth in the use of GIS in recent years include: (i) accessibility, power and diminishing cost of microcomputers and workstations; (ii) accessibility and ease of use of modern GIS software (specifically the command-line interface replaced by the graphical user interface); (iii) more seamless links between GIS, relational databases and statistical packages; (iv) broad availability of public domain information, largely via the internet (e.g. data on human and livestock populations, disease, administrative units, landscape, elevation, hydrology, climate and remotely sensed imagery); and (v) accessibility and cost of global positioning systems (GPS) to facilitate georegistration of field data. Reviews of the application of GIS to epidemiology and public health include Sanson *et al.* (1991), Elliott *et al.* (1992), Mott *et al.* (1995), Clarke *et al.* (1996), Croner *et al.* (1996), McGinn *et al.* (1996), Openshaw (1996) and Vine *et al.* (1997).

Advances in remote sensing have had a massive impact on the ways in which we can explore epidemiological and public health problems, since frequent temporal and wide spatial coverage of environmental correlates of diseases and their vectors can be obtained through remotely sensed data (Hay *et al.*, 1996, 1997). There are certain issues specific to the analysis of remotely sensed data, and image processing software provides specialized spatial analysis tools designed to deal with these. Increasingly, remotely sensed data are combined with other types of thematic data within GIS, as is described in this paper. Reviews on the application of remote sensing to epidemiology and public health include Hugh-Jones (1989); Washino and Wood (1994); Hay (1997); Hay *et al,* (1997) and the four chapters following this one (Rogers, this volume; Hay *et al.*, this volume; Randolph, this volume; Brooker and Michael, this volume).

Many proprietary geographic information systems and image processing systems are reviewed in Thrall (1995) and Vanderzee and Singh (1995). Although some proprietary programmes are available to analyse spatial data, most applications are bespoke and developed by specific research groups. An introductory selection of these programmes is provided in Table 1, with reference to internet sites, publications and company names, but this list is by no means exhaustive.

Table 1 An introductory sample of geographic information (GIS), image processing and spatial statistics software. Company names, internet sites, and references are provided where available. Legendre (1993) provides a further list of spatial analysis programmes.

GIS and image processing software	Spatial statistics software
MapInfo MapInfo Corporation, Troy, NY http://www.mapinfo.com	S Plus MathSoft, Inc., Cambridge, MA http://www.mathsoft.com
GIS+ Caliper Corporation, Newton, MA http://www.caliper.com	SpaceStat Anselling *et al.* (1993) http://www.spacestat.com
AtlasGIS ESRI (formerly Strategic Mapping Inc.) http://www.esri.com	Geographical analysis machine (GAM) Openshaw *et al.* (1987)
ARC/INFO ESRI http://www.esri.com	Spatial scan statistic Kulldorff and Nagarwalla (1995)
ArcView ESRI http://www.esri.com	Spatial analysis by distances indices (SADIE) http://www.iacr.bbsrc.ac.uk/res/depts/ entnem/research/wilfgrp/joeperry/ tjoeperry.html
IDRISI Clarke Laboratories http://www.clarklabs.org	Spatial autocorrelation analysis program (SAAP) http://www.exetersoftware.com/cat/ saap.html
ERDAS Imagine ERDAS http://www.erdas.com	C2D Two-dimensional spatial autocorrelation http://www.exetersoftware.com/cat/ c2d.html
EpiMap USD Inc., Stone Mountain, GA http://www.math.yorku.ca/Who/ Faculty/Monette/Ed-stat/0036.html	The R Package http://www.fas.umontreal.ca/biol/ casgrain/en/ labo/R/index.html
Maptitude Caliper Corporation, Newton, MA http://www.caliper.com	SIMPLE Walker and Moore (1988)
SPANS Tydac Technologies, Canada http://www.tydac.com	
TNTmips MicroImages Inc., Lincoln, Nebraska http://www.microimages.com	

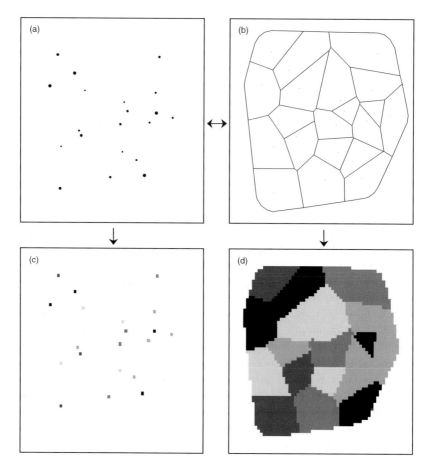

Figure 1 Different data types used in geographic information systems (GIS) and their inter-conversion. In (a) we start with point data, represented here by closed circles whose size is determined by the value of an attribute. Point data can be converted to polygon data, as in (b), using the Thiessen polygon method, and polygons can be converted to points by taking their geometric centroids (note that the points within the Thiessen polygons in (b) are not the polygon centroids). Both point and polygon data can be converted to a raster format, as in (c) and (d)

1.3. Data Types and Interchange

Spatial data generally refer to points or areas (regions, polygons) that may be spaced in a regular or irregular manner, although linear features such as roads and rivers may also be important in spatial data analysis. Such features are referred to as vector data and comprise series of pairs of coordinates. Raster data comprise regular arrays of pixels (picture elements) and have characteristics of both point and area data. Each pixel is related to a location on the Earth's surface through a map projection, and the value of that pixel summarizes the value of a variable on the ground for an area specified by the resolution of the raster array.

Typical examples of point data include health facilities, schools, habitations and inspection sites for livestock diseases. Examples of area data include administrative units, a level at which many population data and health statistics are summarized, and elements of land cover and land use designation. Linear data are less commonly used in a public health context, although transport routes can be important elements in decision support systems, for example in predicting the spread of contagious diseases or estimating the costs of health service delivery.

Different analytical tools are often appropriate to different data types, and methods exist to convert between data types, e.g. raster to vector conversions, enabling the integration of data from different sources. Van der Knaap (1992) discusses some of the potential pitfalls in converting from vector to raster data and compares the performances of a number of GIS packages in this respect.

Figure 1 summarizes the main data types available to spatial analysis and GIS, and illustrates the conversions between data types that are possible and the tools available to achieve these. Point data can be converted into area data using methods such as Thiessen polygons, where polygon boundaries are defined such that each part of the study area is assigned to the point to which it is closest. Point data can also be converted to raster data, at a specified resolution, and there are various ways in which the values of the resulting pixels are defined. For example, the pixel value might reflect presence or absence of one or more points, the number of points falling into each pixel, or some summary of an attribute of the points falling into each pixel, such as the average annual incidence of a disease at all points falling within the bounds of a pixel. Similarly, area data can be converted to point data, for example by assigning the value of a polygon to the point at its geographic mean centre, and area data can also be converted to raster data, at a specified resolution. Again, there are various ways in which the values of the resulting pixels can be defined, for example to reflect the area of a polygon or the value of some attribute of it.

A wide range of tools exists for interpolating point data to generate surfaces or contour maps. Some of these methods strictly interpolate, requiring no information other than the values of points and their positions, for example triangulation, kriging, cubic splines and polynomial smoothing methods. Other methods, such as discriminant analysis and regression analysis (and adaptations of kriging, cubic splines), use ancillary data to derive predictions of values between sample points. These approaches are the subject of Section 4.

2. DISEASE DISTRIBUTION MAPPING

2.1. Overview

The first step in analysing the distribution of disease, is to produce a map. The advent of GIS has greatly facilitated the mapping process and, in particular, enables maps to be updated rapidly and presented in different ways. In this section some of the issues relevant to recording and managing data for mapping diseases are discussed. Methods for mapping diseases are then reviewed, followed by some examples ranging from small-scale studies to disease atlas projects. Finally some more dynamic aspects of disease mapping are considered.

2.2. Data Collection and Recording

Disease data are generally derived from standard record-keeping activities or survey data collected to address specific questions (Myers *et al.*, this volume). Cliff and Haggett (1988) and Cliff *et al.* (1998) discuss, with examples, some some of the issues relating to disease statistics. They also discuss some of the problems, and ways in which these can be overcome, including missing data values, data from multiple sources, changing boundaries, geographical distribution and population structure.

Standardization of disease data is essential for making realistic comparisons, particularly for individuals of different sexes and age groups, though other factors such as ethnic group or income may also be important. Age–sex pyramids, which may be used to standardize data, show age cohorts plotted on the vertical axis and the demographic variable of interest (e.g. mortality) plotted on the horizontal axis. The score of each of the age cohorts on the demographic variable is plotted as a horizontal bar, with males to the left of zero and females to the right (Cliff and Haggett, 1988). From these pyramids standardized estimates of mortality rates and disease incidence can be made.

Certain aspects of data collection and recording are critical. These include accurate georegistration of samples, data quality control and information management. The increasing accuracy and availability of GPS represent major advances towards precise georegistration of field-collected data. Various database management systems have been developed to manage disease data. Some, such as EpiInfo (USD Inc., Stone Mountain, GA), are generic and can be used for a range of diseases. Other bespoke database management systems are specific to particular diseases and surveillance systems, and are provided with standardized operating procedures to ensure quality control of data collection. One example of such a system is the TB99 database (Veterinary Laboratory Administration, Weighbridge, UK) designed specifically to record data on the incidence of bovine tuberculosis in the UK. Another is the Disease And Vector Integrated Database (DAVID), used to record data from tsetse and trypanosomiasis surveys and surveillance, and to produce reports in the form of tables, graphs and GIS maps (Robinson and Hopkins, 1999).

2.3. Methods for Mapping Disease

The two most common types of disease map are: (i) those based on standardized rates, and (ii) those based on significant deviations from the area average. Rate maps are generated from estimates of incidence or prevalence, while significance maps are based on the use of statistical distributions, such as the normal or more usually Poisson, to map significant differences between the observed and expected occurrence of disease. Clayton and Bernardinelli (1992) point out some of the limitations of both methods of disease representation and propose methods that are based on Bayesian statistics, providing, as examples, maps of lip cancer in Scotland and breast cancer mortality in Sardinia. Choynowski (1959) further discusses the advantages of maps based on probabilities, using a study of brain tumours in southern Poland as an example.

There are various ways in which disease data can be represented on maps (Cliff and Haggett, 1988). These include dot-density maps, proportional circles, proportional spheres, grey-scale (chloropleth) maps, contour (isopleth) maps, cartograms and 3-D surface plots. Methods of map generalization (Cliff and Haggett, 1988) include smoothing filters based on weighted moving averages on cellular maps, row and column averages leading to trend surfaces (expected values), and calculation of residuals (observed–expected values) which reveal local deviations from broad patterns in disease distributions.

2.4. Examples from Small-scale Studies

There are a number of examples of small-scale disease mapping projects (see also Brooker and Michael, this volume). Andes and Davis (1995) used a GIS to link vital statistics and census data to provide a contour map of infant mortality rates in the state of Alaska, between 1982 and 1991. This was then used to explore possible causes of increased infant mortality.

Yamagata *et al.* (1986) mapped manually the prevalence of nodules of *Onchocerca volvulus* in Guatemala, using data collected during extensive nodulectomy surveys conducted over 4 years. There appeared to be clustering of onchocerciasis foci associated with particular geological formations, conducive to the existence of the numerous small streams in which the main black-fly vector, *Simulium ochraeceum*, breeds. These data are ideal for a rigorous statistical analysis of disease clustering and underlying causation, and therefore risk mapping, within a GIS, although this was not done.

In a detailed mapping exercise, Beyers *et al.* (1996) produced a GIS map and database of the 1991 incidence of human tuberculosis in two suburbs of the Western Cape, South Africa, using census data and official TB records from the Cape Metropolitan Council. This study suggested certain sub-districts to have a disproportionately high incidence. This mapping exercise facilitated targeting of prophylactic therapy to high-risk areas and provided the base for exploring possible causes, such as overcrowding.

Popovich and Tatham (1997) describe the use of automated mapping techniques to target vaccination programmes against pertussis in Yuna County, Arizona. Baseline maps, including zip codes to locate patients, were combined with patient immunization records, pertussis outbreak data, census data and information on the distribution of health facilities such as schools and clinics. This provided a simple management tool that was used to target public health outreach programmes rationally.

Hightower *et al.* (1998) used more accurate differential GPS to obtain positional data for (amongst other relevant features) houses, schools, health care centres, water bodies and other major mosquito breeding sites. Entomological and parasitological data were collected from selected schools. Whilst they used a GIS to produce digital maps of weekly trap catches of mosquitoes, they found its analytical capabilities insufficient to estimate proximity of households to nearest potential mosquito breeding sites, and conducted this analysis in a statistical package. Though not conducted, a more sophisticated spatial analysis would have enabled them to produce seasonal risk maps for malaria.

2.5. Examples from Disease Atlas Projects

Cliff and Haggett (1988) provide an excellent review of disease atlas data. Through a very thorough analysis of the distribution of cholera cases in metropolitan London in the middle of the nineteenth century, they describe many of the analytical procedures involved in atlas generation and the spatial and temporal analysis of disease data. They use these techniques to describe in detail the spatial dynamics of four contrasting epidemic diseases: AIDS, smallpox, influenza and measles.

Brooker *et al.* (2000) describe the progress made in developing a GIS atlas of the distribution of schistosomiasis and intestinal helminth infections in

Africa. This work has highlighted the complexity of assessing the distribution of infection and disease and emphasized the need for improved approaches (Brooker and Michael, this volume).

2.6. Animations of Modelling Predictions

A further development in mapping disease distributions is to add a temporal dimension. At its simplest this is the visualization of the results from disease surveillance. Cliff and Haggett (1988) used this method to demonstrate the spread of a major measles epidemic in southwest England in 1969/70. They also show a more analytical example that stems from monitoring levels of iodine-131 resulting from the explosion of the Chernobyl nuclear power station in the former USSR in 1986. To simulate dispersal of the radiation cloud, ground measurements of I-131 were interpolated spatially and temporally according to the pattern of surface atmospheric pressure.

Gravity models of flows are one method by which to model changes in disease distribution and the spread of disease epidemics, described by Cliff and Haggett (1988) using the example of the spread of measles in the United States. The movement (flow) of disease is modelled as being directly proportional to the 'masses' of the origin, generating the flow, and the destination, receiving the flow of disease, and inversely proportional to some function of the distance between these points. The 'mass' of a place might be based on some estimate of the population size, adjusted, for example, for demographic structure, risk and infrastructure, whilst the distance may be estimated from some defined cost surface.

At a still higher level of sophistication, process-based models of disease transmission can be produced that model the way in which disease is passed from one person to the next. Such models take into account features like the infectious period of the disease and the sizes and locations of the population at risk. Cliff and Haggett (1988) describe the Hamer–Soper model for epidemic disease transmission and illustrate its application to measles epidemics in four Icelandic medical districts in the middle twentieth century.

3. SPATIAL PATTERNS IN DISEASE DISTRIBUTIONS

3.1. Overview

Having mapped the distribution of a disease the next analytical step is to identify patterns in that distribution. Spatial patterns are highly dependent on scale and extent and some of the issues are highlighted in this section.

Moreover, most patterns in disease data involve some kind of clustering or spatial autocorrelation.

3.2. Issues of Scale and Extent

Since different processes operate at different scales and over different areas in ecology and epidemiology, issues of scale and extent are fundamental to spatial statistical analysis. Exploratory analysis of spatial data sets may help detect patterns at different scales. Spatial analyses are, however, sensitive to (i) variations in zoning systems used, (ii) the scale at which data are recorded, (iii) the sampling systems used to collect data and (iv) the extent of the area to be considered in the analysis. These are reviewed in detail by Fotheringham and Rogerson (1993) and, from a more ecological perspective, by Wiens (1989) and Liebhold et al. (1993).

There are ways in which it is possible to establish the relative importance of different geographical scales in a spatial process. Moellering and Tobler (1972) and Cliff and Haggett (1988), for example, describe the method of 'blocking' or 'continuous quadrat analysis' in which a study area is divided into quadrats that are combined into larger quadrats in a systematic way. A hierarchical analysis of variance then reveals the spatial scales that are important to the processes involved. As will be seen in later sections, analytical procedures such as spatial autocorrelation and second-order neighbourhood analysis can also indicate what scales are involved in spatial processes.

3.3. Spatial Clustering

Given the underlying distribution of a population at risk, clustering occurs if a disease occurs more often in a particular area than would be expected by chance. Whether these cases have a common relationship to some other environmental, biological or social variable, i.e. are the direct result of the distribution of risk factors, is the subject of Section 4. Alexander and Cuzick (1992) make the distinction between clustering in a small area, or around a particular point source, and a general tendency for clustering within the population. They provide a statistical definition of clustering, review models that have been developed and give some examples. Urquhart (1992) emphasizes this distinction and highlights some of the problems in interpreting the results from geographical cluster analysis. Douven and Scholten (1995) distinguish three ways to characterize the properties of patterns in Boolean point data: (i) measures based on density, (ii) measures based on distance, and (iii) measures based on inter-point distance and direction.

Many statistical tests have been developed to explore patterns of geographical clustering in disease data. Examples are provided in Moran (1948), Ohno *et al.* (1979), Cliff and Ord (1981), Whittemore *et al.* (1987), Cuzick and Edwards (1990), Turnbull *et al.* (1990), Alexander and Cuzick (1992) and Kulldorff and Nagarwalla (1995). Here I review three methods.

3.3.1. *Join Counting*

Join counting (Moran, 1948) is a measure of clustering appropriate to nominal area data, where the actual and expected (i.e. if the pattern were random) number of joins between areas with dissimilar values are estimated (Ohno *et al.*, 1979; Cliff and Ord, 1981). The smaller the ratio of dissimilar to total joins the more clustered the disease pattern. This method has the advantage that it is not necessary to convert the area data to point representations, but does not lend itself to elucidating the size or pattern of clustering, nor does it yield different results upon major changes in the size and shape of the areal units involved.

Grimson *et al.* (1981) discuss the issues of defining adjacencies in regions such as county boundaries and adapt the method of join counting to demonstrate an interesting spatial pattern in the incidence of sudden infant death syndrome (SIDS) in North Carolina in the late 1970s.

Cliff and Haggett (1988) use join count statistics to demonstrate a very high degree of clustering of cholera cases in metropolitan London in the middle of the nineteenth century. They provide a further example to explore spatial clustering in the incidence of bronchitis in the middle of the twentieth century in England and Wales.

Ohno *et al.* (1979) used join counting to identify clusters in the geographical pattern of bladder cancer mortality, and of oesophageal cancer mortality, between 1969 and 1971, in Japan. They also developed a chi-square test to estimate the significance of detected clusters.

3.3.2. *The Geographical Analysis Machine*

The Geographical Analysis Machine (GAM) was developed to explore disease data sets for evidence of spatial patterns (Openshaw *et al.*, 1987). The GAM avoids the risk of obtaining spurious results arising from arbitrary administrative boundaries by assigning population census data to the position of the centroid of each area unit (see Section 1.3. and Figure 1). The GAM uses a series of overlapping circles in which observed and expected numbers of cases are computed for circles, of a variety of radii, with centres at every point of a fine grid. Statistical limitations of this methodology are discussed by Cuzick and Edwards (1990).

The GAM has been used to investigate geographical correlates of cancer in children in northern England (Openshaw et al., 1987, 1988), and also clustering of leukaemia incidence in northern New York (Turnbull et al., 1990).

3.3.3. Spatial Scan Statistic

The GAM inspired the development of a similar method to identify clusters in disease distributions, the spatial scan statistic (Kulldorff and Nagarwalla, 1995), which is a generalization of the cluster analysis method developed by Turnbull et al. (1990). In essence, a circle of variable size scans the study area to detect clusters. Under the null hypothesis of no clustering, i.e. random distribution of points, the probability of falsely detecting a cluster of points is equal to the specified significance level.

Kulldorff and Nagarwalla (1995) used the spatial scan statistic to measure the level of clustering of leukaemia incidence in northern New York, and they used this analysis to compare the spatial scan statistic to a number of other cluster detection algorithms, including those of Turnbull et al. (1990) and Openshaw et al. (1987).

The spatial scan statistic was also applied to age-standardized data on the annual incidence of childhood leukaemia in Sweden by Hjalmars et al. (1996). Contrary to at least one previous suggestion of clustering in childhood leukaemia, they demonstrated no significant clustering.

3.4. Spatial Autocorrelation

Spatial autocorrelation refers to the interdependence of values of a variable at different geographic locations. This can be used to elucidate disease patterns, or can be estimated and used to account for the effects of non-independence of closely positioned data points, when exploring relationships between variables. This second application of spatial autocorrelation will be discussed in Section 4. In the following sub-sections, a number of methods of measuring spatial autocorrelation, and their applications, are discussed.

3.4.1. Moran's I and Spatial Correlograms

Moran's I coefficient of autocorrelation (Moran, 1950) measures the degree of similarity between each areal unit and its contiguous neighbours. A similar estimate of spatial autocorrelation is Geary's c (Geary, 1954). Moran's I and Geary's c usually produce very similar results (Getis and Ord, 1996), though

Moran's I is more widely used in the analysis of public health data. Details and formulae for these and other spatial autocorrelation statistics are given in Cliff and Ord (1981); Smans and Esteve (1992) and Kitron and Kazmierczak (1997).

These are known as global statistics since they are based on simultaneous measurements from many locations. The major limitation with these statistics is the assumption that the influence of two areal units on each other is the same regardless of how they are joined, i.e. irrespective of the length of common boundary that they share. Cliff and Ord (1981) overcame this limitation by developing a generalized, or weighted, modification of Moran's I in which a matrix of weights could be used to define the hypothesized relationships between areal units, thus enabling a flexible approach to the definition of neighbours. In such a matrix, each pair of observations (points or areas) is related by a weight that refers to a measure of neighbourhood. Dependent on the application, weights can be assigned in a number of ways. For example, in 'row-standardization' weights are determined by the number of adjacent areas, and must add up to one for each area (e.g. weights would be 0.2 for each of 5 adjoining areas). The weight matrix usually contains simply the reciprocal of the square of the distance between points, though more sophisticated estimates of weights can be derived from distance functions, e.g. distance-decay correlation functions, or from boundary functions such as the length of common boundary between two areas.

A limitation with simple spatial autocorrelation analysis is that it assumes that correlation or covariance isotropic is the same in all directions, i.e. isotropic. There are, however, ways to overcome the problems posed by anisotropic data in which the weight matrix is manipulated to reflect directional inequalities. Furthermore, by varying the structure of the weight matrix different hypotheses concerning the nature of a variable's spatial distribution can be tested.

Moran's I is approximately normally distributed, ranges from -1 to $+1$ and has a value of zero under the null hypothesis of no spatial autocorrelation. The higher the absolute value of I, the stronger the level of spatial autocorrelation. A statistical significance test for Moran's I is based on $Z(I)$, a standardized normal variable (Upton and Fingleton, 1985; Getis and Ord, 1992).

A single measurement of Moran's I at a given distance is of limited use. A spatial correlogram is a series of estimates of Moran's I, evaluated at increasing distances (Glick, 1979; Upton and Fingleton, 1985; Liebhold et al., 1993) and is calculated by means of the weight matrix. On the y-axis of the correlogram (Figure 2) is an estimate of I, ranging from -1 to $+1$, and distance, lag (h), is plotted along the x-axis. The correlogram can be used to determine at what distance on average the spatial autocorrelation is maximized. Various tests exist, some reviewed by Cliff and Ord (1981), to determine whether a spatial autocorrelation value in a correlogram is statistically significant.

Kitron and Kazmierczak (1997) used Moran's I to investigate the spatial distribution of the incidence of Lyme disease by county, between 1991 and 1994, in Wisconsin State. Furthermore, they analysed clustering in two candidate covariates of Lyme disease: distribution of the tick vector *Ixodes scapularis* (recorded from surveys of ticks on deer harvested from annual hunts in 1981, 1989 and 1992) and normalized difference vegetation index (NDVI) values (derived from the National Oceanic and Atmospheric Administrations (NOAA) Advanced Very High Resolution Radiometer (AVHRR)) during the spring and fall when the contrast between agricultural land and woodland is maximized (April and September 1993). Assigning data for each county to its centroid, they estimated Moran's I statistic, the significance test value $Z(I)$, and used spatial correlograms to evaluate the distances where spatial effects were maximized. This analysis demonstrated

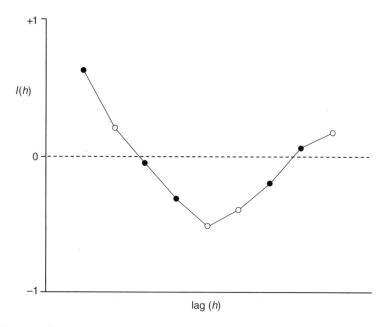

Figure 2 Generalized correlogram. Values of Moran's I coefficient are estimated for different values of lag (h). The further the points are from the dashed line ($I(h) = 0$)), the more the estimates of Moran's I coefficient differ from the null hypothesis of no spatial autocorrelation, at that lag (h). Significance levels of Moran's I coefficient, based on $Z(I)$, are usually indicated on correlograms by using filled circles to represent lags at which values of Moran's I are significant at a specified level, and open circles to represent lags at which they are not.

significant clustering in the mean annual incidence of Lyme disease, tick endemicit, and NDVI values in spring and fall.

Kitron *et al.* (1996) used Moran's *I*, and *Z(I)*, to identify a significant positive spatial autocorrelation in tsetse fly catches in the Ruma National Park in Kenya. Using correlograms they showed that the distance beyond which spatial association had no further influence varied in different parts of the park and at different times of the year.

Glick (1979) used Moran's *I* and correlograms to study the pattern of some cancers in Pennsylvania. He demonstrated the differences in spatial autocorrelation for cancers of the stomach, lung and bladder and, for lung and stomach cancer, and compared the degree of spatial autocorrelation in males and females.

3.4.2. *Semivariograms and Kriging*

A convenient way to explore spatial autocorrelation is to plot a scattergram of $z(x)$, the value of a variable at location x, against $z(x + h)$, the value of the same variable at some distance (or lag) h away. Such a plot is known as an h-scattergram and a tight scatter of points along the 45° line indicates spatial autocorrelation at the specified lag. Different patterns of scatter are typically seen when plotting h-scattergrams over different ranges of h; typically tight when h is small and diffuse when h is large. The 'moment of inertia', defined quantitatively in Liebhold *et al.* (1993), describes the pattern of scatter in an h-scattergram.

Rather than plotting vast numbers of h-scattergrams to describe how autocorrelation varies with distance, we can plot a semivariogram, which summarizes all h-scattergrams over all possible pairings of data for all significant values of h. Quantitative definitions of semivariograms are provided in Liebhold *et al.* (1993) and Maurer (1994).

The semivariogram can be computed either as an average of the semivariance, $\gamma(h)$, in all directions (scalar) or can be specific to a particular direction or range of directions (vector), in order to identify anisotropic processes.

Figure 3 shows a generic semivariogram with the variance between values steadily increasing as points become farther apart, and then levelling off at a distance above which data values are independent of one another. In general, semivariogram values should only be plotted to about half the width of the sampling space in any direction, to avoid boundary effects, and the greater the number of pairs of samples (minimum 30–50), the greater the statistical reliability (Liebhold *et al.*, 1993).

The semivariogram has a number of features that can be used to define its structure, that are illustrated in Figure 3. The *nugget* variance is the value of

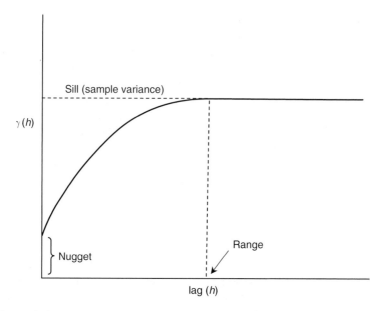

Figure 3 Generalized semivariogram. The *nugget* is the value of the semivariance for which $h = 0$; the maximum value of the semivariance is called the *sill*; and the *range* is the distance, lag (h), at which the *sill* is reached.

the semivariance for which $h = 0$; this represents variability within sample units, or measurement error. The maximum value of the semivariance is called the *sill*; this represents the local variance of the random component of the data. The *range* is the distance, lag (h), at which the *sill* is reached; this is the distance (if any) at which data are no longer spatially autocorrelated. Liebhold *et al.* (1993) discusses at length the alternative ways in which semivariograms can be estimated, ways in which semivariograms can be smoothed and the relationships between correlograms, covariance and semivariograms.

Given a semivariogram that explains the relationship between point values, based on some function of their relative position and separation, it follows that we should be able to predict values at unsampled points, given the values and relative positions of sampled points. Kriging (Delfiner and Delhomme, 1975) is a linear interpolation method that predicts the values of a variable, at unsampled points, based on observations at known locations, using a model of the covariance of a random function. The covariance function is based on a theoretical semivariogram for a specified model such as exponential, spherical, Gaussian, linear or power. To perform kriging in statistics programmes such as the S+ Spatial Stats Package (MarkSoft Inc.,

Cambridge, MA), estimates of the nugget, sill and range are made, and a theoretical semivariogram model is suggested.

Kriging is a method that has been widely used in meteorological work, to interpolate values of climate data from weather stations, but has not been extensively utilised in epidemiology and public health. Nicholson and Mather (1996), however, have used kriging to interpolate densities of the tick *Ixodes scapularis*, vector of Lyme disease, at unsampled locations on Rhode Island, USA. A semivariogram of tick densities at sample points indicated a high level of spatial autocorrelation. A Gaussian model was fitted, and this explained 97.7% of the variation in semivariance. Kriging was then used to make predictions over the whole area, based on this modelled semivariogram.

3.4.3. Gi(d) *Local Statistic*

With very large GIS data sets, and particularly with large raster data sets such as remotely sensed images, global statistics such as Moran's I run the risk of losing information on spatial autocorrelation since they summarize an enormous number of possible disparate spatial relationships. Local statistics overcome this problem by measuring dependence in only a limited portion of the study area. Unlike Moran's I statistic, which measures the correlation between attribute values and location, the $Gi(d)$ local statistic (Getis and Ord, 1992, 1996) is an indicator of local clustering, measuring the 'concentration' of a spatially distributed attribute variable. By comparing local estimates of spatial autocorrelation with global averages, the $Gi(d)$ statistic identifies 'hot spots' in spatial data. Details of computation of the $Gi(d)$ statistic are given in Ding and Fotheringham (1992), Getis and Ord (1996) and Kitron *et al.* (1997), and some of the problems and shortcomings of local statistics are discussed in Getis and Ord (1996).

Kitron *et al.* (1997) used the $Gi(d)$ local statistic to identify significant clustering of cases of La Crosse encephalitis, over distances of 5 and 10 km, around particular towns in the Peoria area of Illinois. This analysis revealed a number of towns to be hot spots for La Crosse encephalitis.

Ding and Fotheringham (1992) used the $Gi(d)$ local statistic to explore clustering of population growth in China between 1982 and 1990, and demonstrated that high rates of population growth were concentrated in the south, which they attributed to less severe restrictions on family size and greater proportions of ethnic minorities with larger family sizes.

Getis and Ord (1992) used the $Gi(d)$ local statistic to explore the spatial pattern of SIDS by county in North Carolina between 1979 an 1984. In addition to a number of small clusters, they identified a major hotspot in the south central portion of the state, which they later associated with the distribution of health care facilities (Getis and Ord, 1996).

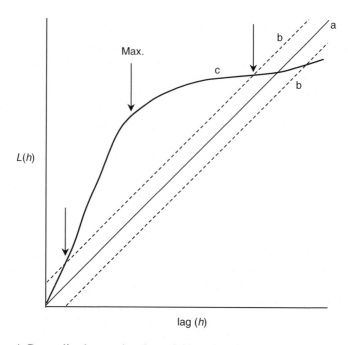

Figure 4 Generalized second-order neighbourhood analysis plot. The plot given by a random Poisson distribution is shown by the straight diagonal line, a. Confidence limits for the Poisson line are indicated by the dashed lines, b. A theoretical data series is shown by line c. The point of maximum deviation of the plot from the Poisson line, over which clustering is maximized, is indicated with an arrow and labelled 'Max.'. The range of distances over which clustering is significant is indicated by the other two arrows.

Local statistics can also be used to quantify the pattern and intensity of spread of a disease away from the core of a hot spot. Getis and Ord (1996) used the $Gi(d)$ local statistic in this way to trace the spread of AIDS away from San Francisco.

3.4.4. *Second-order Neighbourhood Analysis*

Second-order analysis (Ripley, 1976, 1981; Getis, 1984) and second-order neighbourhood analysis (Getis and Franklin, 1987) compare the actual and expected distances between points of similar value, to give estimates of the magnitude and significance of clustering in patterns. A quantitative review of a number of second-order analysis functions is provided in Getis (1984).

Second-order analysis can identify both the distance at which clustering becomes significant and the distance at which maximum clustering is observed. Results are represented by plotting the expected distance between points of similar value against the actual distances. A random Poisson distribution gives a straight diagonal line whilst for clustered data this plot deviates from the $y = x$ line. Confidence limits can be calculated for the Poisson line. The point of maximum deviation of the plot from the Poisson line gives the distance over which clustering is maximized, and the range of distances over which clustering is significant is given by the points at which this line crosses the line denoting the confidence intervals of the Poisson line. This is illustrated in Figure 4, which provides a generalized example of a second-order analysis plot.

Hungerford (1991) demonstrates this technique using data from 1986 on seroprevalence rates of anaplasmosis in cattle, by county, in Illinois. The measured distance between points of similar value was shown to be significantly smaller than would be expected if all points and values were randomly distributed, demonstrating a significant degree of clustering of anaplasmosis incidence.

Kitron *et al.* (1997) used second-order neighbourhood analysis to investigate clustering of 61 cases of La Crosse encephalitis in the Peoria area of Illinois. They found that positive clustering peaked at a distance of 10 km, and that clustering was significant at distances of between 0.5 and 35 km.

4. SPATIAL RELATIONSHIPS AND CAUSAL FACTORS IN DISEASE DISTRIBUTIONS

4.1. Overview

Having mapped the distribution of a disease and quantified the spatial patterns therein, the next task is to establish the causal factors of the disease. In many cases the spatial pattern of the disease will itself provide insights as to the causal mechanisms, and below it is shown how second-order neighbourhood analysis can be used to look for causation. In other cases, however, we need to elucidate sometimes subtle correlates of the disease in order to derive models that can be used to predict disease distributions. Any prediction that is made needs to be to be tested in some way and included in this section are some comments on measuring the fit of predictive models.

Univariate approaches to exploring the correlates and causes of disease are very often useful and have the advantages that results are easily interpreted. Optimal threshold analysis is described as a method for looking at correlates of disease distributions individually that is described here. The regression and correlation methods described later in the section can also be used with single variables.

Univariate methods however, have the major disadvantage that interactions between causal agents of disease cannot be modelled, thus a multivariate approach is often desirable. James and McCullock (1990) review a range of multivariate statistical methods used in ecology and systematics, all of which have potential applications, and some of which have been used in public health and epidemiology. Some of the techniques, for example principal components analysis and supervised and unsupervised classification techniques, are available in GIS and image processing packages, but usually do not have the desired level of flexibility. For most of the techniques described below, specialized programmes need to be developed or the analysis needs to be performed in statistical packages, and the results displayed using GIS.

The first group of multivariate methods that is described in this section is discriminant analysis, making the distinction between linear and non-linear methods (Rogers, this volume). The second group of multivariate methods that is reviewed includes various adaptations of regression and correlation analyses.

A common source of error in analyses that look for correlates of a disease pattern is not to allow for spatial autocorrelation in the disease pattern itself, or in the candidate predictor variables. This violates the rules of independence for parametric data analysis and may lead to underestimates of the errors of model parameters and spuriously high levels of significance, largely due to an effective reduction in sample size. It will be seen in the following discussion that few attempts have been made to allow for spatial autocorrelation.

Two classes of disease distribution modelling are not covered in this review. The first is a set of techniques based on climate matching. Most of these studies are concerned with predicting tick habitat suitability (Randolph, this volume). Reviews and some specific examples are provided by Baker (1978), Sutherst *et al.* (1989), Lessard *et al.* (1990), Perry *et al.* (1990), Norval *et al.* (1991a,b) and Perry *et al.* (1991).

The second set of methods, not covered here, is that based on either supervised or unsupervised classification of high resolution satellite images, followed by some statistical correlation (e.g. discriminant analysis, regression or analysis of variance) between the identified land classes and the incidence of a disease or vector. Although these techniques have been used for other species, most pertain to mapping breeding sites for various species of mosquito (Hay *et al.*, this chapter). Specific examples are given in Hayes *et al.* (1985), Daniel and Kolar (1990), Cooper and Umland Houle (1991), Hugh-Jones (1991), Hugh-Jones *et al.* (1992), Beck *et al.* (1994,1997), Pope *et al.* (1994) and Rejmankova *et al.* (1995).

4.2. Second-order Neighbourhood Analysis

Second-order neighbourhood analysis (see above) can also be used to explore relationships between two different data sets if the points from one data set, for example potential sources of infection, are analysed with respect to points in a different spatial data set, for example the locations of disease cases.

Kitron *et al.* (1992) used second-order neighbourhood analysis in this way to explore clustering of white-tailed deer, recorded from hunting reports, infested and not infested with *Ixodes dammini*, the vector of the Lyme disease bacterium, around Castle Rock State Park in Ogle County, Illinois. Their analysis showed that infested deer were significantly more clustered around the park than non-infested deer.

4.3. Measuring the Fit of Predictive Models

Geostatistical analysis of disease or vector data very often involves making predictions of distribution or abundance, an important aspect of which is to assess prediction errors. Measuring model fit and predictive power is a challenging problem and there is not space within this chapter to do it justice. The subject is well reviewed by Fielding and Bell (1997), who discuss many of the issues and provide guidelines for appropriate tests under different conditions from a conservation perspective.

The problem stems from the fact that, in isolation, assessing success as the percentage of correct predictions is nonsense. To illustrate this point, suppose we have 100 sites, ten with some disease vector present and 90 where it is absent. Suppose also that we fit a model that chooses ten sites at random and defines them as occupied by this vector. By chance we expect to predict correctly one site as present (10%), and 89% (90–1)/100 as absent. Whilst for correctly predicted absence this appears to be a reasonable model, it clearly is not. In reality, the spatial autocorrelation of the data will also effectively reduce the sample size making these error assessments even more optimistic.

There are essentially two issues with error assessment of this kind. First, which data to use to develop the model and which to use to test it, and second, which test to use to assess it. Fielding and Bell (1997) describe various ways to partition data for error assessment, including re-substitution, bootstrapping and jackknifing. They also compare measures of classification accuracy, some derived entirely from information in the confusion matrix, a matrix of actual against predicted presence and absence, including correct classification rate, sensitivity (proportion of positive cases, or presence, correctly predicted), specificity (proportion of negative cases, or absence, correctly predicted) and the kappa statistic. They also describe tests that allow correct presence and correct absence to be weighted differently (cost matrices) and propose two approaches to spatial weighting in error assessment.

In the examples below a number of tests have been used: correct classification rate, sensitivity, specificity and the kappa statistic. The problems with the first three are illustrated above. The kappa statistic (Cohen, 1960; Rosenfield and Fitzpatrick-Lins, 1986; Carstensen, 1987), however, is a useful method to assess classification accuracy (Ma and Redmond, 1995) and has been widely used in epidemiology and public health (e.g. Robinson *et al.*, 1995; Rogers *et al.*, 1996; Hendrickx, 1999).

Kappa can be used for area or raster data sets and estimates the coincidence of two variables, taking into account the degree of overlap that would be expected by chance alone. Values of kappa range from 0 to 1, 1 indicating a perfect coincidence of distributions. Landis and Koch (1977) suggest the following ranges of agreement for the kappa statistic: poor where kappa < 0.4; good where $0.4 <$ kappa < 0.75; and excellent where kappa > 0.75. Tests like the kappa statistic may also be useful for exploring relationships between disease patterns and potential causal factors since confidence intervals can be attached (Ma and Redmond, 1995), making them suitable for hypothesis testing.

4.4. Optimal Threshold Analysis

If we consider candidate predictor variables for disease or vector distributions individually then we can estimate probability density functions by assigning observations to appropriate bin ranges, and dividing the frequency of observations in each bin by the total number of observations (Williams, 1993). The probability density function of presence or absence, per unit of measurement on the x-axis, can then be plotted against each variable.

An example of a probability density function plot is shown in Figure 5, taken from Robinson *et al.* (1997b). The figure shows the probability density function of average maximum temperature, considered for presence and absence of a species of tsetse fly, *Glossina morsitans morsitans*. In its simplest interpretation, the lines for presence and absence cross at the temperature above which the habitat is suitable and below which the habitat is not suitable for this species of fly.

For a single variable, x, we can calculate the probability that x takes the value $\tilde{x}$, given that the disease or vector is present or absent, as $P(\tilde{x} \mid p)$ and $P(\tilde{x} \mid a)$ respectively. From Bayes' theorem (Bonham-Carter, 1994) the probability of presence for this value of x can then be estimated, taking into account the *a priori* probability that it is present. Furthermore, we can estimate the

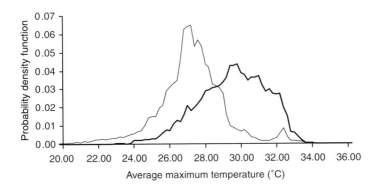

Figure 5 Probability density function of the average of the maximum temperature, considered for presence (bold line) and absence (fine line) of *Glossina morsitans morsitans* in Zambia. Adapted from Robinson *et al.*(1997b).

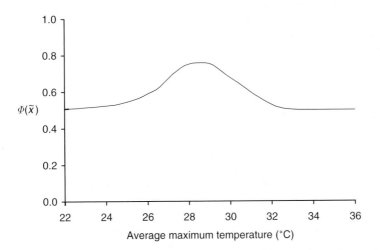

Figure 6 Optimal threshold distribution function, $\Phi(\tilde{x})$, for *Glossina morsitans morsitans* in Zambia, for the average of the maximum temperature. Adapted from Robinson *et al.* (1997b).

cumulative distribution functions of presence and absence for x. This has been termed the optimal threshold distribution function, $\Phi(\tilde{x})$ (Robinson *et al.*, 1997b), and gives the probability that the disease or vector is present above $\tilde{x}$ and absent below $\tilde{x}$. Figure 6 shows a plot of the optimal threshold distribution function for the same data shown in Figure 5. In these plots, flies should be present where the optimal threshold distribution function is falling; where it is rising, flies should be absent. Plotting the curves in this form gives an immediate estimate of the quality of the prediction: if the optimal threshold distribution function reaches either one or zero the prediction is perfect, if it is flat at 0.5 the variable has zero predictive power. Formulae for the optimal threshold distribution function are given in Robinson *et al.* (1997b).

The use of the optimal threshold distribution functions to explore the environmental characteristics underlying the distributions of three species of tsetse, *Glossina morsitans morsitans*, *G. m. centralis* and *G. pallidipes*, in southern Africa (Robinson *et al.*, 1997b) revealed a number of environmental correlates of the distributions of these species. For example, the optimal threshold for *G. m. morsitans*, using the average maximum temperature as a predictor variable, was shown to be 28.6°C (Figures 5 and 6).

4.5. Discriminant Analysis

While in some areas there may be a critical factor or constraint that determines the distribution of a disease or vector, individual variables are unlikely to act on their own and habitat suitability is more likely to be determined by subtle interactions of several factors. When variables are examined in isolation there is no need to make any assumption about the distribution of the data, and there is no ambiguity as to the biological interpretation of the results. There is a risk, however, of missing important interactions between variables, which may be exploited to improve the accuracy of the predictions overall. The next few sections describe a range of multivariate statistical techniques that can be used to explore the distribution of diseases and vectors.

4.5.1. *Classical Linear Discriminant Analysis*

Classical linear discriminant analysis can be used to derive a model that predicts the likelihood of presence or absence of a disease or vector. Points of presence and absence are plotted in multivariate parameter space, as illustrated by the hypothetical set of data points shown in Figure 7, where solid dots indicate presence and open circles absence.

Application of classical linear discriminant analytical methods involves the assumption of a common covariance matrix for sites of presence and absence, and this matrix would be calculated directly from the within-sample sums of squares of the deviations of each point from its sample mean, and the cross-products of these differences. Rogers *et al.* (1996) provides the formulae for classical linear discriminant analysis, in the context of its application to predicting tsetse distributions from remotely sensed data in West Africa (see also Rogers, this volume).

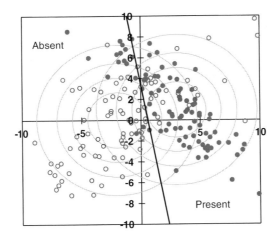

Figure 7 Classical linear discriminant analysis. Theoretical points of presence (solid dots) and absence (open circles) of a disease or vector plotted in bivariate parameter space. Contours are shown that contain 20%, 40%, 60% and 80% of the distributions. Adapted from Robinson *et al.* (1997a).

Linear discriminant analysis has now been widely applied to the problem of predicting the distribution of suitable habitat for a number of species of tsetse flies, based largely on remotely sensed data. For example predictions have been made with varying levels of accuracy in Zambia (Robinson *et al.*, 1997a); in Zimbabwe (Rogers and Williams, 1993, 1994; Williams *et al.*, 1994); in Kenya and Tanzania (Rogers and Randolph, 1993; Rogers and Williams, 1993, 1994); and in Togo (Hendrickx *et al.*, 1993).

Cross *et al.* (1996) used a stepwise discriminant analysis procedure to model presence and absence of *Phlebotomus papatasi*, the vector of sandfly fever and leishmaniasis in south-west Asia. Four hundred and fourteen records of presence and 416 records of absence of sandfly were used to develop a discriminant model based on a series of meteorological variables. Seventy four per cent of the absence sites were correctly predicted as unsuitable for the vector, and 67% of the presence sites were correctly predicted as suitable. Seasonal changes in climatic suitability for the vector were demonstrated.

Wood *et al.* (1991) and Wood *et al.* (1992) used linear discriminant analysis to distinguish high from low mosquito-producing rice fields in California. Using a combination of predictor variables based on Landsat images and estimates of distance from pasture (associated with a source of blood meals), they found that they could distinguish high from low mosquito-producing fields with an accuracy of up to 85% nearly two months before anopheline larval populations peaked in the rice fields. Their results were concomitant with field observations that greater mosquito larval populations were found in rice fields with rapid development of early season vegetation canopy, and located near livestock pastures.

4.5.2. Non-linear Discriminant Analysis (Maximum Likelihood Classification)

The discriminant line shown in Figure 7 is clearly inadequate for distinguishing sites of presence from those of absence, because the assumption of a common covariance matrix is violated by the data. In fact, the data shown in Figure 7 were generated randomly from the two different covariance matrices:

$$C(\text{absent}) = \begin{pmatrix} 20 & 15 \\ 15 & 20 \end{pmatrix} \quad C(\text{present}) = \begin{pmatrix} 15 & -15 \\ -15 & 20 \end{pmatrix}$$

The discriminant function illustrated in Figure 7 is based on the average of these matrices, whilst the appropriate confidence interval contours for these different matrices are shown in Figure 8.

Non–linear discriminant analysis, equivalent to the maximum likelihood classifiers available in most image processing software, allows us to select a discriminating function that follows lines of equi-probability between the two bi-variate distributions for presence and absence. There can be more than one line and these need no longer be straight, but can be curved as they are in Figure 8.

Given that it is possible to estimate the position of any sample point on each of the two bi-variate plots in Figure 5, and to express this in terms of either the probability of presence ($P(p)$: the probability with which the point belongs to the population of points defining presence), or the probability of absence ($P(a)$), we can define a maximum likelihood criterion by which points are assigned to the category of suitability for flies when $P(p) > P(a)$. Normalisation is often required so that $P(p) + P(a) = 1$: hence the likelihood of presence is usually expressed as $P(p)/(P(p) + P(a))$ and of absence as $P(a)/(P(p) + P(a))$ (this normalization does not, of course, affect the decision outcome since each probability is divided by the same total). Formulae for such analyses with un-equal covariance matrices are given in Rogers et al. (1996) (see also Rogers, this volume).

Non-linear discriminant analysis has been used to predict the distributions of tsetse flies in West Africa (Rogers et al., 1996) and in Southern Africa (Robinson et al., 1997a). The increased predictive power conferred by relaxing the constraint of common covariance is evident from these analyses: Robinson et al. (1997a), predicting the distribution G. m. centralis in Zambia, achieved a kappa index of agreement of 0.58 using classical linear

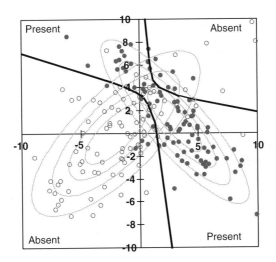

Figure 8 Non-linear discrimminant analysis (maximum likelihood classification). The same data as plotted in Figure 7, but the covariance matrices given in the text are used to plot the 20%, 40%, 60% and 80% contour lines. Adapted from Robinson *et al.* (1997a).

discriminant analysis, but this was increased to 0.68 when maximum likelihood classification was used.

Through these analyses it is becoming increasingly clear that classification accuracy can be further increased by subdividing the habitat into ecological zones prior to performing discriminant analysis. In this case presence or absence within each ecological zone is considered to be a separate class, such that, for example, if the area were divided into three ecological zones, each containing areas of presence and absence, there would be six classes to be distinguished. In the analysis described above, Robinson *et al.* (1997a) further increased the kappa index of agreement to 0.73 by dividing the habitat into two ecological zones prior to analysis.

4.6. Regression and Correlation Analysis

A wide range of adaptations of regression and correlation analysis has also been applied to the problem of modelling the distributions of diseases and their vectors. The methodologies will not be described as they are covered adequately in standard statistical texts such as Sokal and Rohlf (1981), but attention will be drawn to where these techniques have been adapted to analyse spatial data.

4.6.1. *Correlation Analysis*

A variety of rank correlation coefficients can be used to determine whether two sets of rank-ordered data are related, and the Spearman rank correlation coefficient, described in Sokal and Rohlf (1981), has been applied to a number of public health and epidemiological problems.

Spearman rank correlation was used by Kitron and Kazmierczak (1997), in the analysis described above, to explore the relationship between various measures of the incidence of Lyme disease by county in Wisconsin, and spring and fall NDVI estimates and estimates of tick abundance. Lyme disease incidence was significantly correlated with tick distribution, and both were positively correlated with the NDVI in the spring and fall. They produced a map of Lyme disease endemicity for Wisconsin based on a combination of incidence of the disease in humans and the distribution of the tick vector.

In endemic areas, schistosomiasis typically has a patchy distribution, attributed in part to environmental effects. Malone *et al.* (1994) and Malone *et al.* (1997) created diurnal temperature difference images (dT) from the NOAA AVHRR channel 4 brightness temperature data, which they proposed would be related to moisture domains reflecting relative risk from schistosomiasis. Using Spearman rank correlation they demonstrated an inverse relationship between dT values and prevalence of *Schistosoma mansoni*, and proposed that AVHRR thermal difference maps could be used to predict environmental risk of schistosomiasis.

4.6.2. *Spatial Correlation Analysis*

Clifford *et al.* (1989) have developed modified tests of association for spatially autocorrelated processes that are based on the correlation coefficient. For positively autocorrelated processes the effective sample size is reduced, and their methods are based on the evaluation of an effective sample size that takes into account the spatial structure of the data. The Pearson correlation coefficient between two spatial processes is calculated and the correct significance of the correlation is given, taking spatial autocorrelation into account.

Clifford *et al.* (1989) demonstrate this method, exploring the relationships between age and sex-standardised lung cancer, smoking and industrial factors in France between 1968 and 1969. Unadjusted correlation coefficients showed highly significant positive correlations between lung cancer incidence and both smoking and working in the metal industry, and a less significant positive correlation between lung cancer incidence and working in the textile industry. Their alternative correlation coefficients, adjusted for spatial autocorrelation, reduced the effective sample size by some 20% and, although at a lower level of significance, there was still a positive correlation between lung cancer incidence and both smoking and working in the metal industry. The correlation between lung cancer incidence and working in the textile industry dropped from significance, however.

4.6.3. *Regression Analysis*

Regression analysis has been used widely to model disease and vector distributions. As an extension to regression analysis, it is often useful to map the residual values from the regression model. If spatial patterns can be seen in the mapped residuals (geostatistical tools could be used to investigate the distribution of residuals formally), then candidate predictor variables that might match that pattern, i.e. explain some of that unexplained variance, should be sought.

Hay *et al.* (1998) applied linear regression analysis to the seasonal characteristics of malaria transmission in Kenya using Fourier-processed meteorological satellite data. They determined an NDVI threshold for malaria transmission, and used this to make predictive maps of the average number of months for which malaria transmission could occur across Kenya (see also Hay *et al.*, this volume).

Baylis and Rawlings (1998) explored 54 climatic variables in terms of their relationship to catches of *Culicoides imicola*, vector of the African horse sickness (AHS) virus. They found that with just two variables, wind speed and minimum NDVI, they could explain 50% of the variation in abundance of *C. imicola* from 22 sites in Morocco. They applied this model to a set of data from Iberia but with limited success, demonstrating the problems associated with trying to extrapolate empirically derived models outside the range from which their training data originated.

4.6.4. *Spatial Regression Analysis*

The above examples make no attempt to deal with the issues of spatial autocorrelation, as has been discussed throughout this chapter, and thus run the risk of violating assumptions of independence in samples and are likely to overestimate the confidence with which these predictions can be made. In spatial regression analysis we look for two levels of variation: large scale effects caused by a set of independent variables, which may include spatial information, and small-scale effects caused by spatial autocorrelation. Spatial regression analysis allows the incorporation of the spatial structure in both dependent and independent variables into our regression models. Cliff and Ord (1981), Getis (1990), Haining (1990), Richardson (1992) and Cressie (1993) provide details of statistical methods for adjusting for spatial autocorrelation in regression analysis, and discuss the relative merits and appropriate usage of the different methods. Very few examples in public health and epidemiology, however, have made use of these methods.

Kitron *et al.* (1996) have used spatial multiple regression analysis to explore the distribution of tsetse flies (*Glossina pallidipes*) in the Lambwe Valley, Kenya using Landsat TM satellite imagery. In regression analysis they found that values of TM band 7, associated with moisture content of the soil and vegetation, were most highly correlated with fly catches. To separate the spatial component from the association of fly density with the satellite data they used a spatial filtering technique developed by Getis (1990). The spatial component contributed more to the association but there remained a significant spatially independent association between the fly density and the remotely sensed data.

4.6.5. *Logistic Regression Analysis*

Logistic regression is a technique that can be used to model a dichotomous (e.g. presence/absence) dependent variable, as a function of a series of independent variables that can be either continuous or categorical. The inclusion of categorical predictor variables gives logistic regression an advantage over standard multiple regression and discriminant analysis, though the other methods are more efficient for continuous data (James and McCullock, 1990).

Thompson *et al.* (1996) used stepwise logistic regression to model prevalence of Bancroftian filariasis, transmitted by *Culex pipens* mosquitoes, from 299 villages in the southern Nile delta, using the single predictor variable of diurnal temperature differences (dT), derived from NOAA AVHRR data.

Duchateau *et al.* (1997) used a reduced set of climate variables (derived through varimax-rotated principal component analysis of a larger set of temperature, rainfall and evaporation data), NDVI-based vegetation variables and land use as independent predictor variables in a logistic regression model for theileriosis outbreaks in Zimbabwe. They used the final logistic regression model, to which wet-season rainfall and temperature contributed most, to produce a map of probability of theileriosis outbreaks in Zimbabwe.

4.6.6. *Spatial Logistic Regression Analysis*

The problems posed by spatial autocorrelation to logistic regression analysis are identical to those for normal regression analysis, and similar techniques exist to account for them (Cressie, 1993). In ecological studies spatial autocorrelation has been accounted for in logistic regression analysis (e.g. Augustin *et al.*, 1996) and examples are now emerging in the field of public health and epidemiology.

Thomson *et al.* (1999) explored the association between malaria infection in children in the Gambia and age, remotely sensed seasonal vegetation indices, bed net use and possession of a health card. They used logistic regression but explicitly addressed the issues of spatial autocorrelation in the interpretation of their results. This was achieved by computing the semivariogram of the standardized residuals of the fitted logistic regression model, which showed that correlation between pairs of measurements declined exponentially with distance. They then used the method of generalized estimating equations (Liang and Zeger, 1986) to adjust the standard errors of the logistic regression model to account for spatial autocorrelation. When accounting for spatial autocorrelation in the logistic regression model, the significance of all predictor variables was reduced, except for child age (indicating that there is no spatial difference in age structure in children in the Gambia). Some variables dropped from significance, although bed net use and vegetation index remained significant.

5. DECISION SUPPORT FOR DISEASE INTERVENTION

5.1. Overview

The main objective of applying geostatistical tools and GIS to disease data is, ultimately, to assist informed decision making for disease intervention. Once we have mapped the distribution of disease, analysed the patterns of that distribution and attempted to elucidate the causal factors, we are in a stronger position to decide how we can change these patterns to improve the health of people and their livestock. In this section I review some methods that have been applied to decision support for epidemiology and public health. The section starts with a review of some standard GIS overlay functions.

5.2. Standard GIS Functions

Standard GIS overlay functions include a wide range of mathematical and logical expressions, as well as a range of distance operators such as buffering and cost surface estimation. These are easily used to combine data in simple decision rules.

Tempalski (1994) describes the use of a GIS for decision support in the control of dracunculiasis in the Zou Province of Benin. Dracunculiasis is transmitted to humans by drinking water contaminated by the water flea *Cyclops* spp., which carries the immature Guinea worm larvae. Standard GIS buffering and overlay techniques were used first to map the prevalence of dracunculiasis, then to demonstrate a clustering of high prevalence along major roads, and finally to reveal a general reduction in dracunculiasis following the installation of wells. They used the GIS to generate maps of villages that were at risk through lack of access to clean water. They could also target villages where health education was required further to reduce the risk of Guinea worm infection.

In Nadidad *taluka* (administrative area) of Kheda district, Gujarat, comprising 100 villages with a total population of *c*. 0.5 million, malaria is on the increase, the principal vector being *Anopheles culicifacies*. Sharma and Srivastava (1997) digitized map layers for Nadidad that were relevant to mosquito breeding: soil, water table, water quality, relief and hydro-geomorphology. Each of these they classified into high, medium or low, depending on the suitability for waterlogging and mosquito breeding. By a series of overlay and union operations they produced a malaria risk map that divided the *taluka* into 13 zones, each classified as high, medium or low malaria risk. They found a good correlation between this risk map and the average annual parasite incidence in each zone, although not with the individual layers, emphasizing the need for a multivariate approach. They were thus able to suggest control strategies appropriate for each zone. (See also Hay *et al*., this volume.)

Porter (1999) and Tanser and Wilkinson (1999) discuss the use of GIS and GPS in the implementation of the DOTS (directly observed therapy, short course) control strategy for tuberculosis in northern KwaZulu/Natal, South Africa. They use standard GIS distance and buffering functions to demonstrate and quantify how much a community-based programme increases access to tuberculosis treatment, and promote its further use for decision support and programme management in developing countries.

5.3. Weighted Linear Combination

Weighted linear combination (Eastman *et al.*, 1993a,b) is a decision support tool that involves identifying a set of criteria relevant to a decision outcome, standardizing these such that they are comparable in the context of the analysis, assigning weights to these criteria that reflect their relative importance in the decision outcome and summing the product of the criteria, having multiplied them by their assigned weights.

In eastern Zambia, socio-economic surveys demonstrate that trypanosomiasis significantly reduces stocking rates, and the agricultural benefits associated with mixed farming systems. Robinson *et al.* (1999) used a GIS to combine data for six environmental variables: cattle density, human population density, land designation, relative arable potential, crop use intensity and proximity to existing control operations. The distribution of tsetse in the area was predicted using the multivariate methods described earlier in the chapter (Robinson *et al.*, 1997a). Experienced veterinarians and biologists working in the region established criteria weights for the input variables and the data were integrated using weighted linear combination to prioritize areas for trypanosomiasis control. The results of this analysis were used to delimit the optimal location of a community-based tsetse control programme in the Eastern Province of Zambia.

The US/Mexico border area is experiencing a rapid growth in industrialization through multinational corporations due, on the one hand, to proximity to the US market, and on the other to access to cheap levels of Mexican labour and low levels of legislation to protect the environment. This growth is presenting potential health risks to the citizens of the cross-border area, and appropriate planning needs to be undertaken to prepare for emergencies resulting from the release of hazardous materials. Lowry *et al.* (1995) have used weighted linear combinations to identify areas most at risk from potential hazards. They created digital data layers of a series of hazard variables comprising the locations of industrial facilities, risk of surface transmission and risk of sewage transmission of effluents. They also generated data layers pertaining to human risk: overall population, vulnerable component of the population (below the age of 18 years and above the age of 65 years), economic infrastructure and the locations of sensitive institutions (schools, clinics and hospitals). The groups of factors pertaining to hazard risk and human vulnerability were combined independently using weighted linear combination, and the results were amalgamated to produce a map of people that were most at risk. Under different weighting regimes certain sites were consistently ranked as high risk, and these were then analysed individually to ascertain the main causes of risk, and the most appropriate ways to avoid or prepare for hazards.

5.4. Probabilistic Layer Analysis

Probabilistic layer analysis (Njemanze *et al.*, 1999) is somewhat similar to weighted linear combination. It is a method of risk assessment where a level of disease risk is defined and factors contributing to that risk are sequentially weighted, and their probabilities of contributing to that level of disease risk are combined.

Among the poor in developing countries, diarrhoea is a serious health hazard in infants. Njemanze *et al.* (1999) used a GIS to collate a database of geology, hydrology, electricity supply, water sources, population, environmental pollution, towns and villages and 11 537 recorded cases of diarrhoea. Thirty-nine water sources were analysed and probabilistic layer analysis was used to categorize these into three significantly different diarrhoea risk classes. Based on this analysis, solutions to eliminate this risk were proposed.

5.5. Decision Tree Analysis

Decision tree analysis involves the construction of a sequential dichotomous key to classify areas as suitable or not for a particular decision outcome. At each level, a variable is considered and the decision path chosen depending on whether that variable takes a value of above or below a specified threshold.

Robinson (1998) uses GIS to implement decision tree analysis to prioritize areas for control of tsetse-transmitted trypanosomiasis intervention in eastern Zambia. The distribution of tsetse was combined with data on land designation, percentage agriculture, stocking rates and arable potential in a GIS. Two approaches were adopted. In the first, the relevant criteria were combined to prioritize areas where the disease itself was a constraint. The tsetse-infested target areas had medium/high arable potential, relatively low levels of agriculture, and intermediate stocking rates. The second approach addressed the problem of land pressure, where very high densities of livestock and agriculture result from tsetse infestation preventing the movement of people and livestock into adjacent areas that would otherwise be suitable for agricultural development. Areas of high land pressure, and those with a high absorption capacity, were identified using the GIS. High land pressure areas were identified as having intermediate/high levels of agriculture and stocking rates; areas of high absorption capacity that were infested with tsetse, were identified as having low/intermediate stocking rates but a relatively high arable potential. These areas were mapped together to show where they lay adjacent to each other, and thus where trypanosomiasis intervention might contribute to the relief of land pressure.

Using similar methods, Hendrickx (1999) developed a decision support system for trypanosomiasis control in Togo. He used predicted data sets on tsetse abundance, trypanosomiasis prevalence and average herd packed cell volume (an estimate of anaemia in cattle), and combined these with data on cattle breed composition (proportion of trypanotolerance), human population density, agricultural intensity and cattle density. He developed two models: the first aimed at setting priorities for commercial systems, addressing the direct clinical aspects of trypanosomiasis and measuring veterinary needs. The second model prioritized areas for intervention for rural traditional husbandry systems; indicating areas where benefits were likely to be highest in terms of enhanced integration of cattle and crop production.

5.6. Regression Analysis

Aujeszky's disease in swine herds, caused by the pseudorabies virus (PRV), can be transmitted by direct or indirect contact between individuals, and by aerosol suspensions of virus. Understanding the risk of infection posed by these different methods of transmittance could facilitate management and planning for PRV control or eradication strategies. Marsh *et al.* (1991) created a GIS database for 115 swine herds of known PRV status (out of 280) in Blue Earth County, Minnesota. They used the GIS to estimate proximity coefficients for the herds and then extracted the information for further analysis in a statistical package. They performed Cox regression analysis to establish that there was no significant association between PRV status and distance to the nearest county, road, or known quarantine herd. The best fitting model included density of swine herds within a 5 km radius, operation type, housing type, and proximity (<1 km) to a river or lake. Further analysis, with more complete data on swine herd status, should allow a better understanding of these risk factors to produce maps to inform intervention strategies.

5.7. Logistic Regression Analysis

Lyme disease is caused by the bacterium *Borrelia burgdorferi*, which is transmitted by ticks belonging to the genus *Ixodes*. Woodland and forest edge habitats, which support large numbers of the ticks and their wild mammalian hosts, are considered particularly at risk. Glass *et al.* (1995) generated forest distribution and land cover maps for Baltimore County, Maryland, from Landsat Thematic Mapper (TM) imagery from 1990 and planimetric maps of the urban portions of the county. These were combined in a GIS with databases on soils, geology, elevation and watershed, and with the residential locations of 47 Lyme disease cases, reported between 1989 and 1990, and 492 randomly selected control residences. Forty-eight Lyme disease cases reported in 1991, and 495 randomly selected control residences were used to validate the model. The study variables were extracted from the GIS and logistic regression was carried out using statistical software. The resulting logistic function was then run on the maps included in the final model, in the GIS, to generate a risk map. This risk map correctly classified 85.8% of the 1991 data set, although, as mentioned in Section 4.3, this method of accuracy assessment has its limitations.

6. CONCLUSIONS

It is clear from this review that, whilst much has been done in the application of geostatistical analysis and GIS techniques to issues of public health and epidemiology, there is enormous scope for future developments. We are moving rapidly from the purely descriptive mapping of diseases into a far more analytical phase of research towards disease intervention.

We have seen that, although some analyses do take account of spatial autocorrelation in the data, this issue is often not addressed. Discriminant analysis is undoubtedly one of the most powerful tools available for analysing the distribution of diseases and vectors, and elucidating the causal factors behind these. Methods to account for spatial autocorrelation in discriminant analysis however remain to be developed and applied.

In addition to the techniques described in this chapter there are new analytical methods emerging that are likely to have a large impact in this field. Examples include modelling natural spatial structures using fractal geometry (e.g. Maurer, 1994) and the application of wavelets to spatial processes, which allow analyses at multiple spatial scales simultaneously.

Geographical patterns of disease arise from underlying ecological and sociological processes. Whilst providing reasonable grounds on which to base hypotheses, the presence of co-incident patterns does not demonstrate cause and effect. One of the main advances in this field will be in the development of process-based models for disease epidemiology and vector ecology. Significant advances have been made in this direction, for example Rogers and Randolph (1991) have correlated tsetse fly mortality rates with vegetation indices derived from satellite data, and of Randolph and Rogers (1997) have developed a generic population model for the tick *Rhipicephalus appendiculatus*, driven by parameters that can be estimated through remote sensing. When predictions are based on deterministic rather than empirical models, we shall be able to predict more accurately the changes in disease patterns that are likely to occur under conditions of environmental change.

In the area of decision support, the literature includes many kinds of multi-criteria and multi-objective models that can be applied to public health issues, within a spatial context. Furthermore, the inclusion of economic models into decision support tools for public health is a field that has been vastly under-explored and needs to be addressed. Tools are available for estimating the costs of disease and their 'burden' on people, but these remain to be applied in spatial contexts. Malone and Zukowski (1992), in a discussion on the use of GIS to assist the control of cattle liver flukes in southern USA, present some ideas on how estimates of economic risk could be incorporated into the management system, and thus evaluate the costs and benefits of proposed interventions.

Given the extensive literature on spatial statistics, spanning back to the early 1940s, it is quite clear that the evolution of this discipline has been dependent neither on GIS nor on modern computing capabilities. It is also clear that even the most sophisticated GIS packages provide very little in the way of tools for geostatistical analysis. The disciplines of spatial analysis and GIS have developed quite independently of one another and it seems unlikely in the near future that sophisticated spatial analysis tools will be incorporated directly into proprietary GIS programmes. Goodchild (1987) suggests that the full integration of spatial analysis and GIS is currently limited by inadequacy of existing GIS data structures, and Openshaw (1991) further proposes that many of the existing spatial analysis methods are not even suitable for a data-rich GIS environment, advocating the development of new, specifically designed tools.

The conceptual issues of bringing together spatial data analysis and GIS are discussed by Goodchild et al. (1992) who points out the advantages and disadvantages of four levels of integration between spatial data analysis software and GIS: (i) stand-alone spatial analysis software, (ii) loose coupling, (iii) close coupling and (iv) full integration. Some of the practical issues involved in linking GIS and statistical packages are reviewed by Anseling et al. (1993). Bailey (1994) provides a classification of spatial statistical methods and reviews the potential and progress that has been made in integrating these into GIS. With the tendency towards modular geographic information systems, the opportunity arises for 'add-ons' to be developed that provide spatial statistics facilities in a way that is accessible to ecologists and epidemiologists.

The main effect of linking software for geostatistical analysis with GIS will be to increase the accessibility of these powerful analytical tools to a research community that is becoming better versed in increasingly user-friendly GIS software. Furthermore, coupled with current advances in computing power, it will make these tools available for use on large raster data sets, as derived from remote sensing. Tools for geostatistical analysis will need to be advanced to resolve the complex issues that will be posed by vast multi-temporal and/or, multi-spectral data sets.

Considerable progress could be made by drawing together the strengths of spatial analysis and GIS. For example, in spatial autocorrelation analysis, standard GIS functions such as buffering and cost surface analysis could be used to generate the distance matrices used in semivariogram analysis and kriging. Taking this a step further, vector habitat suitability maps derived from multivariate analysis of remotely sensed data could also be used to generate distance matrices for semivariogram analysis, which, combined with genetic analysis, may create a new and potentially powerful tool for analysing the population structure of disease vectors.

In a recent review on the use of GIS in health district management, Boelaert *et al.* (1998) warn of a 'new wave of technology-driven priority setting within the health sector', drawing attention to some of the risks associated with applying inappropriate statistical methods to data, and presenting results to naïve decision-makers in the seductive manner made possible with modern GIS.

We need to be both cautious and rigorous in our application of the analytical power that will arise through closer integration of spatial analysis and GIS. Used appropriately, the sorts of analyses reviewed in this chapter can provide us with an unprecedented insight into biological patterns and processes, and could result in significant savings of resources, through data-driven priority setting and planning for public health.

ACKNOWLEDGEMENTS

I am grateful to the editors and also to Jack Lennon and Brian Williams for their helpful comments on this manuscript. TPR was partially funded for this work under IFAD grant TAG 284.

REFERENCES

Alexander, F.E. and Cuzick, J. (1992). Methods for the assessment of disease clusters. In: *Geographical and Environmental Epidemiology* (P. Elliott, J. Cuzick, D. English and R. Stern, eds), pp. 238–250. Oxford: Oxford University Press.

Andes, N. and Davis, J.E. (1995). Linking public health data using geographic information system techniques: Alaskan community characteristics and infant mortality. *Statistics in Medicine* **14**, 481–490.

Anseling, L., Dodson, R.F. and Hudak, S. (1993). Linking GIS and spatial data analysis in practice. *Geographical Systems* **1**, 3–23.

Augustin, N.H., Mugglestone, M.A. and Buckland, S.T. (1996). An autologistic model for the spatial distribution of wildlife. *Journal of Applied Ecology* **33**, 339–347.

Bailey, T. (1994). A review of statistical spatial analysis in geographical information systems. In: *Spatial Analysis and GIS* (S. Fotheringham and P. Rogerson, eds), pp. 13–44. London: Taylor and Francis.

Baker, G.L. (1978). An outbreak of *Spodoptera exempta* (Walker) (Lepidoptera: Noctuidae) in the highlands of Papua New Guinea. *Papua New Guinea Agricultural Journal* **29**, 11–25.

Baylis, M. and Rawlings, P. (1998). Modelling the distribution and abundance of *Culicoides imicola* in Morocco and Iberia using climatic data and satellite imagery. *Archives of Virology* **14**, 137–153.

Beck, L.R., Rodriguez, M.H., Dister, S.W. *et al.* (1994). Remote sensing as a landscape epidemiologic tool to identify villages at high risk for malaria transmission. *American Journal of Tropical Medicine and Hygiene* **51**, 271–280.

Beck, L.R., Rodriguez, M.H., Dister, S.W. *et al.* (1997). Assessment of a remote sensing-based model for predicting malaria transmission risk in villages of Chiapas, Mexico. *American Journal of Tropical Medicine and Hygiene* **56**, 99–106.

Beyers, N., Gie, R.P., Zietsman, H.L. *et al.* (1996). The use of a geographical information system (GIS) to evaluate the distribution of tuberculosis in a high-incidence community. *South African Medical Journal* **86**, 40–44.

Boelaert, M., Arbyn, M. and Van der Stuyft, P. (1998). Geographical information systems (GIS), gimmick or tool for health district management. *Tropical Medicine & International Health* **3**, 163–165.

Bonham-Carter, G.F. (1994). *Geographic Information Systems for Geoscientists: Modelling with GIS.* Oxford: Pergamon/Elsevier.

Brooker, S., Rowlands, M., Haller, L., Savioli, L. and Bundy D.A.P. (2000). Towards an atlas of human helminth infection in sub-Saharan Africa: the use of geographical information systems (GIS). *Parasitology Today* **16**, 303–307.

Burrough, P.A. and McDonnell, R.A. (1998). *Principles of Geographical Information Systems.* Oxford: Oxford University Press.

Carstensen, L.W. (1987). A measure of similarity for cellular maps. *The American Cartographer* **14**, 345–358.

Choynowski, M. (1959). Maps based upon probabilities. *Journal of the American Statistical Society* **54**, 385–388.

Clarke, K.C., Mclafferty, S.L. and Tempalski, B.J. (1996). On epidemiology and geographic information systems: a review and discussion of future directions. *Emerging Infectious Diseases* **2**, 85–92.

Clayton, D. and Bernardinelli, L. (1992). Bayesian methods for mapping disease risk. In: *Geographical and Environmental Epidemiology.* (P. Elliott, J. Cuzick, D. English and R. Stern, eds), pp. 205–220. Oxford: Oxford University Press.

Cliff, A.D. and Haggett, P. (1988). *Atlas of the Distribution of Diseases: Analytical Approaches to Epidemiological Data.* Oxford: Blackwell.

Cliff, A.D. and Ord, J.K. (1981). *Spatial Processes: Models and Applications.* London: Pion.

Cliff, A., Haggett, P. and Smallman-Raynor, M. (1998). *Deciphering Global Epidemics—Analytical Approaches to the Disease Records of World Cities, 1888–1912.* Cambridge: Cambridge University Press.

Clifford, P., Richardson, S. and Hemon, D. (1989). Assessing the significance of the correlation between two spatial processes. *Biometrics* **45**, 1230–1234.

Cohen, J. (1960). A coefficient of agreement for nominal scale. *Educational and Psychological Measurements* **20**, 37–46.

Cooper, J.W. and Umland Houle, J. (1991). Modelling disease vector habitats using thematic mapper data: identifying *Deramcentor variabilis* habitats in Orange County, North Carolina. *Preventive Veterinary Medicine* **11**, 353–354.

Cressie, N.A.C. (1993). *Statistics for Spatial Data.* New York: Wiley and Sons.

Croner, C.M., Sperling, J. and Broome, F.R. (1996). Geographic information systems (GIS): new perspectives in understanding human health and environmental relationships. *Statistics in Medicine* **15**, 1961–1977.

Cross, E.R., Newcomb, W.W. and Tucker, C.J. (1996). Use of weather data and remote sensing to predict the geographic and seasonal distribution of *Phlebotomus papatasi* in southwest Asia. *American Journal of Tropical Medicine and Hygiene* **54**, 530–536.

Cuzick, J. and Edwards, R. (1990). Spatial clustering for inhomogeneous populations. *Journal of the Royal Statistical Society B* **52**, 73–104.

Daniel, M. and Kolar, J. (1990). Using satellite data to forecast the occurrence of the common tick *Ixodes ricinus* (L.). *Journal of Hygiene, Epidemiology, Microbiology and Immunology* **34**, 243–252.

Delfiner, P. and Delhomme, J.P. (1975). Optimum interpolation by kriging. In: *Display and Analysis of Spatial Data* (J.P. Davis and M.J. McCullagh, eds), pp. 96–114. New York: Wiley & Sons.

Ding, Y. and Fotheringham, A.S. (1992). The integration of spatial analysis and GIS. *Computers in Environmental and Urban Systems* **16**, 3–19.

Douven, W. and Scholten, H.J. (1995). Spatial analysis in health research. In: *The Added Value of Geograhical Information Systems in Public and Environmental Health* (M.J.C. de Lepper, H.J. Scholten and R.M. Stern, eds), pp. 117–133. The Netherlands: Kluwer Academic Press.

Duchateau, L., Kruska, R.L. and Perry, B.D. (1997). Reducing a spatial database to its effective dimensionality for logistic-regression analysis of incidence of livestock disease. *Preventive Veterinary Medicine* **32**, 207–218.

Eastman, J.R., Kymen, P.A.K. and Toledano, J. (1993a). A procedure for multi-objective decision making in GIS under conditions of conflicting objectives. European Conference on Geographical Information Systems, pp. 438–447.

Eastman, J.R., Jin, W., Kyem, P.A.K. and Toledano, J. (1993b). Participatory multi-objective decision-making in GIS. Proceedings of AUTOCARTO XI, pp. 33–43.

Elliott, P., Cuzick, J., English, D. and Stern, R. (1992). *Geographical and Environmental Epidemiology.* Oxford: Oxford University Press.

Fielding, A.H. and Bell, J.F. (1997). A review of methods for the assessment of prediction errors in conservation presence/absence models. *Environmental Conservation* **24**, 38–49.

Fotheringham, A.S. and Rogerson, P.A. (1993). GIS and spatial analytical problems. *International Journal of Geographical Information Systems* **7**, 3–19.

Geary, R. (1954). The contiguity ratio and statistical mapping. *Incorporated Statistician* **5**, 115–145.

Gesler, W. (1986). The uses of spatial analysis in medical geography: a review. *Society for Science in Medicine* **23**, 963–973.

Getis, A. (1984). Interaction modelling using second-order analysis. *Environmental Planning A* **16**, 173–183.

Getis, A. (1990). Screening for spatial dependence in regression analysis. *Papers of the Regional Science Association* **69**, 69–81.

Getis, A. and Franklin, J. (1987). Second-order neighbourhood analysis of mapped point patterns. *Ecology* **68**, 473–477.

Getis, A. and Ord, J.K. (1992). The analysis of spatial association by use of distance statistics. *Geographical Analysis* **24**, 189–206.

Getis, A. and Ord, J.K. (1996). Local spatial statistics: an overview. In: *Spatial Analysis: Modelling in a GIS Environment* (P. Longley and M. Batty, eds), pp. 261–277. Cambridge: Geoinformation International.

Glass, G.E., Schwartz, B.S., Morgan, J.M.I., Johnson, D.T., Noy, P.M. and Israel, E. (1995). Environmental risk factors for Lyme disease identified with geographic information systems. *American Journal of Public Health* **85**, 944–948.

Glick, B. (1979). The spatial autocorrelation of cancer mortality. *Society for Science in Medicine* **13D**, 123–130.

Goodchild, M.F. (1987). A spatial analytical perspective on geographical information systems. *International Journal of Geographical Information Systems* **1**, 327–334.

Goodchild, M., Haining, R. and Wise, S.E.A. (1992). Integrating GIS and spatial data analysis: problems and possibilities. *International Journal of Geographic Information Systems* **6**, 407–423.

Grimson, R., Wang, K. and Johnson, P. (1981). Searching for hierarchical clusters of disease: spatial patterns of sudden infant death syndrome. *Society for Science in Medicine* **15D**, 287–293.

Haining, R. (1990). *Spatial Data Analysis in the Social and Environmental Sciences.* Cambridge: Cambridge University Press.

Hay, S.I. (1997). Remote sensing and disease control: past, present and future. *Transactions of the Royal Society of Tropical Medicine and Hygiene* **91**, 105–106.

Hay, S.I., Tucker, C.J., Rogers, D.J. and Packer, M.J. (1996). Remotely sensed surrogates of meteorological data for the study of the distribution and abundance of arthropod vectors of disease. *Annals of Tropical Medicine and Parasitology* **90**, 1–19.

Hay, S.I., Packer, M.J. and Rogers, D.J. (1997). The impact of remote sensing on the study and control of invertebrate intermediate host and vectors for disease. *International Journal of Remote Sensing* **18**, 2899–2930.

Hay, S.I., Snow, R.W. and Rogers, D.J. (1998). Predicting malaria seasons in Kenya using multitemporal meteorological satellite sensor data. *Transactions of the Royal Society of Tropical Medicine and Hygiene* **92**, 12–20.

Hayes, R.O., Maxwell, E.L., Mitchell, C.J. and Woodzick, T.L. (1985). Detection, identification, and classification of mosquito larval habitats using remote sensing scanners in earth-orbiting satellites. *Bulletin of the World Health Organization* **63**, 361–374.

Hendrickx, G. (1999). Georeferenced decision support methodology towards trypanosomosis management in West Africa. Ph.D. thesis, Universiteit Gent, Gent.

Hendrickx, G., Rogers, D.J., Napala, A. and Slingenbergh, J.H.W. (1993). Predicting the distribution of riverine tsetse and the prevalence of bovine trypanosomiasis in Togo using ground-based and satellite data. International Scientific Council for Trypanosomiasis Research and Control, 22nd Meeting, Kampala.

Hightower, A.W., Ombock, M., Otieno, R., Odhiambo, R. *et al.* (1998). A geographic information system applied to a malaria field study in western Kenya. *American Journal of Tropical Medicine and Hygiene* **58**, 266–272.

Hjalmars, U., Kulldorff, M., Gustafsson, G. and Nagarawalla, N. (1996). Childhood leukaemia in Sweden: using GIS and a spatial scan statistic for cluster detection. *Statistics in Medicine* **15**, 707–715.

Hugh-Jones, M. (1989). Applications of remote sensing to the identification of the habitats of parasites and disease vectors. *Parasitology Today* **5**, 244–251.

Hugh-Jones, M. (1991). LANDSAT-TM identification of the habitats of the cattle tick, *Amblyomma variegatum*, in Guadeloupe, French Windward Islands. *Preventive Veterinary Medicine* **11**, 355–356.

Hugh-Jones, M., Barre, N., Nelson, G. *et al.* (1992). Landsat-TM identification of *Amblyomma variegatum* (Acari: Ixodidae) habitats in Guadeloupe. *Remote Sensing of Environment* **40**, 35–55.

Hungerford, L.L. (1991). Use of spatial statistics to identify and test significance in geographic disease patterns. *Preventive Veterinary Medicine* **11**, 237–242.

James, F.C. and McCullock, C.E. (1990). Multivariate analysis in ecology and systematics: panacea or Pandora's box? *Annual Review of Ecology and Systematics* **21**, 129–166.

Kitron, U. and Kazmierczak, J.J. (1997). Spatial analysis of the distribution of Lyme disease in Wisconsin. *American Journal of Epidemiology* **145**, 558–566.

Kitron, U., Jones, C.J., Bouseman, J.K., Nelson, J.A. and Baumgartner, D.L. (1992). Spatial analysis of the distribution of *Ixodes dammini* (Acari: Ixodidae) on white-tailed deer in Ogle County, Illinois. *Journal of Medical Entomology* **29**, 259–266.

Kitron, U., Otieno, L.H., Hungerford, L.L. *et al.* (1996). Spatial analysis of the distribution of tsetse flies in the Lambwe Valley, Kenya, using Landsat TM satellite imagery and GIS. *Journal of Animal Ecology* **65**, 371–380.

Kitron, U., Michael, J., Swanson, J. and Haramis, L. (1997). Spatial analysis of the distribution of Lacrosse encephalitis in Illinois, using a geographic information system and local and global spatial statistics. *American Journal of Tropical Medicine and Hygiene* **57**, 469–475.

Kulldorff, M. and Nagarwalla, N. (1995). Spatial disease clusters: detection and inference. *Statistics in Medicine* **14**, 799–819.

Landis, J.R. and Koch, G.C. (1977). The measurement of observer agreement for categorical data. *Biometrics* **33**, 159–174.

Legendre, P. (1993). Spatial autocorrelation: trouble or new paradigm? *Ecology* **74**, 1615–1673.

Lessard, P., L'Eplattenier, R., Norval, R.A.I. *et al.*(1990). Geographical information systems for studying the epidemiology of cattle diseases caused by *Theileria parva*. *Veterinary Record* **126**, 255–262.

Liang, K.Y. and Zeger, S.L. (1986). Longitudinal data analysis using generalized linear models. *Biometrika* **73**, 13–22.

Liebhold, A.M., Rossi, R.E. and Kemp, W.P. (1993). Geostatistics and geographic information systems in applied insect ecology. *Annual Review of Entomology* **38**, 303–327.

Lowry, J.H.J., Miller, H.J. and Hepner, G.F. (1995). A GIS-based sensitivity analysis of community vulnerability to hazardous contaminants on the Mexico/US Border. *Photogrammetric Engineering & Remote Sensing* **61**, 1347–1359.

Ma, Z. and Redmond, R.L. (1995). Tau coefficients for accuracy of assessment of classification of remote sensing data. *Photogrammetric Engineering & Remote Sensing* **61**, 435–439.

Malone, J.B. and Zukowski, S.H. (1992). Geographic models and control of cattle liver flukes in the southern USA. *Parasitology Today* **8**, 266–270.

Malone, J.B., Huh, O.K., Fehler, D.P. *et al.* (1994). Temperature data from satellite imagery and the distribution of schistosomiasis in Egypt. *American Journal of Tropical Medicine and Hygiene* **50**, 714–722.

Malone, J.B., Abdel-Rahman, M.S., El Bahy, M.M., Huh, O.K., Shafik, M. and Bavia, M. (1997). Geographic information systems and the distribution of *Schistosoma mansoni* in the Nile delta. *Parasitology Today* **13**, 112–119.

Marsh, W.E., Damrongwatanapokin, T., Larntz, K. and Morrison, R.B. (1991). The use of a geographic information system in an epidemiological study of pseudorabies (Aujeszky's disease) in Minnesota swine herds. *Preventive Veterinary Medicine* **11**, 249–254.

Maurer, B.A. (1994). *Geographical Population Analysis: Tools for the Analysis of Biodiversity.* London: Blackwell Scientific.

McGinn, T.J.I., Cowen, P. and Wray, D.W. (1996). Geographic information systems for animal health management and disease control. *Journal of the American Veterinary Medicine Association* **209**, 1917–1921.

Moellering, H. and Tobler, W.R. (1972). Geographical variances. *Geographical Analysis* **4**, 34–50.

Moran, P.A.P. (1948). The interpretation of statistical maps. *Journal of the Royal Statistical Society B* **10**, 243–251.

Moran, P.A.P. (1950). Notes on continuous stochastic phenomenon. *Biometrika* **37**, 17–23.

Mott, K.E., Nuttall, I., Desjeux, P. and Cattand, P. (1995). New geographical approaches to control of some parasitic zoonoses. *Bulletin of the World Health Organization* **73**, 247–257.

Nicholson, M.C. and Mather, T.N. (1996). Methods for evaluating Lyme disease risks using geographic information systems and geospatial analysis. *Journal of Medical Entomology* **33**, 711–720.

Njemanze, P.C., Anozie, J., Ihenacho, J.O., Russell, M.J. and Uwaeziozi, A.B. (1999). Application of risk analysis and geographic information system technologies to the prevention of diarrheal diseases in Nigeria. *American Journal of Tropical Medicine and Hygiene* **61**, 356–360.

Norval, R.A.I., Perry, B.D., Gebreab, F. and Lessard, P. (1991a). East Coast fever: a problem of the future for the horn of Africa? *Preventive Veterinary Medicine* **10**, 163–172.

Norval, R.A.I., Perry, B.D., Kruska, R. and Kundert, K. (1991b). The use of climate data interpolation in estimating the distribution of *Amblyomma variegatum* in Africa. *Preventative Veterinary Medicine* **11**, 365–366.

Ohno, Y., Aoki, K. and Aoki, N. (1979). A test of significance for geographical clusters of disease. *International Journal of Epidemiology* **8**, 273–281.

Openshaw, S. (1991). A spatial analysis research agenda. In: *Handling Geographical Information: Methodology and Potential Applications* (I. Masser and M.J. Blakemore, eds), pp. 18–37. Harlow: Longman.

Openshaw, S. (1996). Geographical information systems and tropical diseases. *Transactions of the Royal Society of Tropical Medicine and Hygiene* **90**, 337–339.

Openshaw, S., Charlton, M., Wymer, C. and Craft, A. (1987). A mark I geographical analysis machine for the automated analysis of point data sets. *International Journal of Geographical Information Systems* **1**, 335–358.

Openshaw, S., Craft, A., Charlton, M. and Birch, J.M. (1988). Investigation of leukaemia clusters by use of a geographical analysis machine. *Lancet* **1**, 272–273.

Perry, B.D., Lessard, P., Norval, R.A.I., Kundert, K. and Kruska, R. (1990). Climate, vegetation, and the distribution of *Rhipicephalus appendiculatus* in Africa. *Parasitology Today* **6**, 100–137.

Perry, B.D., Kruska, R., Lessard, P., Norval, R.A.I. and Kundert, K. (1991). Estimating the distribution and abundance of *Rhipicephalus appendiculatus* in Africa. *Preventive Veterinary Medicine* **11**, 261–268.

Pope, K.O., Rejmankova, E., Savage, H.M., Arredondojimenez, J.I., Rodriguez, M.H. and Roberts, D.R. (1994). Remote sensing of tropical wetlands for malaria control in Chiapas, Mexico. *Ecological Applications* **4**, 81–90.

Popovich, M.L. and Tatham, B. (1997). Use of immunization data and automated mapping techniques to target public health outreach programs. *American Journal of Preventive Medicine* **13**, Supplement: Developing Immunization Registries, 102–107.

Porter, J.D.H. (1999). Geographical information systems (GIS) and the tuberculosis DOTS strategy. *Tropical Medicine & International Health* **4**, 631–633.

Randolph, S.E. and Rogers, D.J. (1997). A generic population model for the African tick *Rhipicephalus appendiculatus*. *Parasitology* **115**, 265–279.

Rejmankova, E., Roberts, D.R., Pawley, A., Manguin, S. and Polanco, J. (1995). Predictions of adult *Anopheles albimanus* densities in villages based on distances to remotely sensed larval habitats. *American Journal of Tropical Medicine and Hygiene* **53**, 482–488.

Richardson, S. (1992). Statistical methods for geographical correlation studies. In: *Geographical and Environmental Epidemiology* (P. Elliott, J. Cuzick, D. English and R. Stern, eds), pp. 181–204. Oxford: Oxford University Press.

Ripley, B.D. (1976). The second-order analysis of secondary point processes. *Journal of Applied Probability* **13**, 255–266.

Ripley, B.D. (1981). *Spatial statistics.* New York: Wiley & Sons.

Robinson, T.P. (1998). Geographic information systems and the selection of priority areas for control of tsetse-transmitted trypanosomiasis in Africa. *Parasitology Today* **14**, 457–461.

Robinson, T.P. and Hopkins, J.S. (1999). Managing livestock disease data: the Disease And Vector Integrated Database (DAVID). Annual Conference of the Society of Veterinary Epidemiology and Preventative Medicine, Bristol, pp. 24–26.

Robinson, T.P., Rogers, D.J. and Williams, B. (1997a). Mapping tsetse habitat suitability in the common fly belt of Southern Africa using multivariate analysis of climate and remotely sensed vegation data. *Medical and Veterinary Entomology* **11**, 235–245.

Robinson, T.P., Rogers, D.J. and Williams, B. (1997b). Univariate analysis of tsetse habitat in the common fly belt of Southern Africa using climate and remotely sensed vegetation data. *Medical and Veterinary Entomology* **11**, 223–234.

Robinson, T.P., Hopkins, J.S., Williams, B.G. and Harris, R.S. (1999). Decision support for trypanosomiasis control: An example using a geographic information system in eastern Zambia. XXV Meeting of the International Scientific Council for Trypanosomiasis Research and Control (ISCTRC). Mombassa, Kenya.

Rogers, D.J. and Randolph, S.E. (1991). Mortality rates and population density of tsetse flies correlated with satellite imagery. *Nature* **351**, 739–741.

Rogers, D.J. and Randolph, S.E. (1993). Distribution of tsetse and ticks in Africa: past, present, and future. *Parasitology Today* **9**, 266–271.

Rogers, D.J. and Williams, B.G. (1993). Monitoring trypanosomiasis in space and time. *Parasitology* **106**, S77-S92.

Rogers, D.J. and Williams, B.G. (1994). Tsetse distribution in Africa: seeing the wood and the trees. In: *Large-Scale Ecology and Conservation Biology*, British Ecological Society Symposium XXXV (P.J. Edwards, R.M. May and N.Webb, eds), pp. 247–271. Oxford: Blackwell Scientific.

Rogers, D.J., Hay, S.I. and Packer, M.J. (1996). Predicting the distribution of tsetse flies in West Africa using temporal Fourier processed meteorological satellite data. *Annals of Tropical Medicine and Parasitology* **90**, 225–241.

Rosenfield, G.H. and Fitzpatrick-Lins, K. (1986). A coefficient of agreement as a measure of thematic classification accuracy. *Photogrammetric Engineering & Remote Sensing* **52**, 223–227.

Sanson, R.L., Pfeiffer, D.U. and Morris, R.S. (1991). Geographic information systems: their application in animal disease control. *Revue Scientifique et Technique de l'Office International des Epizooties* **10**, 179–195.

Sharma, V.P. and Srivastava, A. (1997). Role of geographic information system in malaria control. *Indian Journal of Medical Research* **106**, 198–204.

Smans, M. and Esteve, J (1992). Practical approaches to disease mapping. In: *Geographical and Environmental Epidemiology* (P. Elliott, J. Cuzick, D. English and R. Stern, eds), pp. 141–150. Oxford: Oxford University Press.

Sokal, R.R. and Rohlf, F.J. (1981). *Biometry. The Principles and Practice of Statistics in Biological Research.* New York: W.H. Freeman.

Sutherst, R.W., Spradbery, J.P. and Maywald, G.F. (1989). The potential geographic distribution of the Old World screw-worm fly, *Chrysomya bezziana*. *Medical and Veterinary Entomology* **3**, 273–280.

Tanser, F. and Wilkinson, D. (1999). Spatial implications of the tuberculosis DOTS strategy in rural South Africa: a novel application of geographical information system and global positioning system technologies. *Tropical Medicine & International Health* **4**, 634–638.

Tempalski, B.J. (1994). The case of Guinea worm: GIS as a tool for the analysis of disease control policy. *Geographical Information Systems* **4**, 32–39.

Thompson, D.F., Malone, J.B., Harb, M. *et al.* (1996). Bancroftian filariasis distribution and diurnal temperature differences in the southern Nile data. *Emerging Infectious Diseases* **2**, 234–235.

Thomson, M.C., Connor, S.J., Dalessandro, U. *et al.* (1999). Predicting malaria infection in Gambian children from satellite data and bed net use surveys: the importance of spatial correlation in the interpretation of results. *American Journal of Tropical Medicine and Hygiene* **61**, 2–8.

Thrall, I.G. (1995). New generation of mass-market GIS software: a commentary. *Geographical Information Systems* **5**, 58–60.

Turnbull, B.W., Iwano, E.J., Burnett, W.S., Howe, H.L. and Clark, L.C. (1990). Monitoring for clusters of disease: applications to leukemia incidence in upstate New York. *American Journal of Epidemiology* **132**, S136-S143.

Upton, G.J.G. and Fingleton, B. (1985). *Spatial Data Analysis by Example. Vol. 1. Point Pattern and Quantitative Data.* New York: Wiley & Sons.

Urquhart, J. (1992). Studies of disease clustering: problems of interpretation. In: *Geographical and Environmental Epidemiology* (P. Elliott, J. Cuzick, D. English and R. Stern, eds), pp. 278–285. Oxford: Oxford University Press.

van der Knaap, W.G.M. (1992). The vector to raster conversion: (mis)use in geographical information systems. *International Journal of Geographical Information Systems* **6**, 159–170.

Vanderzee, D. and Singh, A. (1995). Survey of geographical information system and image processing software. *International Journal of Remote Sensing* **16**, 383–389.

Vine, M.F., Degnan, D. and Hanchette, C. (1997). Geographic information systems: their use in environmental epidemiologic research. *Environmental Health Perspectives* **105**, 598–605.

Walker, P.A. and Moore, D.M. (1988) SIMPLE: an inductive modelling and mapping tool for spatially oriented data. *International Journal of Geographical Information Systems* **2**, 347–364.

Washino, R.K. and Wood, B.L. (1994). Application of remote sensing to arthropod vector surveillance and control. *American Journal of Tropical Medicine and Hygiene* **50**, 134–144.

Whittemore, A., Friend, N., Brown, B.W. *et al.* (1987). A test to detect clusters of disease. *Biometrica* **74**, 631–637.

Wiens, J.A. (1989). Spatial scaling in ecology. *Functional Ecology* **3**, 385–397.

Williams, B. (1993). *Biostatistics: Concepts and Applications for Biologists.* London: Chapman and Hall.

Williams, B., Rogers, D., Staton, G., Ripley, B. and Booth, T. (1994). Statistical modelling of georeferenced data: mapping tsetse distributions in Zimbabwe using climate and vegetation data. Proceedings of a workshop organised by ILRAD in collaboration with the FAO, Nairobi.

Wood, B., Washino, R., Beck, L. *et al.* (1991). Distinguishing high and low anopheline-producing rice fields using remote sensing and GIS technologies. *Preventive Veterinary Medicine* **11**, 277–288.

Wood, B.L., Beck, L.R., Washino, R.K., Hibbard, K.A. and Salute, J.S. (1992). Estimating high mosquito-producing rice fields using spectral and spatial data. *International Journal of Remote Sensing* **13**, 2813–2826.

Yamagata, Y., Suzuki, T. and Garcia Manzo, G.A. (1986). Geographical distribution of the prevalence of nodules of *Onchocerca volvulus* in Guatemala over the last four decades. *Tropical Medicine and Parasitology* **37**, 28–34.

Satellites, Space, Time and the African Trypanosomiases

D.J. Rogers

*Trypanosomiasis and Land Use in Africa (TALA) Research Group,
Department of Zoology, University of Oxford, South Parks Road,
Oxford OX1 3PS, UK*

ADVANCES IN PARASITOLOGY VOL 47
0065-308-X $30.00

ABSTRACT

The human and animal trypanosomiases of Africa provide unique challenges
to epidemiologists because of the spatial and temporal scales over which
variation in transmission takes place. This chapter describes how our
descriptions of the different components of transmission, from the parasites
to the affected hosts, eventually developed to include geographical
dimensions. It then briefly mentions two key analytical techniques used in the
application of multi-temporal remotely sensed imagery to the interpretation
of field data; temporal Fourier analysis for data reduction, and a variety of
discriminant analytical techniques to describe the distribution and abundance
of vectors and diseases. Satellite data may be used both for biological,
process-based models and for statistical descriptions of vector populations
and disease transmission. Examples are given of models for the tsetse
Glossina morsitans in the Yankari Game Reserve, Nigeria, and in The
Gambia. In both sites the satellite derived index of Land Surface
Temperature (LST) is the best correlate of monthly mortality rates and is
used to drive tsetse population models. The Gambia model is then
supplemented with a disease transmission component; the mean infection
rates of the vectors and of local cattle are satisfactorily described by the
model, as are the seasonal variations of infection in the cattle.

High and low spatial resolution satellite data have been used in a number
of statistical studies of land cover types and tsetse habitats. In addition multi-
temporal data may be related to both the incidence and prevalence of
trypanosomiasis. Analysis of past and recent animal and human
trypanosomiasis data from south-east Uganda supports the suggestion of the
importance of cattle as a reservoir of the human disease in this area; mean
infection prevalences in both human and animal hosts rise and fall in a
similar fashion over the same range of increasing vegetation index values.

Monthly sleeping sickness case data from the districts and counties of
south-east Uganda are analysed and often show significant correlations with
local LST. Case numbers increase with LST in areas that are relatively cooler
than average for this part of Uganda, but decrease with LST in areas that are
on average warmer. This indicates different seasonal cycles of risk across the
region, and may be related to the differing vectorial roles of the two local
tsetse, *G. fuscipes* and *G. pallidipes*.

Finally, the increasing pace of change, and the likelihood of new or re-
emerging vector-borne diseases, highlight the need for accurate and timely
information on habitat changes and the impacts these will have on disease
transmission. The next generation of satellites will have significantly more
spectral and spatial resolution than the current satellites, and will enable us
to refine both statistical and biological predictions of trypanosomiasis and
other vector-borne diseases within disease early warning systems.

1. INTRODUCTION

1.1. The Geography of the African Trypanosomiases

The African trypanosomiases affect humans, domestic animals and wildlife over *c.*10 million km^2 of sub-Saharan Africa (Murray and Gray, 1984). The 30 species and subspecies of the tsetse vectors of these diseases are adapted to forest (*Glossina fusca* group), forest and riverine (*G. palpalis* group) and savannah (*G. morsitans* group) vegetation types and each species has unique habitat requirements. Human African trypanosomiasis (HAT), 'sleeping sickness', is a zoonosis with domestic and wild vertebrates recognized as competent hosts for both of the human-infective strains, *Trypanosoma* (*Trypanozoon*) *brucei gambiense* in West Africa and *T. b. rhodesiense* in East Africa. Domestic animal trypanosomiasis, 'nagana', involves mostly *T.* (*Duttonella*) *vivax*, *T.* (*Nannomonas*) *congolense* and, to a much lesser extent, *T. b. brucei*, none of which infects humans (with very rare exceptions) (Hoare, 1972).

   Putting together the pieces of this particular epidemiological jigsaw has exercised generations of scientists and administrators in Africa. If the three steps of science are to describe, to explain and then to predict, we began the first step more than a century ago, with Bruce (1895) first associating the symptoms of nagana with trypanosome infection and tsetse with its transmission and Kleine (1909) later showing the cyclical development of the trypanosomes within these insect vectors (Vickerman, 1997). The second step was taken by colonial administrators, scientists and historians, not always holding the same view of this complex problem (Ford, 1971). The colonial administrators saw tsetse and trypanosomiasis as a long-standing scourge on the continent, and devoted a disproportionate amount of colonial research funding (approximately one quarter) to this single group of diseases (Farley, 1991). Scientists saw a more detailed picture of the interaction between three very different systems, those of the parasites, the vectors and the hosts, but their studies initially emphasized one or other element of this triumvirate and only rarely considered the totality, i.e. epidemiology.

1.2. The Trypanosome Parasites

Once sufficiently precise diagnostic techniques were established, the parasitologists were generally struck by the profusion of trypanosome strains circulating in areas with tsetse vectors, and by the antigenic variation of each strain over time within each individual host's bloodstream (Barry, 1997). Trypanosome strains were initially collected and cryopreserved in parasitological 'stamp collections', which were drawn upon for laboratory experiments, but otherwise unexploited. A question asked in the 1980s of the scientists engaged in the feverish search for yet more strains to add to such unorganized collections was 'When is more, enough?' The answer came when strain variation was fitted into its correct geographical context. We are now able to put geographical pages in our stamp albums; some trypanosome strains are restricted to quite small areas; others are mostly restricted to small areas with occasional outliers hinting at a more widespread distribution, and other strains are already known to be quite widespread (Hide, 1997; Komba *et al.*, 1997).

Individual strains retain their infectivity to host species such as humans over periods of more than 20 years spent in alternative hosts (Ashcroft, 1959), and modern molecular techniques confirm the persistence of relatively homogenous strain groups of trypanosomes in the same geographical areas for periods of at least 35 years (Hide, 1997). This part of the trypanosomiasis story is now near the completion of its descriptive stage: explanations will almost certainly involve elements of space and time that are implicit in the term 'landscape epidemiology' (Kitron, 1998).

1.3. The Tsetse Vectors

Whilst the parasitological work continued, the entomologists were engaged in pioneering work on tsetse ecology. The distribution of the different tsetse species was captured in a series of maps based on point captures of flies and judicious interpolation, often based on elevation that assumed cold temperature limitation of tsetse breeding (Newstead *et al.*, 1924; Ford and Katondo, 1977). Excellent but generally forgotten work in West Africa, following Bodenheimer's pioneering studies of climatic constraints on animal species distributions (Bodenheimer, 1938), began to relate tsetse distributions to a combination of temperature and humidity conditions, captured by climatographs (Gaschen, 1945). Much later this approach was given a quantitative treatment that allowed predictions to be made of continental species' distributions, based on temperature and saturation deficit surfaces (Rogers, 1979; Rogers and Randolph, 1986). Within tsetse distributional limits, a number of studies investigated seasonal variations in fly numbers (Glasgow and Welch, 1962). Jackson's mark–release–recapture studies in the 1930s were advised by Ronald Fisher in Cambridge, UK, and were the first of their kind for any insect species (Jackson, 1930, 1937). It was quickly appreciated that the numbers of tsetse captured by a variety of methods (a relative estimate of population size) was only indirectly related to the numbers of flies present in an area (the absolute population size), and the links between relative and absolute population size estimates continue to be investigated (Hargrove and Borland, 1994). Both fly distribution and fly abundance have environmental correlates and must therefore be studied in a geographical framework.

1.4. Human Trypanosomiasis

Both Jackson and Nash examined the impact of the environment on tsetse populations and Nash was amongst the first to appreciate that the degree of contact between humans and flies ('man–fly contact') could locally be more important than fly abundance in the transmission of HAT (Nash, 1944, 1948). A small population of flies persisting around a watering point can present a significant risk to humans coming to collect water. The link between vectors and the human host was, however, most critically investigated by Morris in both West and East Africa (Morris and Morris, 1949; Morris, 1951, 1952, 1959, 1960a,b, 1963). Morris showed first how the risk of human disease could be related to the distance of human habitation from the nearest tsetse habitat. Second, he showed that a reduction of this habitat, especially by the removal of riverine forest, could bring about a sustained reduction of both the fly population at watering points and the risk of human infection in such places. Thus both the broad geographical distribution of flies and the fine-scale interactions of flies with habitat features are vital ingredients of our present, rather poor, understanding of HAT.

1.5. Animal Trypanosomiasis

Veterinary research on the African trypanosomiases involved pathology, parasitology and epizootiology. It is clear that the major cattle-rearing regions of Africa are constrained by the presence of tsetse, which exclude the relatively susceptible zebu (*Bos indicus*) races from potentially productive areas. The relatively more trypanotolerant (*Bos taurus*) races are widespread in parts of West Africa, especially in areas of low to intermediate tsetse challenge (neither type of animal can survive in the areas of highest challenge). The impact of trypanosomiasis on much of Africa can be appreciated by examining places where it does not occur, such as the tsetse-free highland areas of Ethiopia and Kenya, and the colder regions of southern Africa; in such places, livestock are integrated into farming systems and contribute to significant increases in productivity in terms of meat, milk, ploughing and manuring (Swallow, 1997). These comparisons led major United Nations agencies to estimate that continental meat production could be doubled if tsetse could be eradicated from Africa. Maps may be produced showing the potential productivity of tsetse-infested areas, to inform the debate between scientists, politicians and development specialists about how best to tackle the African trypanosomiasis problem.

1.6. Colonial Impacts and HAT Epidemics

The parasitological, entomological and veterinary research outlined above was prompted by political interest in, and concern about, devastating epidemics of human sleeping sickness that swept through Africa at the end of the nineteenth and beginning of the twentieth centuries. Unlike the ubiquitous animal trypanosomiases, HAT is often confined to relatively small foci that appear to have persisted for many centuries (Ford, 1971). European exploration followed rapidly by colonial expansion in Africa at the end of the nineteenth century appear to have spread the disease from these persistent foci into new areas. Thus, beginning in 1912, a wave of HAT outbreaks began in what is now the Democratic Republic of the Congo and spread through West Africa taking 30 years to reach Sierra Leone about 3000 km away (Scott, 1965); prevalences commonly exceeded 5%, and locally reached 50%. At more or less the same time, East Africa began to experience hitherto unknown epidemics of HAT in Uganda and Tanzania, with peak numbers of cases in 1902 and 1929 respectively. Several major expeditions, including those of H.M. Stanley accompanied by thousands of African porters, passed through known disease foci in the basin of the River Zaire (Dutton and Todd, 1906) before returning to East Africa, perhaps carrying the infections that later caused HAT outbreaks in new areas with many susceptible humans (Ford, 1971; McLynn, 1991).

Whilst modern epidemiological theory can explain why the human disease is so focal (Rogers, 1988), historical analysis is required to provide an explanation for the place and timing of HAT outbreaks.

1.7. Alternative Views of the Tsetse/Trypanosomiasis Problem

The impacts of both human and animal trypanosomiases are undoubtedly severe but, towards the end of the last century, arguments were put forward suggesting a beneficial effect of both tsetse and trypanosomiasis; that of preserving habitats from human destruction by overgrazing and desertification (Ormerod, 1976, 1986, 1990; Ormerod and Rickman, 1988). Such problems are apparent in the drier regions of West Africa, where animal production is limited on its southerly borders by tsetse and in the north by the Sahara desert; such marginal areas appear to show dramatic short-term impacts of heavy use by cattle, although longer term damage is limited, because the local vegetation has a remarkable capacity to recover when rain eventually falls (Hiernaux and Justice, 1986). Once again, therefore, the importance of trypanosomiasis must be seen in a geographical framework.

2. DISEASE TRANSMISSION MODELS

The third step of science, that of prediction, can be made successfully only once all the above influences have been fitted into a quantitative framework. Models for the African trypanosomiases, as for most vector-borne diseases, owe much to Ross's pioneer modelling of malaria, again at the beginning of the twentieth century (Ross, 1909, 1911). Macdonald's development of this model (Macdonald, 1957) provided the stimulus to modelling the more complex African trypanosomiases, with multiple vector and host species (Milligan and Baker, 1988; Rogers, 1988). Because of the complexity of these models, they were generally solved only for their equilibrium predictions, and were used to estimate the basic reproductive number of the trypanosomiases, R_0 (Anderson and May, 1991). The spatial nature of disease risk was not explicitly modelled and the seasonality of transmission was modelled only in a very general way, by imagining simple monthly (sinusoidal) variations in vector numbers (Rogers, 1988). The result of such mathematical exercises was therefore not so much the production of a model for the African trypanosomiases, but rather a model of a model for these diseases. These proto-models told us what we still needed to know about the trypanosomiases, and many of these 'needs' had the dimensions of both space and time.

2.1. Problems of Space and Time

Intensive studies of parasites, vectors or hosts tend to be relatively restricted geographically. Extending the results to other places involves several crucial assumptions, most of which are difficult to justify. For example, will a model for a tsetse population developed for one place apply to another, which has the same tsetse species but a different habitat and probably, therefore, a different set of natural enemies of tsetse? Can models for the natural disease situation be modified to describe the impacts of intervention with trypanocidal drugs or insecticides? These problems of time and space will be reduced by a better understanding of the dynamics of disease transmission that will come from the development of spatially and temporally explicit disease transmission models.

2.2. Problems of Scale

Related to the problems of space and time are the problems of scale, which become especially important when disease control is being considered. Can the techniques developed for fly suppression in a small area be applied without modification to a much larger area? The answer to this question probably depends on habitat structure and whether or not habitats are scale invariant, i.e. have the same fractal dimension at all spatial scales (Bonham-Carter, 1994; Turner *et al.*, 1998). Can a transmission model for the endemic situation of the disease also describe epidemic outbreaks, i.e. longer time-scale events? The answer to this question depends upon the origin of such epidemics, and whether or not they are intrinsic to the transmission being modelled (i.e. can be described by the standard susceptible–infected–recovered/immune models (Anderson and May, 1991)) or are extrinsically driven by variation in climate or habitats. There is evidence for both in the African trypanosomiases (Rogers and Williams, 1993) and no clear idea of which is the more important. For both sorts of questions of scale, however, a system monitoring variations in both space and time will provide information to test the alternative hypotheses, or at least to rule out some of the alternatives. Satellite imagery provides the potential source of such information.

3. RELATING FIELD AND SATELLITE DATA

The two types of modelling approaches adopted here and elsewhere in this volume fall into the general categories of descriptive (statistical, descriptive) or predictive (biological, process-based) (Table 1). Descriptive models are best developed over large areas that show a variety of environmental conditions, and assume that other places are statistically similar to those for which the original models were developed. For this reason they may be called status quo models. Statistical models can give impressive descriptions of spatial variation in risk but are generally poor at predicting temporal variation in risk because the varying time delays that operate in disease transmission systems are not well captured by the statistical approach. Predictive models, on the other hand, are properly based on a description of the biological interactions between species and therefore, in theory, predict variation through time rather well. Such models, however, tend to be

Table 1 The important differences between statistical and biological approaches to vector and disease mapping.

Type	Requirements	Comments
Statistical	Large-area, multiple layer data sets sampling a wide range of ecological conditions	Extensive studies, status quo models, descriptive
Biological	Long-term study of local dynamics and transmission, measuring demographic rates and variables	Intensive studies, dynamic models, explanatory

developed for relatively small areas (because they are based on local, intensive studies), and so are poor at describing large area spatial variation. The relative strengths of statistical and biological descriptions of disease processes through space and time suggest that their combination should provide us with new and powerful tools for combating complex tropical diseases. Such a combination may be brought about by exploiting remotely sensed satellite data that provide both habitat descriptions for the descriptive approach and measures of important driving climatic variables for process-based models and the predictive approach. Very often the same satellite variable appears to be important in the two rather different approaches (Randolph, this volume).

It is clear from the above that the available satellite imagery (Hay, this volume) will find very different applications in the alternative approaches. The high spatial resolution imagery derived from sensors on the Landsat and Satellite Pour l'Observation de la Terre (SPOT) satellites are ideal for local area descriptions of the habitat (the statistical approach), at a level of detail unavailable from meteorological satellites, but it is only the latter that are capable of regularly monitoring the seasonality of the habitat that influences biological transmission variables (the biological approach). Studies at the regional and continental levels would require such large volumes of high spatial resolution Landsat TM images that meteorological satellite data may be preferred, even for statistical descriptions.

3.1. Image Preparation—Fourier Analysis

Multi-temporal data from Advanced Very High Resolution Radiometer (AVHRR) on board the National Oceanographic and Atmospheric Administration's (NOAA) series of oceanographic satellites and the High Resolution Radiometer (HRR) on the Meteosat meteorological satellites show a repeated annual pattern of variation that may be used to describe habitat AVHRR seasonality in the different satellite sensor channels. Once the raw data are processed to provide monthly indices correlated with vegetation abundance and activity (the normalized difference vegetation index, NDVI) or meteorological variables (AVHRR—Channel 3 land surface radiance, split-window land surface temperatures, vapour pressure deficit; HRR—Cold Cloud Duration (CCD); Hay, this volume; Goetz, this volume), the data may be subjected to temporal Fourier analysis (see the Appendix) that extracts information about the seasonal cycles of these indices in terms of their annual, bi-annual, tri-annual etc. cycles (or 'harmonics'), each one described by its phase and amplitude. Once defined, the various Fourier harmonics may be recombined to provide a description of the original signal with as much detail as required by the user. Fourier analysis performs three quite separate and useful functions:

(1) It removes noise from the original satellite signal. Noise is generally at a high frequency relative to seasonal events, so the corresponding harmonics may be omitted from the Fourier description to produce a smoothed picture of seasonal change.

(2) It achieves data reduction (ordination) of monthly data sets that often show strong month-to-month correlations.

(3) It achieves data ordination in a way that has an obvious biological interpretation in terms of cycles of seasonal events. This is in marked contrast to other methods of data reduction such as principal components analysis where seasonal events may contribute in complex ways to any number of the principal component axes and images derived from them (Eastman and Fulk, 1993).

Further details of temporal Fourier processing are given in Rogers *et al.* (1996) and examples of the application of Fourier-processed imagery to studying ecological patterns and processes are given in Rogers and Williams (1994).

3.2. Distribution and Abundance Analysis—Maximum Likelihood Methods

The reduced-dimension data set produced by the methods outlined above form the set of predictor variables used to describe the field observations on vectors or diseases. As detailed elsewhere (Curran, this volume; Robinson, this volume), there is a variety of statistical methods that may be used for such descriptions, ranging from 'black-box' techniques such as neural networking, to simple thresholding on single environmental variables such as temperature or rainfall (reviewed by Williams *et al.*, 1992; see also Manel *et al.*, 1999). In general these alternatives fall into two categories; those that assume an underlying uni- or multi-variate normal distribution of the predictor variables in areas of the presence or absence of vectors or diseases, and those that do not. It is often tempting to opt for the statistical flexibility of the latter techniques but the danger of doing so is that we may, in the process, lose sight of the information about the biology of the vectors that the former methods can provide. Whilst it is unlikely that the continental distribution of widespread species such as the mosquito *Anopheles gambiae*, or the tsetse *G. morsitans,* may be described by a single set of multi-variate normal conditions, it does not seem unreasonable to imagine that their distributions are adapted to a series of multi-variate normal conditions, each differing subtly from the next. If this is the case, the application of modified multi-variate techniques will not only allow us to make more accurate statistical prediction of distributions, but will also provide information useful for regional biological models. One useful multi-variate technique, discriminant analysis, is described in the Appendix. This flexible technique can deal both with multiple categories of distribution and abundance data that arise from clustering environmental data, and with non-linearities of the response of these biological data to the environmental variables.

4. SMALL AREA, BIOLOGICAL MODELS

In this section I develop a generic approach to vector-borne disease modelling, and examine the potential contribution of satellite imagery to such studies.

4.1. Models for Tsetse Populations

Generic models for vectors involve components of birth and death, at least one of which must be density dependent. The temperature and humidity dependence of demographic rates may be investigated in the laboratory (e.g. Buxton and Lewis, 1934), but these studies provide only an approximate guide to such rates in field conditions. Tsetse apparently carefully select micro-habitats that are considerably moister (Bursell, 1959) and, at least in the hottest conditions, cooler than ambient (Hargrove and Packer, 1993). The remarkably slow rate of offspring production by tsetse has been described by simple equations involving only air temperature (Glasgow, 1963; Hargrove, 1994). These equations may be used in predictive models using either meteorological or satellite data. One problem of the latter is that satellite sensors record directly only the thermal radiance (reflected or emitted) of the soil and vegetation cover, which is often much higher than air temperature (Hay *et al.*, 1996). Various manipulations of the satellite data can give air temperature estimates with accuracies of a few degrees Celsius (Prihodko and Goward, 1997; Prince *et al.*, 1998; Hay and Lennon, 1999; Green and Hay, 2000; Goetz *et al.*, this volume), sufficiently good for initial models.

In sharp contrast to birth rates, death rates in tsetse appear to depend on both temperature and atmospheric moisture (Rogers and Randolph, 1986), and there is also strong evidence for density dependence at both the puparial and adult stages (Rogers, 1974; Rogers and Randolph, 1984; Rogers *et al.*, 1984).

A combination of birth and death rates, each described by the locally appropriate meteorological variables, and with a variable amount of density dependence, successfully described tsetse population changes in both West and East Africa (Rogers, 1990; Rogers *et al.*, 1994). It is now possible to use satellite data as a surrogate for the standard meteorological data. For example, in the Yankari Game Reserve in Nigeria the correlation between the bi-monthly mortality rate of *G. morsitans submorsitans* and LST estimates derived from satellites orbiting more than 800 km above the earth's surface is stronger than the correlation between mortality and saturation deficit (the best ground-based correlate) calculated from meteorological data collected about 50 km from the field site (Figure 1).

A satisfactory description of seasonal changes in tsetse populations is achieved by fine tuning several critical parameter values in the biological models. This fitting process can be automated by steepest descent search methods (Hargrove and Williams, 1998), although a careful check must be kept on parameter values since there appear to be many locally stable equilibria when models are fitted to population data. One example, using the satellite data to predict the monthly mortality rate of *G. m. submorsitans* in Nigeria, is shown in Figure 2. The model also requires some estimate of air temperature for predicting inter-larval and puparial developmental periods. Rather than predicting temperature from published formulae relating land surface radiance to air temperature, the model in Figure 2 included an additional fitted parameter to relate satellite (LST) and air temperature directly. This parameter was varied along with all the others to achieve a least-squares fit of the model to the field data.

This and other fitted tsetse population models suggest that substantial density dependence operates on these vectors (Rogers and Randolph, 1984), although the agents of these mortalities have never been sufficiently investigated. Fly abundance is a product of both the density independent, abiotic mortalities (which may be predicted from satellite data) and density dependent biotic ones, and hence models developed for one area may not be extended to others unless the density dependent components are in some way described by satellite data.

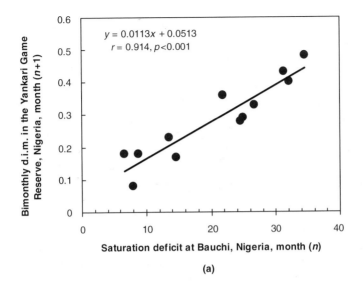

(a)

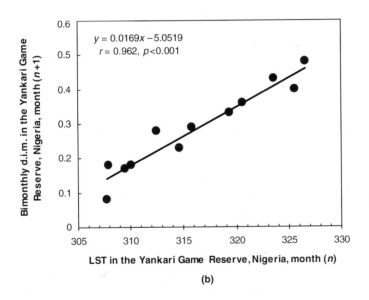

(b)

Figure 1 Comparison of the relationship between the bimonthly density independent mortality (d.i.m.) of the tsetse *Glossina morsitans submorsitans* in the Yankari Game Reserve in Nigeria and (a) saturation deficit of the previous month derived from meteorological records from Bauchi, 50 km away and (b) land surface temperature (LST) of the previous month from the NOAA series of satellites orbiting >800 km above.

4.2. Models for Trypanosome Transmission

Once the fit of the tsetse model is validated, the disease transmission component may be added. A simple transmission model for the African trypanosomiases, based on the standard susceptible–infected—recovered/immune model (Anderson and May, 1991) is described in the Appendix. This model contains equations describing changes in the proportions of vectors and hosts that are currently incubating infections, and of hosts that have recovered and are immune to re-infection for a period of time (these proportions are usually set to their equilibrium values in models that predict only equilibrium disease prevalences (Rogers, 1988)).

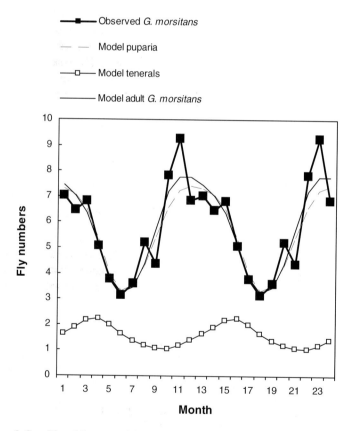

Figure 2 Satellite driven model for the tsetse *Glossina morsitans submorsitans* in the Yankari Game Reserve, Nigeria. The linear relationship shown in Figure 1(b) is used to predict monthly fly mortality rates within a generic population model for tsetse, fitted by least squares methods.

Figure 3 shows the result first of fitting the tsetse model and then of adding the trypanosomiasis component to field observations from The Gambia (data from Rawlings *et al.*, 1991). This model applies to locations where the major vector species is *G. morsitans submorsitans* and the hosts are the local trypanotolerant N'Dama cattle. The various parameters of the transmission equation can be varied to provide an excellent fit to the field data both in terms of the average level of infections in both the vectors and hosts and in the seasonal changes in infection rates. Fly infection rates are highest when

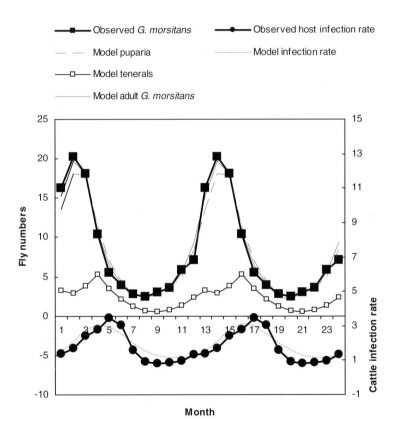

Figure 3 In The Gambia the best satellite correlate of the monthly mortality rate of the tsetse *Glossina morsitans submorsitans* is again the LST. This relationship is used in a least squares-fitted model for this species, to which are later added disease transmission equations from Appendix, C. The resulting trypanosome transmission model describes adequately the mean and seasonal variations in N'Dama cattle infection rates (lower panel; observed mean 1.77% and model mean 1.96%) and the mean fly infection rates (observed mean 2.61%, model mean 2.02%, not shown) (original fly and cattle infection data from Rawlings *et al.*, 1991, 'late dry season' data).

large proportions of the vectors are old; this does not occur at peak tsetse population levels, which include many young flies. Figure 3 shows that the model fly population reaches a peak in February, as do the flies in the field. Fly infection rates show two peaks in the model (not shown in Figure 3), in February and August, whilst field infection rates were variable and showed no significant month-to-month differences. Model host infection rates are highest in April, compared with May in the field, and show similar seasonal changes.

The surprising and encouraging result of this modelling exercise is that remotely sensed satellite data, selected on the basis of current understanding of tsetse dynamics in the field, may be used to drive a fully integrated disease transmission model. The challenge for the future is to see if such models give realistic predictions when extended to other areas, and when subjected to variations mimicking those of natural (climate) and anthropogenic (intervention) changes.

5. LARGE AREA, STATISTICAL MODELS

5.1. High Spatial Resolution Studies

5.1.1. *Vegetation Mapping*

Tsetse live in habitats that provide shade for developing puparia and resting sites for adults. If these habitats can be identified in remotely sensed imagery, not only may tsetse distributions be mapped over very wide areas, but the impact of tsetse on land use over time may be followed in a series of images taken every few years.

Giddings was apparently the first to suggest the use of Landsat Multispectral Scanner (MSC) derived imagery to identify tsetse habitats, in a feasibility exercise that was never followed up (Giddings and Naumann, 1976). A few years later, Bourn in Nigeria showed a significant positive relationship between *G. palpalis* abundance and the size of riverine forest patches as determined from Landsat MSS data (Bourn, 1983).

A recent study in Zimbabwe, using Landsat Thematic Mapper (TM) images taken between 1972 and 1993, concluded that changes in the natural vegetation cover, and in agricultural areas, could not be quantified easily because of spectral differences between similar vegetation types in the different images (Pender and Rosenberg, 1995; Pender *et al.*, 1997). The study concentrated on agricultural areas (Human Dominated Land Use, HDLU) and these were best revealed by applying an edge filter to Landsat TM band 7 and then overlaying the result on a band 3 image. This emphasized boundaries, and agricultural areas could be relatively easily distinguished and mapped. The study concluded that tsetse-borne trypanosomiasis was only one of several reasons for changing land use patterns in the area, and seldom the dominant one. This study showed that Landsat imagery, even when taken in the same month but in different years, may not be comparable at least partly because the precise timing of seasonal events differs from one year to the next. Rainfall causes a rapid and dramatic change in habitat appearance in many tropical savannahs and is notoriously unpredictable from one year to the next.

Recently, supervised classification of a single SPOT image in Burkina Faso, West Africa, using field observations of vegetation types, was able to produce a detailed classified habitat image, to which were related the mean trap catches of both *G. palpalis* and *G. tachinoides* (La Rocque, 1997). In all, 13 separate vegetation types were involved and the resulting producer's error (i.e. the error in classifying known vegetation types; Ma and Redmond, 1995) ranged from 1% to 61% whilst the consumer's error (i.e. the error in pixels classified as a particular vegetation type) ranged from 1% to 78%.

5.1.2. Tsetse Flies

Kitron and colleagues (1996) carried out a detailed spatial analysis of the numbers of *G. pallidipes* caught in several blocks of traps around the Ruma National Park in the Lambwe Valley, Kenya, an area of intermittent HAT transmission. Each trap location was described in terms of vegetation type and canopy cover. Spatial autocorrelation methods were applied and demonstrated strong spatial associations between the trap catches of tsetse. Landsat TM band 7, associated with vegetation and the moisture content of the soil, was the most consistent satellite correlate of fly numbers. Spatial filtering determined the relative contribution to this relationship of spatial and aspatial effects and showed a strong spatially independent association for some of the monthly catches, and for the average catches. This suggests that a large component of the association was the result of other determinants underlying the spatial distribution of both fly numbers and spectral values.

5.2. Low Spatial Resolution Studies

The ability of low resolution multi-temporal advanced very high resolution (AVHRR) data to capture habitat seasonality makes such data ideal for vector-borne disease studies not requiring the full spatial resolution of Landsat and SPOT imagery. AVHRR data have been used in a number of statistical studies to map vegetation, tsetse and disease.

5.2.1. *Vegetation Mapping*

Aerial surveys of Nigeria were carried out in the 1990 wet and dry seasons; every 20 km grid square was surveyed with a 5% sampling intensity at the end of the dry season, and 80% of these squares were re-surveyed at the end of the wet season (Bourn *et al.*, 1994). Digital elevation and the means, maxima, minima and standard deviations of selected raw channel AVHRR data and also processed vegetation indices and thermal data were used in a step-wise, non-linear discriminant analysis of eight main land-cover types (Rogers *et al.*, 1997). At the spatial resolution of the sample grid, many grid squares contained a mixture of vegetation types (and therefore mixed spectral signals for the satellites to detect), so a training set was selected that included only grid squares containing between 50% and 70% of a single vegetation type (sample sizes were too small for higher thresholds). At the 60% threshold, producer's accuracy ranged from 48% to 100%, and consumer's accuracy from 39% to 100% (both figures omitting a small sample of 'bare ground' squares that were misclassified as 'scrub' or 'woodland'). These figures were good enough for the classification signatures to be applied to the whole imagery, and the resulting vegetation map was supported by expert opinion. Many of the classification errors could easily be explained. For example, 'open woodland' was often misclassified as 'dense woodland', and the latter in turn as 'forest'. Cultivated areas were frequently misclassified as 'scrubland', 'open woodland' or 'dense woodland' and more rarely as 'forest'. Humans exploit these land-cover types for their agricultural activities, and trees are often left in place to provide shade for young crops. Although the producer's accuracy for cultivated areas was only 48%, the consumer's accuracy was 92%. Hence we conclude that areas mapped as cultivated by supervised methods are correctly identified, but some areas identified as woodland types may be under at least partial cultivation. As with the high resolution imagery, the identification of cultivated areas is crucial in determining the impact of tsetse and trypanosomiasis on development. Unfortunately, different crops can have very different spectral signatures, making identification of a single agricultural class from satellite imagery very difficult, and it may therefore be necessary to divide study areas into zones according to one or more determinants of crop type (e.g. rainfall) before carrying out supervised classifications on the zoned imagery.

5.2.2. *Tsetse Flies*

Methods for analysing tsetse distribution using AVHRR data were developed in a series of studies of *G. morsitans* and *G. pallidipes* in Kenya, Tanzania and Zimbabwe (Rogers, 1993; Rogers and Randolph, 1993; Rogers and Williams,

Table 2 The relative importance of meteorological satellite variables for predicting the distribution of tsetse flies in Africa.

	Africa (all)	West Africa	East Africa
NDVI	22	8	14
Thermal	8	15	12
Cold cloud duration	16	6	8
Elevation	4	1	3

The table records the number of times each type of variable appeared in the top 10 predictor data sets during discriminant analysis of the distributions of *G. morsitans*, *G. longipalpis*, *G. palpalis*, *G. f. fuscipes*, *G. pallidipes* and *G. tachinoides* either in West or East Africa, or for the whole continent. Thermal variables were AVHRR Channel 4 only. Variables which were a combination of two types (e.g. thermal/NDVI) are not included.

1993) and of the two subspecies of *G. morsitans* in Zambia (Robinson *et al.*, 1997a,b). Linear discriminant analysis, using a single cluster each for presence and absence data, and equal prior probabilities during classification (see Appendix, B), gave predictive accuracies of 80% or greater. These studies revealed several important features of local and regional fly distribution. First, only a single climatic variable appears to determine tsetse distribution at the edge of its continental range (e.g. the maximum of the mean monthly temperature for *G. morsitans* in Zimbabwe). Second, more than one climatic variable is required to describe distributions well within the continental range (e.g. in Kenya and Tanzania). In such areas, one variable excludes the flies from some places and others are more important elsewhere. Third, the average temperature difference between areas of fly presence and absence may be less than 1°C (Rogers and Randolph, 1993). Fourth, the two southern African subspecies of *G. morsitans* appear to respond very differently to climatic variables, but the distribution of each is described well by multivariate methods (Robinson *et al.*, 1997a,b). The studies were later extended to cover the entire continental distribution of tsetse, and the analytical methods adapted to allow clustering of the environmental variables in areas of both presence and absence (see Plate 2). The analyses also allowed the use of different covariance matrices for each cluster, and variable *a priori* probabilities (see Appendix, B), and thus provided the maximum likelihood solutions to the problem of defining tsetse distributions. These further studies suggested that flies show adaptations to regional climate (deduced from the increased accuracy obtained with clustered data) and that in different regions, different sets of variables appear to be important in determining fly distributions. This is illustrated in Table 2 which records the frequency of occurrence of elevation, CCD, temperature or NDVI data in the top ten selected variables used to define the distribution of key species (*G.*

tachinoides in West and *G. pallidipes* in East Africa), or closely related subspecies (*G. morsitans* ssp. and *G. palpalis* ssp. or *G. f. fuscipes*) either at regional or continental levels. In West Africa, temperature variables are almost twice as frequently selected as are NDVI variables (West Africa is on average 2–3°C warmer than East and central southern Africa), while in East Africa NDVI variables are more important than are thermal variables, and elevation also frequently plays a role. At the continental level NDVI dominates, followed by CCD. These statistical analyses in many ways confirm and extend our biological knowledge of tsetse. Regional differences in the genetics and behaviour of *G. pallidipes* were already known (Langley *et al.*, 1984) and population analyses had also suggested that the same species of tsetse showed a different tolerance to atmospheric dryness in the Lambwe Valley of Kenya (where conditions are near to ideal for this species) than in Somalia (Rogers, 1990). A challenge for future research is to investigate whether the environmental variable clusters of a single species' range correspond geographically to genetic differences between flies from the different areas.

Endemic HAT in much of West Africa is especially common in moist savannah regions, but is rare both in the more forested South and in the drier north. In an attempt to explain this distribution, the World Health Organization carried out detailed studies in villages in the affected zone in Côte d'Ivoire and supported a transect study running from the coast more or less directly inland to Bobo Dioulasso, Burkina Faso, about 700 km away. At about 100 km intervals the local populations of *G. palpalis* (which occurs throughout the transect) were sampled in both wet and dry seasons, and various measures were made of fly population behaviour and condition (map and imagery, see Plate 3). One striking result from this study was the demonstration that the sizes of the flies along the transect were similar in the wet season but significantly different in the dry season, when flies in the north were smaller than flies in the south. This result was shown separately for both male and female tsetse and the satellite images confirmed both a lower mean NDVI in the dry season and variations in NDVI values that related to changes in fly size (Figure 4). Tsetse fly size is determined by the parent female at the time of larval development *in utero*; environmentally and nutritionally stressed flies suffer a higher death rate, and also produce smaller offspring about one month later (hence the best correlate of fly size was with satellite data of the previous month). The apparently very small, but significant, variation in fly size recorded along this transect has been associated in other studies on *G. pallidipes* with a four-fold change in fly mortality rate seasonally (Dransfield *et al.*, 1989).

Our tentative explanation for the restriction of HAT to the central zone of the transect is that only in this region are there sufficient numbers of seasonally stressed flies biting humans to maintain the human disease. Flies in the south are not as seasonally stressed and so avoid biting humans (not a favoured host); to the north, although seasonal stress causes flies to bite humans, there are insufficient vectors to maintain the human disease endemically. This explanation is consistent with the results of large area surveys of this vector species in Côte d'Ivoire that showed the largest numbers of tsetse at intermediate NDVI levels (Rogers and Randolph, 1991).

Habitat signatures in the thermal (LST), vegetation (NDVI) and rainfall (CCD) variables were used in conjunction with ground data in the interpretation of a unique data set from Togo, that included contemporary observations on flies, disease and cattle at a spatial resolution of 0.125 degrees across the entire country (Hendrickx, 1999; Hendrickx et al., 1999a,b; 2000a,b). In selected sites, fly populations and cattle disease were monitored monthly, so that the project recorded both spatial and temporal patterns of vectors and disease. The project confirmed and extended the relationships previously found elsewhere in West Africa, between mean trap catches of G. tachinoides and G. palpalis and remotely sensed AVHRR NDVI (Rogers and Randolph, 1991), with further relationships between fly abundance and both CCD and LST estimates. The country-wide distributions of these two species and of G. m. submorsitans and G. longipalpis were subjected to non-linear discriminant analysis involving Fourier processed AVHRR data and were described with accuracies exceeding 90%. Fly abundance, divided into three classes of low, medium and high, were described with accuracies greater than 70% for G. tachinoides, but only 56% for G. palpalis (the only two species for which abundance data were available).

The study also investigated the effects on accuracy of using a sub-sample of the entire data-set and of changing the numbers of predictor variables used. As might be expected, accuracy tended to diminish with a smaller proportion (and therefore number) of observations in the training set but, within limits, increased with an increasing number of predictor variables. The accuracy of predicting samples not included in the training set was maximized with fewer predictor variables than were required to maximize the predictions of the training set itself; this suggests there is a danger of 'over-fitting' a training set so that the results are less generalizable to predict conditions in unsampled areas (Hendrickx et al., 2000a). The study also compared the fly situation in Togo with predictions previously made for Togo on the basis of less accurate tsetse distribution maps for Côte d'Ivoire and Burkina Faso and satellite data for all three countries (Rogers et al., 1996). These previous predictions for Togo were rather poor; whether this was a problem of a rather unsatisfactory training set (old and possibly out-dated maps), or represents a genuine problem of extending such analyses from one place to another requires urgent investigation.

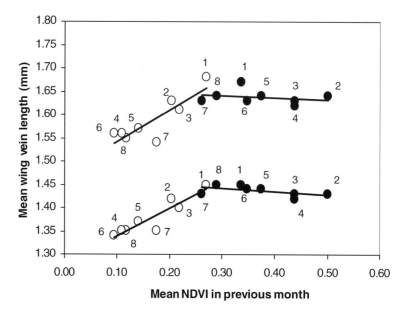

Figure 4 Relationship between the mean vein length of the hatchet cell of female (above) and male (below) *G. palpalis* collected along the transect shown in Plate 3 and satellite derived NDVI of the previous month for the dry (left, open circles) and wet (right, filled circles) season samples. All sites appear to be equally suitable in the wet season, but fly size is significantly smaller in the northern, drier sites in the dry season. Site numbers (see Plate 3 for map and images) are 1, Pauli Brousse; 2, Antonihio; 3, Degbézéré; 4, Bo'Pri; 5, River N'zi; 6, Komborodougou; 7, Oulokoussou; 8, La Guingette. Regression details: wet season, males $y = 1.463–0.072x$, $r = 0.584$, n.s.; females $y = 1.651–0.037x$, $r = 0.201$, n.s.: dry season, males $y = 1.2794 + 0.598x$, $r = 0.915$, $P < 0.01$; females $y = 1.476 + 0.670x$, $r = 0.852$, $P < 0.01$. (Redrawn from Rogers and Randolph, 1991.)

One criticism of the environmental envelope approach to mapping distributions is that it ignores the influence of biotic factors that are so important in determining the abundance of flies within their geographical limits (Davis *et al.*, 1998). We suspect that in most cases there are very few flies at the edge of distributions, and therefore that biotic factors are much less important there than abiotic ones. The fact that we can produce such good fits without explicitly modelling biotic factors supports this idea, although it is also possible that some of the satellite data channels are acting as surrogates for unquantified biotic factors.

Several problems remain for local, regional and continental mapping of tsetse distributions using satellite data, the most important being the ultimate degree of accuracy that such an approach can provide. Tsetse control personnel want a map that is 100% reliable, but this degree of accuracy is essential only when tsetse eradication is the ultimate aim. Even ground surveys fail to detect some tsetse species (e.g. *G. austeni* which is not attracted to conventional sampling techniques) and no method is particularly efficient at detecting the presence of any all tsetse species at very low densities. The prohibitive costs of total eradication and the difficulty of maintaining this situation over time have caused many to think more realistically of fly control. Control, rather than eradication, requires the identification of tsetse 'hot-spots' which must be targeted by control measures, and of ecological corridors along which flies might move to re-invade controlled areas. Each of these tasks should be more easy to accomplish using satellite data and carefully prepared predictive risk maps.

Combining such information in a GIS and decision making framework can also reveal the desirability of such operations in many locations (Robinson, this volume).

5.2.3. *Animal Trypanosomiasis*

Relatively few studies provide area-wide estimates of the prevalence of trypanosomiasis in domestic animals. The Togo study, mentioned above, collected cattle infection data from sedentary herds in 92% of the grid squares that contained sufficient cattle (approximately two-thirds of the total squares) and thus provides another unique data set for analysis (Hendrickx, 1999; Hendrickx *et al.*, 1999a,b). Overall infection rates, divided into low, medium and high categories, were described with an accuracy of 80%; this figure improved when infections of *T. congolense* (83%) and *T. vivax* (89%) were analysed separately. In discussing these results Hendrickx emphasized that this degree of accuracy was obtained only by using as a predictor variables both the satellite data and anthropogenically determined factors such as herd management and nutrition, breed of cattle, and percentage of land under cultivation. The link between satellites and animal disease is clearly less direct than that between satellites and the tsetse vectors, because the chain of causation is longer.

The Togo study revealed close links between trypanosomiasis prevalence and an easily obtained field measure of anaemia, the packed cell volume, which future area-wide studies could use. It also showed how genes from zebu cattle (i.e. more trypanosusceptible) are introgressing into the trypanotolerant stock, especially in areas where tsetse are disappearing under human pressure on the land. This reveals an elegant degree of adaptation of local farming practices (which may of course arise through trial-and-error processes) to local epizootiological situations. Furthermore, it suggests that development should concentrate on sensible encouragement of the evolution of the current situation rather than revolutionary, and probably inappropriate, changes to it. As satellite data improve with the next generations of satellites (Goetz *et al.*, this volume) and analytical methods are refined, landscape changes brought about by human activity will be much more easily monitored from space.

5.2.4. *Human African Trypanosomiasis in Uganda*

As in the case of animal trypanosomiasis, there are few extensive data sets for the human disease. Uganda's long history of HAT outbreaks, however, provides us with data that are both historical and contemporary with the satellite imagery. Although other areas in the country have been affected by HAT, the Busoga region that borders the north-eastern shores of Lake Victoria has had the most persistent and well documented HAT focus, with records from the early years of the twentieth century until the present day (Ford, 1971). In the 1960s, Robertson suggested that HAT in this area was an occupational disease, occurring among fishermen in the dry season (with peak incidence in January) and others, mostly agriculturalists, in the wet season (with a peak in May/June; Robertson, 1963). Robertson was never quite able to explain the differences he found, since the seasonal activities of the fishermen did not coincide with their peak risk of infection.

Soon after this, Mwambu and Odhiambo surveyed trypanosomiasis in cattle in the Tororo District of Uganda along a transect from Mjanji, about 8 km from Lake Victoria, inland to about 45 km from the lake (Mwambu and Odhiambo, 1967). *T. vivax* and *T. congolense* were found at all distances from the lake, although *T. congolense* infections fell rapidly beyond about 13 km from the lake, coinciding with the local limits of *G. pallidipes*. *T. brucei* infections (not further identified) were restricted to the lowland forests and thickets favoured by this species of tsetse, and thus did not occur beyond the 13 km limit. These observations tended to confirm the earlier impressions that *G. pallidipes*, a species that readily feeds on domestic and wild animals, is the major source of introduction of *T. b. rhodesiense* into human populations from these alternative hosts. Once introduced into the human population, infections are then rapidly transmitted by *G. fuscipes*, which more readily bites humans and other primates.

Although there are no contemporary satellite data, we can relate the point prevalences of Mwambu and Odhiambo's cattle survey to satellite NDVI for the same sites recorded in 1984–88 (Figure 5a). These relationships were confirmed by two Cambridge undergraduate expeditions to the area in 1991 and 1992, which examined infection rates in cattle along transects determined by the NDVI satellite imagery (J. Hodgson, *pers. comm.*). A new epidemic of HAT began in this area in the 1970s. In many

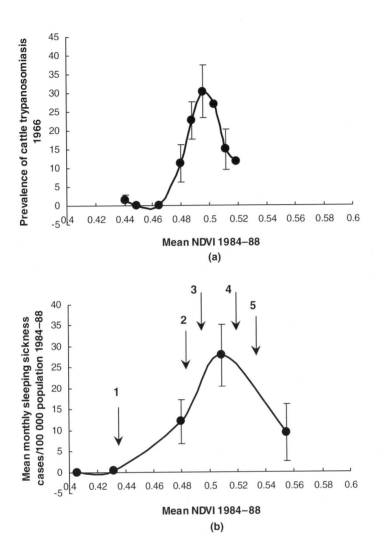

Figure 5 (a) Relationship between the mean trypanosome prevalence in groups of village cattle in the Busoga region of south-eastern Uganda in 1960s (data from Mwambu and Odhiambo, 1967) and mean annual satellite NDVI 1984–88 for the sample sites. (b) Relationship between the mean monthly prevalence of human sleeping sickness (cases per 100 000) in the same region of Uganda between 1984 and 1988 and mean NDVI from each county, for the same period. The similarity of these two curves lends support to the idea that cattle are quantitative reservoirs of the human disease in this area (Hide, 1997). The arrows in (b) indicate the mean NDVI of the local districts; 1, Tororo; 2, Kamuli; 3, Iganga; 4, Jinja; 5, Mukono. The figures also show ±1 standard error of the means.

ways it repeated the pattern shown during the previous outbreak in the 1940s (Ford, 1971). Starting near Jinja and the source of the White Nile, the epidemic spread slowly east, towards Tororo and the border with Kenya. In the 5 year period 1984–88, the mean prevalence of HAT in the sub-counties of the region showed a remarkably similar pattern of variation to that of the cattle surveys, across approximately the same range of NDVI values (Figure 5b). This is evidence, albeit indirect, that cattle are a potentially important reservoir of the human disease in this region; this idea originated with the isolation of human-infective *T. brucei* from cattle in an ecologically similar lake-side site in Kenya in 1966 (Onyango *et al.*, 1966) and was more recently confirmed by parasitological surveys in this area (Hide, 1997). It is likely that cattle are particularly important as a reservoir of HAT in this part of Uganda because they are relatively resistant to the local strains of *T. vivax* (*T. congolense* is rare in the area). In the absence of any pathogenic infections in their animals, local cattle owners rarely use trypanocides, which would otherwise kill off both the pathogenic trypanosomes and the (to cattle) non-pathogenic *T. brucei* spp. among which are human-infective forms.

The relatively fine temporal and spatial resolution of the HAT data from the Busoga region (cases recorded monthly at sub-county level) allows us to investigate the relationships between the incidence of disease in humans and the satellite data. Before analysis, the monthly case numbers were expressed as a percentage of each year's total, to remove the effects of variation in the annual totals as the epidemic moved slowly through the area. Counties were excluded from the analysis in years when they had fewer than 12 cases. Finally the monthly percentage figures were averaged for the 5 years 1984–88 inclusive, for comparison with the AVHRR data for the same years. Correlations of case numbers with LST were significant at the 5% level or better in 12 out of 20 counties, with a lag of 0, 1 or 2 months (the approximate delay between a person receiving an infective fly bite and presenting with HAT at a local clinic) (Figure 6). The slopes of the significant relationships were significantly inversely related to the mean land surface temperatures for each county, and positively related to the respective mean NDVIs; they could therefore be positive in some areas and negative in others (Figure 7). Where temperatures are relatively cool, and NDVIs relatively high, HAT case numbers increase with monthly temperatures and are most frequent in the hottest months, at the start of the year. Where temperatures are higher, by an average of only 1 or 2°C, in the east of the area near Tororo, HAT case numbers decrease with monthly temperatures and cases peak in June, the coolest time of the year. We can thus reinterpret Robertson's observations by suggesting that HAT in this area is not so much a disease of occupation (fishermen vs. non-fishermen) as a disease of location.

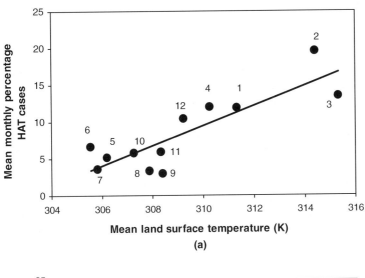

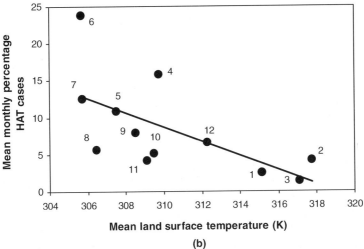

Figure 6 Relationship between the mean monthly numbers of sleeping sickness cases in the Busoga region of Uganda and satellite derived land surface temperature (LST) for the same month and district, 1984–88, (a) Kamuli District, Bulamogi County ($y = 1.361x{-}412.3, r = 0.847, P < 0.001$); (b) Tororo District, Tororo County ($y = -0.972x + 310.05$, $r = 0.641$, $P < 0.05$). Case numbers are expressed as a percentage of the annual totals for each county; months are indicated by numbers, 1 = January etc.

Analysis of tsetse fly catches made in this area show that the mean monthly mortality of one of the local vector species, *G. fuscipes*, is highest in the locally hotter season, whilst that of the other, *G. pallidipes*, is highest in the locally cooler season (D.J.Rogers, unpublished observation). The different responses of these two species to effectively the same climate, suggests that their relative

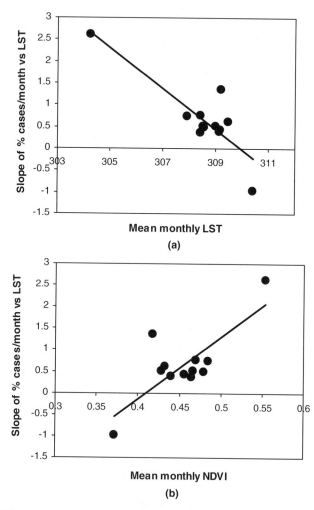

Figure 7 The regression coefficients shown in Figure 6 are significant in 12 of the 20 counties of south-eastern Uganda affected by HAT during the epidemic of the 1970s and 80s. Here these significant coefficients are related to (a) the mean land surface temperature (LST) ($y = -0.467x + 144.69, r = 0.855, P < 0.001$) and (b) the mean NDVI ($y = 14.344x - 5.876, r = 0.777, P < 0.01$) for each county for the period 1984–88.

importance as HAT vectors will change seasonally, and perhaps spatially. As in the case of the West African study (see above), tsetse tend to bite humans only when especially stressed and the seasonal risk to humans in this part of Uganda will therefore depend upon the local mix of these two vector species. There is anecdotal evidence that *G. pallidipes* numbers in the eastern end of this fly belt have fallen considerably in recent years, as a result of human impacts, although it survives in the lakeside vegetation nearer to Jinja.

Epidemics have raged through this area of Uganda for more than 100 years and sequential disease management strategies (resettlement, aerial spraying with insecticides, or trapping) have complicated the interpretation of the entomological, parasitological and clinical data from the region. Probably the only certainty is that there will be further epidemics in the future, and satellite imagery may be used to monitor and perhaps predict where the next epidemic is likely to begin, and to generate predictive maps to highlight populations at greatest risk of infection.

6. SATELLITE IMAGERY AND DISEASE RISK MAPPING: THE WAY FORWARD

This chapter has illustrated how remotely sensed data may be used to map habitats, vector distribution and abundance, and disease incidence and prevalence over time and space. The unprecedented rates of change imposed by dynamic and expanding human communities upon their natural habitats have several consequences. First it is less and less likely that static paper maps will be an acceptable way to store information about the distribution of natural resources, or disease risk (Myers *et al.*, this volume). Secondly, it becomes more likely that old diseases will occur in new areas, or new diseases will emerge from zoonotic reservoirs. Thirdly, more problems will be transnational, or will originate from travellers returning from further afield. Fourthly, the ability of traditional veterinary and medical services to cope with such problems will be severely limited; every delay in the response to a new disease situation will result in more people or animals eventually becoming infected, and greater strains will be placed on resource providers. Vector-borne diseases have basic reproductive numbers which are often orders of magnitude greater than those of directly transmitted diseases (Rogers and Packer, 1993) and this will result in rapid increases in case numbers when conditions are suitable. Finally, the tragedies of civil conflicts are often accompanied by epidemic disease outbreaks, most recently in Uganda, the Democratic Republic of Congo and Sudan, placing additional burdens on local health services. For these and many other reasons the world community needs to establish disease forecasting and early warning systems

(Myers *et al*, this volume.). These must couple the best statistical analyses of current situations with our developing ability to produce robust biological models (Randolph and Rogers, 1997; Randolph, this volume) in which a variety of intervention strategies may be tested before they are applied in the field. As emphasized in this chapter, remotely sensed data may be used in both the statistical and biological models and provide a vital link between the two approaches. The benefit is a more rapid advance in our understanding of variations of disease systems in both space and time.

Several problems remain to keep research agendas full. First we need to establish the levels of accuracy of the predictions of both statistical and biological models. This is difficult to do when the field data themselves contain unquantified errors: vector distribution maps drawn by interpolation using elevation contours (as seems to have been the case for Zimbabwe's historical tsetse maps) are likely to be inaccurate, at least in detail, but will be well described by environmental variables that vary with elevation. Statistical predictions of distributions using satellite and other data are probably most useful where they are least accurate; investigation of both false positive and false negative areas will give us more insight into the real determinants of local distributions, of both vectors and diseases, and thus lead to better maps.

Secondly, we need to establish generic ways of collecting field data that will be more useful for the modelling approaches. Given limited resources, is it best to distribute them thinly to produce an extensive picture of the disease situation (ideal for statistical modelling), or to concentrate them into a single area, thus producing data for detailed biological models? Satellite data may be used both to select where the extensive data should be collected (to sample as wide a variety of habitats as possible) and to identify high risk areas that would benefit most from the intensive, biological studies.

Thirdly, we urgently need to increase the spatial resolution of our multi-temporal data. It is remarkable that we can achieve quite so much with the present poor spatial resolution (generally 8×8 km pixel size) of these data: tropical habitats are extremely heterogeneous spatially, and many vectors are effectively spatial opportunists, inhabiting a very small, but precise, fraction of the entire habitat. In the short term we can use a variety of wavelet techniques (Stollnitz *et al.*, 1996; Mallat, 1999) to increase the spatial resolution of past data sets artificially: these techniques offer the best way to integrate the high spatial resolution of the Landsat and SPOT series of satellites with the high temporal resolution of the NOAA and Meteosat satellites. In the longer term, the increased spatial (and spectral) resolution of the next generation of satellites (Hay, this volume; Wood *et al.*, this volume) will improve our discrimination of vector and disease habitats.

Fourthly, temporal variation in disease risk is least amenable to analysis using the present generations of both geographical information systems and remote sensing software. The research challenge here is to couple the increasingly sophisticated understanding we have of the intrinsic determinants of temporal variation in risk (Anderson and May, 1991) with our still relatively poor understanding of the extrinsic determinants, from seasonal weather patterns to longer time-scale climatic variation. To address this challenge we may need to adopt some of the analytical techniques of time-series analysis for intensive study data (Chatfield, 1980; Hay *et al.*, 2000) and of oceanography and meteorology for extensive study data (Storch and Zwiers, 1999).

Finally, having put all of the above together, we must concentrate on the delivery of the ideas and approaches to the health-care providers who are in the best position to benefit from them. Whilst the habitat sensing may be 'remote', product delivery must be personal, accurate and timely. The problems to be addressed here involve the ownership both of the field data and of the developed methodologies; clearly a mutually beneficial constructive dialogue must be established between the national and international partners. The development of prototype disease early warning systems (Myers *et al.*, this volume) is a first step along this exciting path to the future.

Models of one sort or another are a vital ingredient of all remote sensing activities and are especially important when satellite data are coupled to biological studies. Whilst modellers would do well to adopt as a daily mantra the warning that 'all models are wrong, but some are useful' (Anderson and May, 1991), their inspirational creed should be that of T. S. Eliot (1944):

> We shall not cease from exploration
> And the end of all our exploring
> Will be to arrive where we started
> And know the place for the first time.

ACKNOWLEDGEMENTS

I thank John Townshend, Tim Robinson, and Fred Snijders for the supply of my first AVHRR NDVI imagery in 1986, and the Pathfinder program (NASA—Goddard Space Flight Center), and Compton Tucker and the GIMMS Group at NASA-GSFC for more recent data.

Dr Mbulamberi kindly provided the HAT data for Uganda.

Sarah Randolph and Simon Hay commented on the manuscript, and research support came from the Department for International Development under Schemes R5794 and R6626, administered through the Natural Resources Institute International.

APPENDIX

A. Temporal Fourier Analysis

Temporal Fourier analysis describes variations through time of satellite signals as a series of simple sine curves with different frequencies and amplitudes. It is applicable to regularly collected data such as maximum value composited monthly AVHRR data, $\{x_t\}$.

The Fourier series representation of $\{x_t\}$ is found from the following:

$$x_t = a_0 + \sum_{p=1}^{N/2-1} [a_p \cos(2\pi pt/N) + b_p \sin(2\pi pt / N)] + a_{N/2} \cos \pi t$$

$$(t = 1, 2, ..., N) \qquad (1)$$

with coefficients $\{a_p, b_p\}$ defined as follows:

$$a_0 = \bar{x}$$

$$a_{N/2} = \frac{\sum (-1)^t x_t}{N}$$

$$\left. \begin{array}{l} a_p = \dfrac{2 \left(\sum x_t \cos(2\pi pt/N) \right)}{N} \\[3mm] b_p = \dfrac{2 \left(\sum x_t \sin(2\pi pt/N) \right)}{N} \end{array} \right\} p = 1, ..., N/2-1 \qquad (2)$$

The component at a frequency $\omega p = 2\pi p/N$ is called the pth harmonic, and for all $p \neq N/2$ these harmonics may be written in the equivalent form

$$a_p \cos \omega_p t + b_p \sin \omega_p t = R_p \cos(\omega_p t + \phi_p) \qquad (3)$$

where

R_p = the amplitude of the pth harmonic

$\quad = \sqrt{(a_p^2 + b_p^2)}$

and

ϕ_p = the phase of the pth harmonic

$\quad = \tan^{-1}(-b_p/a_p)$

Thus the important elements of these equations are a_p and b_p that describe the relative mix of sine and cosine terms that together determine the timing of the peak of each harmonic, p. Further details of time series analysis are given by Chatfield (1980) and Diggle (1990).

B. Discriminant Analysis

In its simplest form, discriminant analysis assumes both multivariate normality and a common within-group covariance of the variables for all points defining vector or disease presence and absence. Covariances are estimated from representative samples from reliable distribution maps, the 'training sets'. Means of multivariate distributions are referred to as centroids and are defined by mathematical vectors $[\mathbf{x}_n]$ where n is the number of dimensions (= variables). The Mahalanobis distance, D^2, is the distance between two multivariate distribution centroids, or between a sample point and a centroid, and is defined as follows:

$$D_{12}^2 = (\bar{\mathbf{x}}_1 - \bar{\mathbf{x}}_2)'\mathbf{C}_w^{-1}(\bar{\mathbf{x}}_1 - \bar{\mathbf{x}}_2)$$
$$= \mathbf{d}'\mathbf{C}_w^{-1}\mathbf{d} \tag{4}$$

where the subscripts refer to groups 1 (e.g. for vector absence) and 2 (e.g. for vector presence), $\mathbf{d} = (\bar{\mathbf{x}}_1 - \bar{\mathbf{x}}_2)$ and $\mathbf{C}_w^{-1}$ is the inverse of the within-groups covariance (= dispersion) matrix (Green, 1978). Thus D^2 is the distance between the sample centroids adjusted for their common covariance. Equation (4) may be used in a number of ways. First it may be used to assign new data points to one or other category (of presence or absence) by examining the value of D^2 between each point and each of the training-set defined centroids. The point is then assigned to the group for which D^2 is a minimum. Secondly, the equation may be used to calculate the probability with which each data point belongs to each of the training set groups. This involves defining the position of the point within each multivariate distribution around each centroid (most easily achieved by calculating D^2 which is distributed as χ^2 with $(g-1)$ degrees of freedom, where g is the number of variables defining each centroid). In general these measures are normalized by dividing each by the sum of all measures (i.e. the sum of the probabilities across all classes in the training set) to give posterior probabilities, defined as follows:

$$P(1\,|\,x) = \frac{p_1 e^{-D_1^2/2}}{\sum\limits_{g=1}^{2} p_g e^{-D_g^2/2}}$$

$$P(2\,|\,x) = \frac{p_2 e^{-D_2^2/2}}{\sum\limits_{g=1}^{2} p_g e^{-D_g^2/2}} \tag{5}$$

where $P(1\,|\,x)$ is the posterior probability that observation $\mathbf{x}$ belongs to group 1 and $P(2\,|\,x)$ the posterior probability that it belongs to group 2 (Green,

1978). In equation (5), p_1 and p_2 are the prior probabilities of belonging to the same two groups respectively, defined as the probabilities with which any observation might belong to either group given prior knowledge or experience of the situation. In the absence of any prior experience it is usual to assume equal prior probability of belonging to any of the groups; in the simple case of two-group discrimination, therefore, $p_1 = p_2 = 0.5$. The normalization invoked by summing the probabilities in equation (5) is based on the assumption that observation x must come from one or other of the classes defined in the training-set data (this important assumption is frequently not tested by examining the absolute value of the Mahalanobis distances that are used in the calculations of $P(1|x)$ and $P(2|x)$, although image processing systems are usually capable of optional output of an image of these distances). Other terms of the multivariate normal distribution in equation (5) then cancel out (Tatsuoka, 1971).

As indicated earlier, equations (4) and (5) should be modified when the assumption of common covariances is obviously invalid. Not only may areas of presence and absence differ in their environmental characteristics, but different parts of a species range may also show more subtle differences, requiring separate multivariate descriptions of their climatic conditions. It is generally convenient to anticipate this need in the case of widespread species, and to define clusters of environmental similarity of areas of presence and absence before statistical analysis. Each cluster (either for presence or absence) is then treated as a separate multi-variate normal distribution, with its own covariance characteristics, and the posterior probabilities are calculated by summing across all distributions. In the case of two groups only (one for presence and one for absence), equation (5) is then modified as follows:

$$P(1|x) = \frac{p_1 \, |\mathbf{C}_1|^{-1/2} \, e^{-D_1^2/2}}{\sum_{g=1}^{2} p_g \, |\mathbf{C}_g|^{-1/2} \, e^{-D_g^2/2}}$$

$$P(2|x) = \frac{p_2 \, |\mathbf{C}_2|^{-1/2} \, e^{-D_2^2/2}}{\sum_{g=1}^{2} p_g \, |\mathbf{C}_g|^{-1/2} \, e^{-D_g^2/2}}$$

(6)

where $|\mathbf{C}_1|$ and $|\mathbf{C}_2|$ are the determinants of the covariance matrices for groups $g = 1$ and 2 respectively. The Mahalanobis distances in equation (6), calculated from equation (4), are evaluated using the separate within-group covariance matrices $\mathbf{C}_1$ and $\mathbf{C}_2$ (Tatsuoka, 1971). When there is more than a single class of presence or absence data (e.g. multiple clusters) the summation in the denominators of equation 6 covers the entire set of $g > 2$ groups and there are as many posterior probability equations as there are groups. With unequal covariance matrices the discriminant axis (strictly speaking a plane) that separates the two groups in multivariate space is no longer linear, and equation (6) then effectively defines the maximum likelihood solution to the problem (Swain, 1978).

Training-set data are generally limited in size, and it is unwise to subdivide them by clustering if this results in too few observations in some of the clusters. This will cause the covariance matrices to be ill-characterized, and their inversions (equation (4)) will either be impossible to calculate or will perform badly in predicting presence and absence through use of equation (6). Similarly there is no obvious rule about the use of expected or observed prior probabilities in equations (5) or (6). Use of observed (generally training-set) prior probabilities shifts the equi-probability contours towards the smaller groups, resulting in a larger proportion of assignments to the classes with larger group sizes. This shift, however, may occasionally be large enough to reduce the accuracy of describing even the training-set data.

Ideally the training set should be divided, with half used to develop the covariance matrices and the other half used to test the accuracy of the predictions. Frequently, however, training data are scarce and the entire data set must be used in the training exercise. The resulting predictions will tend to inflate estimates of the accuracy of the techniques, although perhaps only modestly (Randolph, this volume). A further problem is that environmental variables are often spatially auto-correlated, whilst the above multivariate methods assume no such auto-correlation, i.e. statistical independence of training-set observations. The importance of these effects may be judged by including some measure of spatial covariation in the analyses. This is done by generating new variables of the auto-correlations, and by making these variables available for inclusion during step-wise discriminant analysis. If they are selected, the spatial effects contribute significantly to the distributions being modelled. In some analyses spatial covariation affects the variance of the resulting estimates but not their means (Thomson $et\ al.$, 1999), whilst in others both means and variances appear to be affected (Augustin $et\ al.$, 1996). It is a curious feature of satellite data sets that spatial covariation creates problems for some sorts of analyses but is actively exploited by other methods of analysis, such as co-kriging (refer to Robinson, this volume). Further details of multivariate analysis may be found in Tatsuoka (1971), Green (1978) and Krzanowski and Marriott (1995).

C. A Simple Model for Trypanosome Transmission

Changes in the proportions of the vertebrate hosts that are susceptible (s), infected but not yet infectious (f), infectious (x) and recovered, immune (i) are described by the following set of equations:

$$\frac{ds}{dt} = -abmx's + tN + wi - \mu s$$

$$\frac{df}{dt} = abmx's - vf - \mu f$$

$$\frac{dx}{dt} = vf - rx - \mu x$$

$$\frac{di}{dt} = rx - wi - \mu i$$

(7)

where

a is the biting rate of vectors on hosts,
b is the transmission coefficient from vector to vertebrate,
m is the ratio of vectors to hosts ($= M/N$),
x' is the proportion of infected vectors,
t is the birth rate of the hosts,
N is host population size (the present equations describe proportions, so that $N = s + f + x + i = 1$),
w is the rate of loss of immunity, returning immune animals to the susceptible category, s,
μ is the host's natural death rate,
v is the incubation rate of the disease in the vertebrate hosts, and
r is the rate of recovery of the vertebrates from infection.

The equivalent equations for the vectors, continuing to use prime to indicate vector parameters and variables analagous to those of the hosts, are as follows:

$$\frac{ds'}{dt} = -acxs' + t'M - \mu's'$$

$$\frac{df'}{dt} = acxs' - v'f' - \mu'f'$$

$$\frac{dx'}{dt} = v'f' - \mu'x'$$

(8)

where, in addition,

c is the transmission coefficient from vertebrate to vector,
t' is the birth rate of the tsetse population,
M is the tsetse population size (as in the case of the hosts, $M = s' + f' + x' = 1$), and
μ' is the tsetse mortality rate ($= t'$ at equilibrium).

Vector mortality rate appears explicitly in these equations since it is assumed that fly infections can be lost only when infected flies die. These losses are balanced by births, which introduce new, susceptible flies into s'. In the case of the vertebrate hosts, animals which lose their immunity recycle into the susceptible category.

The sum of each set of equations, for both hosts and vectors, is zero, indicating that although each subpopulation (susceptible, infected etc.) may change over time there is no net change in the summed proportions, which must always be equal to 1.0.

The above equations, which apply to the simple situation of a single-vector, single-host disease may be applied to the African trypanosomiases assuming all other hosts fed upon by flies are negligible sources of infection compared to the modelled hosts. The equations are therefore simplified versions of those that have been written for the African trypanosomiases (Rogers, 1988) and may be taken to apply to the situation of trypanosomiasis in domestic animals in areas with few alternative hosts; this simplification allows us to model disease transmission quickly, and to estimate the need for more complex transmission models in field situations. HAT involving domestic and wild reservoir hosts will not be satisfactorily modelled in this simple way.

In making the output of the tsetse population model one of the inputs into the disease transmission model, additional scaling parameters are required to relate the tsetse numbers to the vertebrate host numbers. Since the vector/host ratio (m) always and only appears in the above equations with the transmission coefficient b, estimates of these two quantities will vary together so that their product remains the same.

REFERENCES

Anderson, R.M. and May, R.M. (1991). *Infectious Diseases of Humans: Dynamics and Control*. Oxford: Oxford University Press.

Ashcroft, M.T. (1959). A critical review of the epidemiology of human trypanosomiasis in Africa. *Tropical Diseases Bulletin* **56**, 1073–1093.

Augustin, N.H., Mugglestone, M.A. and Buckland, S.T. (1996). An autologistic model for the spatial distribution of wildlife. *Journal of Applied Ecology* **33**, 339–347.

Barry, J. (1997). The biology of antigenic variation in African trypanosomes. In: *Trypanosomiasis and Leishmaniasis: Biology and Control* (G. Hide, J.C. Mottram, G.H. Coombs and P.H. Holmes, eds), pp. 89–107. Wallingford: CAB International.

Bodenheimer, F.S. (1938). *Problems of Animal Ecology*. Oxford: Clarendon Press.

Bonham-Carter, G.F. (1994). *Geographical Information Systems for Geoscientists*. Oxford: Pergamon/Elsevier.

Bourn, D. (1983). Tsetse control, agricultural expansion and environmental change in Nigeria. Ph.D. thesis, University of Oxford.

Bourn, D., Wint, W., Blench, R. and Wolley, E. (1994). Nigerian livestock resources survey. *World Animal Review* **78**, 49–58.

Bruce, D. (1895). *Preliminary Report on The Tsetse Fly Disease or Nagana in Zululand.* Durban: Bennet and David.

Bursell, E. (1959). The water balance of tsetse flies. *Transactions of the Royal Entomological Society of London* **3**, 205–235.

Buxton, P.A. and Lewis, D.J. (1934). Climate and tsetse flies: laboratory studies upon *Glossina submorsitans* and *tachinoides. Philosophical Transactions of the Royal Society of London, Series B* **224**, 175–240.

Chatfield, C. (1980). *The Analysis of Time Series: An Introduction.* London: Chapman and Hall.

Davis, A.J., Jenkinson, L.S., Lawton, J.H., Shorrocks, B. and Wood, S. (1998). Making mistakes when predicting shifts in species range in response to global warming. *Nature* **391**, 783–786.

Diggle, P.J. (1990). *Time Series: A Biostatistical Introduction.* Oxford: Clarendon Press.

Dransfield, R.D., Brightwell, R., Kiilu, J., Chaudhury, M.F. and Adabie, D.A. (1989). Size and mortality rates of *Glossina pallidipes* in the semi-arid zone of southwestern Kenya. *Medical and Veterinary Entomology* **3**, 83–95.

Dutton, J.E. and Todd, J.L. (1906). *The Distribution and Spread of Sleeping Sickness in the Congo Free State with Suggestions on Prophylaxis.* Liverpool: Liverpool School of Tropical Medicine.

Eastman, J.R. and Fulk, M. (1993). Long sequence time series evaluation using standard principal components. *Photogrammetric Engineering & Remote Sensing* **59**, 991–996.

Eliot, T.S. (1944). *Four Quartets.* London: Faber and Faber.

Farley, J. (1991). *Bilharzia: A History of Imperial Tropical Medicine.* Cambridge: Cambridge University Press.

Ford, J. (1971). *The role of the Trypanosomiases in African Ecology: A Study of the Tsetse Fly Problem.* Oxford: Clarendon Press.

Ford, J. and Katondo, K.M. (1977). *The Distribution of Tsetse Flies in Africa.* Nairobi, OAU: Cook, Hammond & Kell.

Gaschen, H. (1945). Les glossines de l'Afrique Occidentale Francaise. *Acta Tropica* (Supplement) **2**, 1–131.

Giddings, L.E. and Naumann, A. (1976). Remote sensing for the control of tsetse flies. *Technical Memorandum Prepared by Lockheed Electronics Company Inc. for Bioengineering Systems Division, NASA.* Houston, TX: Johnson Space Flight Center.

Glasgow, J.P. (1963). *The Distribution and Abundance of Tsetse.* Oxford: Pergamon Press.

Glasgow, J.P. and Welch, J.R. (1962). Long-term fluctuations in numbers of tsetse fly *G. swynnertoni* Austen. *Bulletin of Entomological Research* **53**, 129–137.

Green, P.E. (1978). *Analyzing Multivariate Data.* Hinsdale, IL: Dryden Press.

Green, R.M. and Hay, S.I. (2000). Mapping of climate variables across tropical Africa and temperate Europe using meteorological satellite sensor data. *Remote Sensing of Environment,* in press.

Hargrove, J.W. (1994). Reproductive rates of tsetse flies in the field in Zimbabawe. *Physiolgical Entomology* **19**, 307–318.

Hargrove, J.W. and Borland, C.H. (1994). Pooled population parameter estimates from mark–recapture data. *Biometrics* **50**, 1129–1141.

Hargrove, J.W. and Packer, M.J. (1993). Nutritional states of male tsetse flies (*Glossina* spp.) (Diptera: Glossinidae) caught in odour-baited traps and artificial refuges: models for feeding and digestion. *Bulletin of Entomological Research* **83**, 29–46.

Hargrove, J.W. and Williams, B.G. (1998). Optimized simulation as an aid to modelling, with an application to the study of a population of tsetse flies, *Glossina morsitans morsitans* (Diptera: Glossinidae). *Bulletin of Entomological Research* **88**, 425–435.

Hay, S.I. and Lennon, J.J. (1999). Deriving meteorological variables across Africa for the study and control of vector-borne disease: a comparison of remote sensing and spatial interpolation of climate. *Tropical Medicine & International Health* **4**, 58–71.

Hay, S.I., Tucker, C.J., Rogers, D.J. and Packer, M.J. (1996). Remotely sensed surrogates of meteorological data for the study of the distribution and abundance of arthropod vectors of disease. *Annals of Tropical Medicine and Parasitology* **90**, 1–19.

Hay, S.I., Myers, M.F., Burke, D.S. *et al.* (2000). Etiology of mosquito-borne disease epidemics. *Proceedings of the National Academy of Sciences of the USA*, in press.

Hendrickx, G. (1999). *Georeferenced Decision Support Methodology Towards Trypanosomosis Management in West Africa*. Merelbeke: Universiteit Ghent.

Hendrickx, G., Napala, A., Dao, B. *et al.* (1999a). The area-wide epidemiology of bovine trypanosomosis and its impact on mixed farming in subhumid West Africa: a case study in Togo. *Veterinary Parasitology* **84**, 13–31.

Hendrickx, G., Napala, A., Dao, B. *et al.* (1999b). A systematic approach to area-wide tsetse distribution and abundance maps. *Bulletin of Entomological Research*, in press.

Hendrickx, G., Napala, A., Slingenbergh, J.H.W., De Deken, R. and Rogers, D.J. (2000a). The contribution of remote sensing to area-wide tsetse surveys. *Bulletin of Entomological Research*, in press.

Hendrickx, G., Napala, A., Slingenbergh, J.H.W., Dr Deken, R., Vercruysse, J. and Rogers, D.J. (2000b). The spatial patterns of trypanosomosis predicted with the aid of satellite imagery. *Parasitology* **120**, 121–134.

Hide, G. (1997). The molecular epidemiology of trypanosomatids. In: *Trypanosomiasis and Leishmaniasis: Biology and Control* (G. Hide, J.C. Mottram, G.H. Coombs and P.H. Holmes, eds), pp. 289–303. Wallingford: CAB International.

Hiernaux, P.H.Y. and Justice, C.O. (1986). Suivie du développement végétal au cours de l'été 1984 dans le Sahel Malien. *International Journal of Remote Sensing* **7**, 1515–1531.

Hoare, C.A. (1972). *The Trypanosomes of Mammals: A Zoological Monograph*. Oxford: Blackwell Scientific.

Jackson, C.H.N. (1930). Contributions to the bionomics of *Glossina morsitans*. *Bulletin of Entomological Research* **21**, 491–527.

Jackson, C.H.N. (1937). Some new methods in the study of *Glossina morsitans*. *Proceedings of the Zoological Society of London 1936*, 811–895.

Kitron, U. (1998). Landscape ecology and epidemiology of vector-borne diseases: tools for spatial analysis. *Journal of Medical Entomology* **35**, 435–445.

Kitron, U., Otieno, L.H., Hungerford, L.L. *et al.* (1996). Spatial analysis of the distribution of tsetse flies in the Lambwe Valley, Kenya, using Landsat TM satellite imagery and GIS. *Journal of Animal Ecology* **65**, 371–380.

Kleine. (1909). Weitere wissenschafteliche beobachtungen über die Entwicklung von Trypanosomen in Glossinen. *Deutsche Medizinische Wochenschrift*, 469–470.

Komba, E.K., Kibona, S.N., Ambwene, A.K., Stevens, J.R. and Gibson, W.C. (1997). Genetic diversity among *Trypanosoma brucei rhodesiense* isolates from Tanzania. *Parasitology* **115**, 571–579.

Krzanowski, W.J. and Marriott, F.H.C. (1995). *Multivariate Analysis Part 2. Classification, Covariance Structures and Repeated Measurements*. London: Arnold.

Langley, P.A., Maudlin, I. and Leedham, M.P. (1984). Genetic and behavioural differences between *Glossina pallidipes* from Uganda and Zimbabwe. *Entomologia Experimentalis et Applicata* **35**, 55–60.

La Rocque, S. de (1997). *Identification des Facteurs Discriminants Majeurs de la Présence des Glossines dans une Zone Agro-pastorale du Burkina Faso. Intérêt pour la Prévision du Risque Trypanosomien*. Montpellier: Université Montpellier II.

Ma, Z.K. and Redmond, R.L. (1995). TAU coefficients for accuracy assessment of classification of remote sensing data. *Photogrammetric Engineering & Remote Sensing* **61**, 435–439.

Macdonald, G. (1957). *The Epidemiology and Control of Malaria*. London: Oxford University Press.

Manel, S., Dias, J.M., Buckton, S.T. and Ormerod, S.J. (1999). Alternative methods for predicting species distributions: an illustration with Himalayan river birds. *Journal of Applied Ecology* **36**, 734–747.

McLynn, F. (1991). *Stanley. The Making of an African Explorer*. Oxford: Oxford University Press.

Milligan, P.J.M. and Baker, R.D. (1988). A model of tsetse-transmitted animal trypanosomiasis. *Parasitology* **96**, 211–239.

Morris, K.R.S. (1951). The ecology of epidemic sleeping sickness. I. The significance of location. *Bulletin of Entomological Research* **42**, 427–443.

Morris, K.R.S. (1952). The ecology of epidemic sleeping sickness. II. The effects of an epidemic. *Bulletin of Entomological Research* **43**, 375–396.

Morris, K.R.S. (1959). The epidemiology of sleeping sickness in East Africa. Part I. A sleeping sickness outbreak in Uganda in 1957. *Transactions of the Royal Society of Tropical Medicine and Hygiene* **53**, 384–393.

Morris, K.R.S. (1960a). Studies on the epidemiology of sleeping sickness in Africa. II. Sleeping sickness in Kenya. *Transactions of the Royal Society of Tropical Medicine and Hygiene* **54**, 71–86.

Morris, K.R.S. (1960b). Studies on the epidemiology of sleeping sickness in East Africa. III. The endemic area of Lakes Edward and George in Uganda. *Transactions of the Royal Society of Tropical Medicine and Hygiene* **54**, 212–224.

Morris, K.R.S. (1963). The movement of sleeping sickness across central Africa. *Journal of Tropical Medicine and Hygiene* **66**, 59–76.

Morris, K.R.S. and Morris, M.G. (1949). The use of traps against tsetse in West Africa. *Bulletin of Entomological Research* **39**, 491–528.

Murray, M. and Gray, A.R. (1984). The current situation on animal trypanosomiasis in Africa. *Preventive Veterinary Medicine* **2**, 23–30.

Mwambu, P.M. and Odhiambo, J.O. (1967). Cattle trypanosomiasis in the area adjoining the South Busoga fly-belt. In: *East African Trypanosomiasis Research Organization Annual Report 1966*, pp. 56–59. Tororo: EATRO.

Nash, T.A.M. (1944). A low density of tsetse associated with a high incidence of sleeping sickness. *Bulletin of Entomological Research* **35**, 51–59.

Nash, T.A.M. (1948). *Tsetse Flies in British West Africa*. London: HMSO.

Newstead, R., Evans, A.M. and Potts, W.H. (1924). *Guide to the Study of Tsetse Flies*. Liverpool: Liverpool School of Tropical Medicine.

Onyango, R.J., van Hoeve, K. and de Raadt, P. (1966). The epidemiology of *T.*

rhodesiense sleeping sickness in Alego Location, Central Nyanza, Kenya. *Transactions of the Royal Society of Tropical Medicine and Hygiene* **60**, 175–182.

Ormerod, W.E. (1976). Ecological effect of control of African trypanosomiasis. *Science* **191**, 815–821.

Ormerod, W.E. (1986). A critical study of the policy of tsetse eradication. *Land Use Policy* **3**, 85–99.

Ormerod, W.E. (1990). Africa with and without tsetse. *Insect Science and its Applcation* **11**, 455–461.

Ormerod, W.E. and Rickman, L.R. (1988). Sleeping sickness control–how wildlife and man could benefit. *Oryx* **22**, 36–40.

Pender, J. and Rosenberg, L.J. (1995). *Impact of Tsetse Control on Land Use in the Semi-arid Zone of Zimbabwe. Phase 1: Classification of Land Use by Remote Sensing Imagery*. Chatham: Natural Resources Institute.

Pender, J., Mills, A.P. and Rosenberg, J. (1997). *Impact of Tsetse Control on Land Use in the Semi-arid Zone of Zimbabwe. Phase 2: Analysis of Land Use Change by Remote Sensing Imagery*. Chatham: Natural Resources Institute.

Prihodko, L. and Goward, S.N. (1997). Estimation of air temperature from remotely sensed surface observations. *Remote Sensing of the Environment* **60**, 335–346.

Prince, S.D., Goetz, S.J., Dubayah, R.O., Czajkowski, K.P. and Thawley, M. (1998). Inference of surface and air temperature, atmospheric precipitable water and vapour pressure deficit using advanced very high-resolution radiometer satellite observations: comparison with field observations. *Journal of Hydrology* **212–213**, 230–249.

Randolph, S.E. and Rogers, D.J. (1997). A generic population model for the African tick *Rhipicephalus appendiculatus*. *Parasitology* **115**, 265–279.

Rawlings, P., Dwinger, R.H. and Snow, W.F. (1991). An analysis of survey measurements of tsetse challenge to trypanotolerant cattle in relation to aspects of analytical models of transmission. *Parasitology* **102**, 371–377.

Robertson, D.H.H. (1963). Human trypanosomiasis in South-East Uganda: a further study of the epidemiology of the disease among fishermen and peasant cultivators. *Bulletin of the World Health Organization* **28**, 627–643.

Robinson, T., Rogers, D.J. and Williams, B. (1997a). Mapping tsetse habitat suitability in the common fly belt of Southern Africa using multivariate analysis of climate and remotely sensed vegetation data. *Medical and Veterinary Entomology* **11**, 235–245.

Robinson, T., Rogers, D.J. and Williams, B. (1997b). Univariate analysis of tsetse habitat in the common fly belt of Southern Africa using climate and remotely sensed vegetation data. *Medical and Veterinary Entomology* **11**, 223–234.

Rogers, D. (1974). *Ecology of Glossina: Natural Regulation and Movement of Tsetse Fly Populations*. Paris: Institut d'Élevage et de Médecine Vétérinaire des Pays Tropicaux.

Rogers, D. (1979). Tsetse population dynamics and distribution: a new analytical approach. *Journal of Animal Ecology* **48**, 825–849.

Rogers, D.J. (1988). A general model for the African trypanosomiases. *Parasitology* **97**, 193–212.

Rogers, D.J. (1990). A general model for tsetse populations. *Insect Science and its Application* **11**, 331–346.

Rogers, D.J. (1993). Remote sensing and the changing distribution of tsetse flies in Africa. In: *Insects in a Changing Environment* (R. Harrington and N.E. Stork, eds), pp. 167–183. London: Royal Entomological Society.

Rogers, D.J. and Packer, M.J. (1993). Vector-borne diseases, models, and global change. *Lancet* **342**, 1282–1284.

Rogers, D.J. and Randolph, S.E. (1984). A review of the density-dependent processes in tsetse populations. *Insect Science and its Application* **5**, 397–402.

Rogers, D.J. and Randolph, S.E. (1986). Distribution and abundance of tsetse flies (*Glossina* spp.). *Journal of Animal Ecology* **55**, 1007–1025.

Rogers, D.J. and Randolph, S.E. (1991). Mortality rates and population density of tsetse flies correlated with satellite imagery. *Nature* **351**, 739–741.

Rogers, D.J. and Randolph, S.E. (1993). Distribution of tsetse and ticks in Africa: past, present and future. *Parasitology Today* **9**, 266–271.

Rogers, D.J. and Williams, B.G. (1993). Monitoring trypanosomiasis in space and time. *Parasitology* **106**, S77–S92.

Rogers, D.J. and Williams, B.G. (1994). Tsetse distribution in Africa: seeing the wood and the trees. In: *Large-scale Ecology and Conservation Biology* (P.J. Edwards, R.M. May and N.R. Webb, eds), pp. 249–273. Oxford: Blackwell Scientific.

Rogers, D.J., Randolph, S.E. and Kuzoe, F.A.S. (1984). Local variation in the population dynamics of *Glossina palpalis palpalis* (Robineau-Desvoidy) (Diptera: Glossinidae). I. Natural population regulation. *Bulletin of Entomological Research* **74**, 403–423.

Rogers, D.J., Hendrickx, G. and Slingenbergh, J. (1994). Tsetse flies and their control. *Revue Scientifique et Technique de l'Office International des Epizooties* **13**, 1075–1124.

Rogers, D.J., Hay, S.I. and Packer, M.J. (1996). Predicting the distribution of tsetse flies in West Africa using temporal Fourier processed meteorological satellite data. *Annals of Tropical Medicine and Parasitology* **90**, 225–241.

Rogers, D.J., Hay, S.I., Packer, M.J. and Wint, G.R.W. (1997). Mapping land-cover over large areas using multispectral data derived from the NOAA-AVHRR: a case study of Nigeria. *International Journal of Remote Sensing* **18**, 3297–3303.

Ross, R. (1909). *Report on the Prevention of Malaria in Mauritius*. London: Churchill.

Ross, R. (1911). *The Prevention of Malaria*. London: Murray.

Scott, D. (1965). *Epidemic Disease in Ghana 1901–1960*. London: Oxford University Press.

Stollnitz, E.J., Derose, T.D. and Salesin, D.H. (1996). *Wavelets for Computer Graphics*. San Francisco: Morgan Kaufmann.

Storch, H. von and Zwiers, F.W. (1999). *Statistical Analysis in Climate Research*. Cambridge: Cambridge University Press.

Swain, P.H. (1978). Remote sensing: the quantitative approach. In: *Remote Sensing: the Quantitative Approach* (P.H. Swain and S.M. Davis, eds), pp. 136–187 and 221–223. New York: McGraw-Hill.

Swallow, B.M. (1997). Impacts of trypanosomosis on African agriculture. In: 24th Meeting of the International Scientific Council for Trypanosomiasis Research and Control, Maputo, p. 607. Nairobi: Organisation of African Unity.

Tatsuoka, M.M. (1971). *Multivariate Analysis: Techniques for Educational and Psychological Research.* New York: Wiley & Sons.

Thomson, M.C., Connor, S.J., D'Alessandro, U., Rowlinson, B., Diggle, P., Cresswell, M. and Greenwood, B. (1999). Predicting malaria infection in Gambian children from satellite data and bed net use surveys: the importance of spatial correlation in the interpretation of results. *American Journal of Tropical Medicine and Hygiene* **61**, 2–8.

Turner, M.J., Blackledge, J.M. and Andrews, P.R. (1998). *Fractal Geometry in Digital Imaging*. San Diego: Academic Press.

Vickerman, K. (1997). Landmarks in trypanosome research. In: *Trypanosomiasis and Leishmaniasis: Biology and Control*, (G. Hide, J.C. Mottram, G.H. Coombs and P.H. Holmes, eds), pp. 1–37. Willingford: CAB International.

Williams, B., Rogers, D., Staton, G., Ripley, B. and Booth, T. (1992). Statistical modelling of georeferenced data: mapping tsetse distributions in Zimbabwe using climate and vegetation data. In: *Modelling Vector Borne and Other Parasitic Diseases* (B.D. Perry and J.W. Hansen, eds), pp. 267–280. Nairobi: International Laboratory for Research on Animal Diseases.

Earth Observation, Geographic Information Systems and *Plasmodium falciparum* Malaria in Sub-Saharan Africa

S.I. Hay,[1] J.A. Omumbo,[2] M.H. Craig[3] and R. W. Snow[2]

[1] *Trypanosomiasis and Land Use in Africa (TALA) Research Group, Department of Zoology, University of Oxford, South Parks Road, Oxford OX1 3PS, UK;*
[2] *KEMRI/Wellcome Trust Collaborative Programme, PO Box 43640, Nairobi, Kenya;*
[3] *MARA/ARMA Investigating Centre, Medical Research Council, 771 Umbilo Road, PO Box 17120, Congella, 4013, South Africa*

ADVANCES IN PARASITOLOGY VOL 47
0065-308-X $30.00

ABSTRACT

This review highlights the progress and current status of remote sensing (RS) and geographical information systems (GIS) as currently applied to the problem of *Plasmodium falciparum* malaria in sub-Saharan Africa (SSA). The burden of *P. falciparum* malaria in SSA is first summarized and then contrasted with the paucity of accurate and recent information on the nature and extent of the disease. This provides perspective on both the global importance of the pathogen and the potential for contribution of RS and GIS techniques. The ecology of *P. falciparum* malaria and its major anopheline vectors in SSA is then outlined, to provide the epidemiological background for considering disease transmission processes and their environmental correlates. Because RS and GIS are recent techniques in epidemiology, all mosquito-borne diseases are considered in this review in order to convey the range of ideas, insights and innovation provided. To conclude, the impact of these initial studies is assessed and suggestions provided on how these advances could be best used for malaria control in an appropriate and sustainable manner, with key areas for future research highlighted.

1. INTRODUCTION

1.1. The Burden of *Plasmodium falciparum* Malaria in Sub-Saharan Africa

It is recognized that 90% of the global malaria burden is concentrated in sub-Saharan Africa (SSA) and caused by *P. falciparum*. It has been estimated that approximately one million people died from the direct consequences of *P. falciparum* malaria infection in SSA during 1997 (Snow *et al.*, 1999a) and that 75% of these deaths occurred among pre-school children. The malaria parasite is one of the most significant infectious agents African children encounter as they pass through childhood. Malaria not only poses a risk to survival but the repeated clinical consequences of infection during early life place a burden on households, health services and ultimately the economic development of nation states (Bloom and Sachs, 1998). Sachs and Warner (1997) have argued that the persistence of endemic malaria in the tropics, and particularly in Africa, is contributory to a perpetual state of depressed economic growth in these regions. These macro-estimates of burden and economic associations provide clear support for a renewed effort aimed at halving malaria mortality by the year 2010, referred to as the Roll Back Malaria (RBM) initiative (Nabarro and Tayler, 1998; WHO, 1998). This

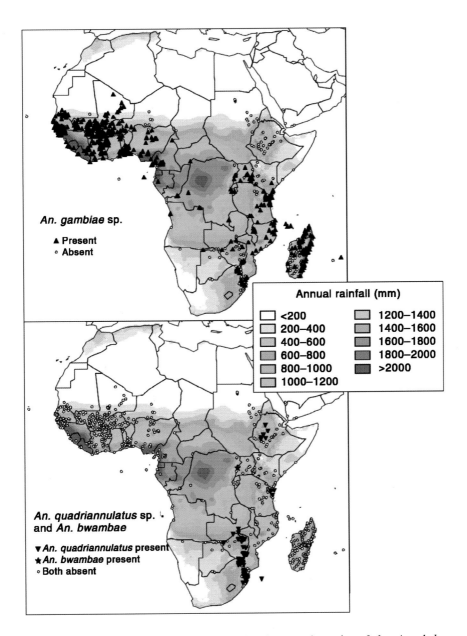

Plate 5 Records of the distribution of the six named species of the *Anopheles gambiae* complex superimposed on a background of total annual rainfall. The addition of 'sp.' to *An. gambiae* and *An. quadriannulatus* indicates that these taxa include more than one species. See Hay *et al*. (this volume).

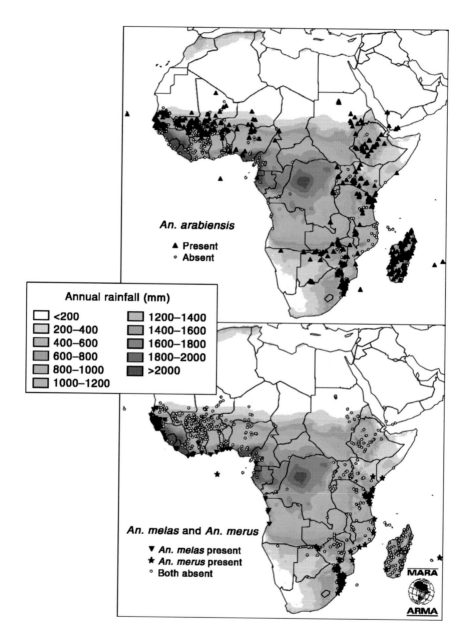

Plate 5 (contd) Records of the distribution of the six named species of the *Anopheles gambiae* complex superimposed on a background of total annual rainfall. The addition of 'sp.' to *An. gambiae* and *An. quadriannulatus* indicates that these taxa include more than one species. See Hay *et al.* (this volume).

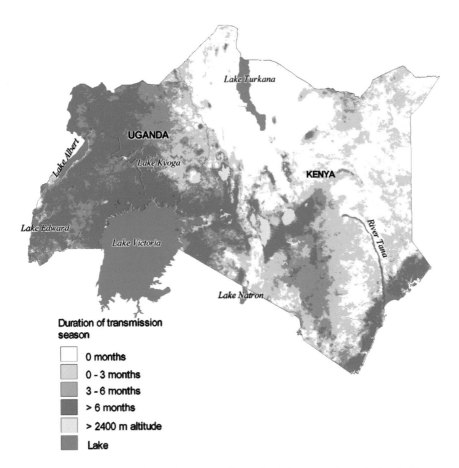

Plate 6 Predictions of malaria seasonality at 1×1 km spatial resolution in Kenya and Uganda (north is to the top of the page). The map shows the number of months for which *Plasmodium falciparum* malaria transmission is possible, determined by a Normalized Difference Vegetation Index (NDVI) threshold of 0.35. See Hay *et al.* (this volume).

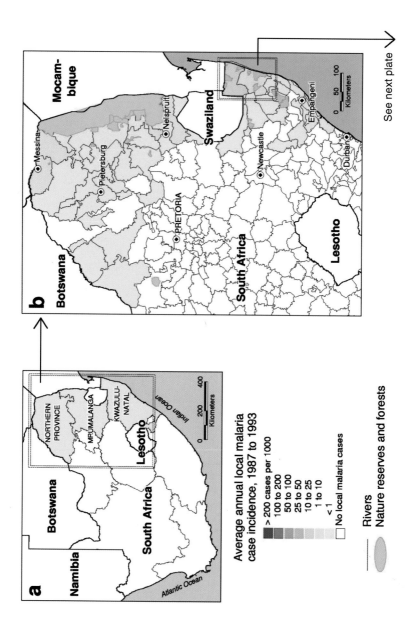

Plate 7 Average annual incidence of locally transmitted malaria in South Africa from 1987 to 1993. Incidence at province level is very low (a); reporting at district level highlights the four high-risk districts along the northern and eastern borders (b). See Hay *et al.* (this volume).

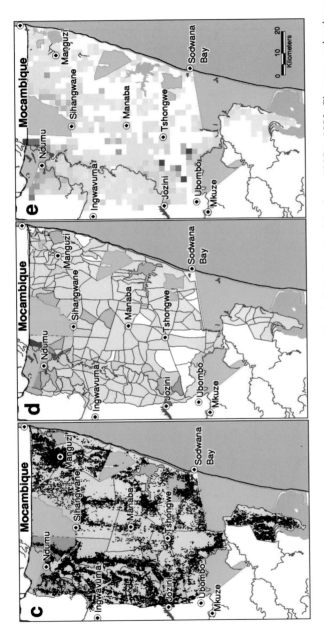

Plate 7 (contd) Average annual incidence of locally transmitted malaria in South Africa from 1987 to 1993. Closer examination of the two high-risk districts in KwaZulu-Natal at malaria control area (c) and sector (d) level reveals sub-district variability, with higher incidence around the towns of Ndumu, Sihangwane and Jozini, and low incidence along the coast; reporting cases at household level (individual households are shown in (c)) allows higher-resolution examination, for instance on a 2.5 × 2.5 km² grid (e). See Hay *et al.* (this volume).

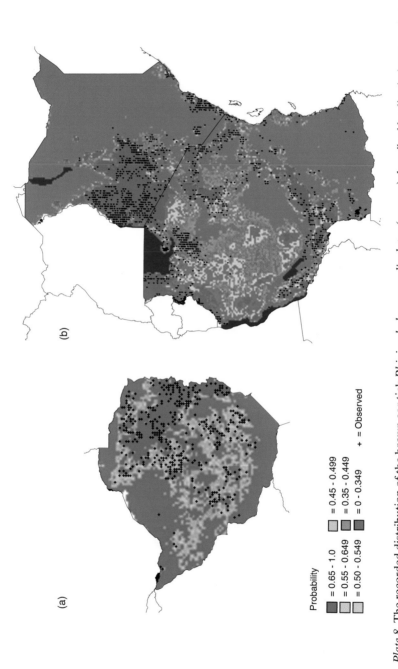

Plate 8 The recorded distribution of the brown ear tick *Rhipicephalus appendiculatus* (crosses) described by discriminant analysis of (a) a single variable, the mean monthly maximum temperature, in Zimbabwe, and (b) two additional variables, the minimum of the monthly NDVI and elevation, in Kenya and Tanzania. See Randolph (this volume). (Reproduced from Rogers and Randolph, 1993.)

Probability

■ = 0.65 - 1.0   □ = 0.45 - 0.499
■ = 0.55 - 0.649   ■ = 0.35 - 0.449
■ = 0.50 - 0.549   ■ = 0 - 0.349

+ = Observed

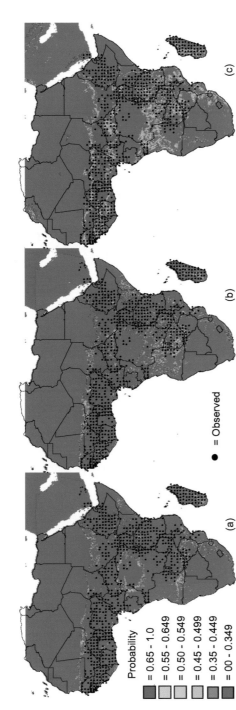

Plate 9 The recorded pan-African distribution of the bont tick *Amblyomma variegatum* (•) described by discriminant analysis of a digital elevation model and remotely sensed variables of NDVI, cold cloud duration and infrared radiation. (a) No clustering, (b) four absence and three presence clusters and (c) five absence and two presence clusters. See Randolph (this volume). (Unpublished maps created by Jonathan Toomer and David Rogers, reproduced from Toomer, 1996.)

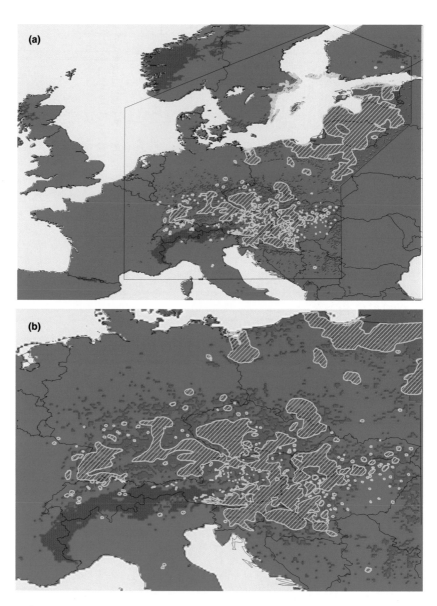

Plate 10 (a) Predicted (red) and observed (yellow hatched) pan-European distributions of foci of tick-borne encephalitis (TBE) virus, based on analysis of remotely sensed environmental variables and elevation within the outlined area. TBE virus occurs extensively to the east of this area, in Russia, Belarus, Ukraine and Romania (Immuno, 1997), but is not yet mapped in any detail. High mountain areas (darker green) were excluded from the analysis as the satellite data for them are less reliable because of more frequent cloud contamination. (b) Detail of central Europe taken from (a).

optimistic goal has been conceived at a time when existing, affordable therapeutics are rapidly failing, health service provision is breaking down, vaccines seem a distant dream, and poverty, conflict and corruption continue to afflict many African states (Desowitz, 1999).

Nevertheless in the wake of this desperate position there is a renewed hope offered by new and old approaches aimed at disease management and the prevention of infection. These include (i) the delivery of services to manage clinical disease through the integrated management of childhood illnesses (IMCI) or engaging mothers and the informal drug sector for home-based management, (ii) new therapeutics and strategies involving drug combinations, (iii) approaches to preventing infection in pregnant women to reduce anaemia and the consequences of placental infection in newborns, (iv) insecticide treated bed nets provided by the informal and commercial sectors and (v) residual house spraying in areas where the interruption of transmission is a realistic objective.

Massive financial investment will be required even to begin to reduce the incidence of malaria mortality in SSA using these tools. The resources targeted at malaria control will always be limited when compared with other social sector investment. Furthermore, not all malaria interventions will be equally appropriate for every setting. The challenge for the public-health sector is to decide which interventions would be appropriate where, and how these may be tailored to the local epidemiology to achieve maximal health impact for minimal investment. Clearly not all resource allocation is evidence-based but there is an increasing recognition among the malaria research and control communities that mapping risk and the projected benefits of intervention is a fundamental monitoring and decision-informing tool. It is in this context the utility of remote sensing (RS) and geographical information systems (GIS) is evaluated.

1.2. Defining Epidemiological Risk for Malaria Control

It is only during the 1990s that the public-health risks of *P. falciparum* malaria in SSA have become better defined. Against a constant risk of infection throughout life, individuals develop functional immunity against the severe and fatal outcomes of infection in infancy. The risk of a self-limiting clinical attack continues through the early years of life before declining later in childhood. The speed of acquired functional immunity depends upon the frequency of parasite exposure from birth. Consequently, complicated clinical disease and mortality from malaria are concentrated amongst the very young children under conditions of intense, high transmission and those at risk include increasingly older children as the intensity of parasite transmission declines (Snow *et al.*, 1997; Snow and Marsh, 1998). Ultimately when parasite transmission is extremely low and subject to unstable epidemic conditions, disease and mortality risks are equally distributed across all age groups. In these conditions the lack of immunity is balanced by the low risk of infection, but any disturbance in the frequency of parasite transmission can lead to devastating effects among the population. The range of environments found in SSA permits the entire spectrum of transmission conditions. Urbanization is a rapidly proliferating demographic feature of the continent and traditionally urban settings represent areas of low parasite exposure. Unstable transmission conditions exist in the south of Africa, the highlands of East Africa and the Horn of Africa and traditionally arid areas of the Sahel and Namibia (see Plate 4). In the remainder of the stable endemic areas in SSA, conditions can vary from one infectious bite from local vectors every 3 years to one every night (Hay *et al.*, 2000). Moreover, seasonal risks of infection vary enormously from a few weeks to constant risk throughout the year.

Facility-based health information systems or national civil registration systems are notoriously inadequate for national disease priority setting or programme planning. Consequently, little information exists on the epidemiological patterns and levels of disease risk in many areas of SSA. The location of populations in relation to health services is fundamental baseline information for health planning in the developed world. Such information, however, is often not available to those responsible for resource allocation in the developing world. Climate and its seasonality are coupled to the temporal variation in vector numbers and hence disease risk. These climate associations led to the development of crude, expert-opinion 'seasonal risk maps' for East Africa during the 1950s. No new map was developed until GIS rekindled interest in these types of information. Information on population distribution, health services, disease risk and seasonality should remove some of the barriers to providing a credible platform upon which to institute selected or targeted malaria control and prevention. GIS and RS provide a framework to develop high-resolution maps of risk, population and service delivery.

2. THE ECOLOGY OF *P. FALCIPARUM* MALARIA IN SUB-SAHARAN AFRICA

2.1. Epidemiology and Control

SSA supports a range of malaria endemicity whose geographical distribution is determined by a complex series of environmental, social, economic and historical patterns and contingencies. A complete epidemiology of *P. falciparum* malaria requires an analysis far beyond the scope of this review. The most important epidemiological aspects are outlined however, so that the rationale for using RS and GIS techniques can be appreciated.

The basic reproductive number, R_0, provides a useful structure for thinking about environmental influences on disease. It describes the average number of new cases of a disease that will arise from the introduction of an infective host into a wholly susceptible population (Anderson and May, 1991a) and, for an insect vector-borne disease such as malaria, it is given by the standard formula:

$$R_0 = \frac{a^2 mcbe^{-\mu T}}{\mu r}$$

where *a* is the vector biting rate, *m* is the ratio of vectors to hosts, *c* is the transmission coefficient from vertebrate to vector (i.e. the proportion of bites by vectors on infected hosts that eventually give rise to mature infections in the vectors), *b* is the transmission coefficient from vector to vertebrate, μ is the vector mortality rate, *T* is the incubation period of the parasite within the vector (sometimes referred to as the extrinsic incubation period) and *r* is the rate of recovery of the vertebrate from infection (Anderson and May, 1991b). Importantly, all of the variables are associated with the vector except *r*, which is a function of the host alone. Due to the dominance of vector biology on the transmission dynamics of pathogens they carry, it follows that the distribution and intensity of such diseases is dependent upon the distribution and abundance of the vector. For the purposes of this discussion we describe both the sensitivity of *P. falciparum* and its vectors to environmental conditions.

2.2. The Life Cycle

2.2.1. Plasmodium falciparum

Mosquitoes of the genus *Anopheles* were first identified as the vectors of human malaria in 1897 by Sir Ronald Ross (Wernsdorfer and McGregor, 1988). Today, four species of *Plasmodium* are known to infect humans namely *P. falciparum, P. malariae, P. vivax* and *P. ovale. P. falciparum* is the dominant malaria parasite found in the stable endemic areas of Africa (Young, 1976; Gilles, 1993). The malaria parasite develops in two stages; a sexual cycle that takes place within the mosquito vector and an asexual cycle in the human host (Fujioka and Aikawa, 1999). The haematophagous adult female *Anopheles* seek vertebrate hosts soon after emergence. The ingested blood is used to support egg production and, following development and subsequent oviposition, the female vector seeks further blood meals to nourish future broods. It is this repeated feeding that facilitates the transmission of parasites between hosts. Infection of the human host begins when sporozoites from an infected mosquito are injected into the blood of a susceptible human during a blood meal. It then takes 0.5–4 h for the sporozoites to invade host liver cells where they multiply and release as many as 30 000 merozoites, which, in turn, invade red blood cells. This asymptomatic period usually lasts about a week in tropical countries. The erythrocytic asexual development stage follows when the parasite develops from a ring form to a trophozoite that then becomes a schizont which multiplies to produce 4–32 merozoites. This intracellular multiplication causes red blood cells to rupture, with the resultant release of toxins into the blood, occurring in synchronized 48 h cycles for *P. falciparum, P. vivax and P. ovale* and in 72 h cycles for *P. malariae*. The bouts of fever associated with malaria correspond with these episodes of toxin release. Continued asexual multiplication with the invasion of further erythrocytes, as well as sexual differentiation, results in the production of macrogametocytes (female) and microgametocytes (male). These are the forms of the parasite infective to the mosquito.

The parasite's sexual cycle begins when gametocytes are ingested by a mosquito vector feeding on an infected individual (Beier, 1998). Fertilization of the gametocytes to form ookinetes takes place in the midgut of the mosquito and these lodge in the midgut outer wall as oocysts. Numerous sporozoites develop within the oocysts and, as the oocysts rupture, migrate to the mosquito's salivary gland from where they are injected into the human host during subsequent blood meals. Various aspects of this complex life cycle are affected by climate and are explored below.

2.2.2. *The* Anopheles gambiae *Complex*

Members of the *Anopheles* genus (subfamily Anophelinae, family Culicidae, order Diptera) are thought to be unique in being competent vectors for human malaria (Gillies and de Meillon, 1968; Service, 1985; Gillies, 1987). *Anopheles gambiae s.l.* is the dominant vector within the Afrotropical and Ethiopian faunal region (Boyd, 1930) and comprises six sibling species (see Plate 5). *An. gambiae s.s.*, *An. arabiensis* and *An. quadriannulatus* are fresh water breeders (White, 1974), *An. melas* and *An. merus* are salt water breeders (White, 1974), while *An. bwambae* breeds in mineral rich waters (Gillies and de Meillon, 1968; Service, 1985; Gillies, 1987).

An. gambiae s.s and *An. arabiensis* tend to occur sympatrically in approximately 70% of SSA (White *et al.*, 1972; Lindsay *et al.*, 1998) with their relative abundance dependent on local ecological conditions. *An. gambiae s.s.* (formerly species A) prefers more humid and vegetated areas. It is predominantly anthropophagic and endophilic and tends to have higher vectorial capacity than other species of the gambiae complex due to its greater longevity.

An. arabiensis (formerly species B) predominates in the Horn of Africa and southern Arabia and is the sole vector of malaria in northern Sudan (Wernsdorfer and McGregor, 1988) (see Plate 5). This species is noted for its ecophenotypic plasticity, being predominantly exophagic and exophilic but facultatively endophagic and endophilic in adverse environmental conditions. Its occasional abundance in arid areas is due to its ability to breed in residual pools of water in dry riverbeds and rapidly increase at the onset of rains.

Anopheles melas and *An. merus* are adapted for salt water breeding and are hence found only along the coastlines of West and East Africa respectively (White, 1974). *An. bwambae* (formerly species D) and *An. quadriannulatus* (formerly species C) are not vectors of human malaria and therefore are not considered further.

2.2.3. Anopheles funestus

An. funestus is the most important vector of malaria in SSA after *An. gambiae s.s.* and *An. arabiensis* (see White *et al.*, 1972; Service, 1993). It prefers shaded, long-lived bodies of water with emergent vegetation for oviposition (Christie, 1959). This species is predominantly anthropophagic and feeds both indoors and outdoors. In areas of rice cultivation it emerges when the rice plants have grown and provide shade and therefore appears later in transmission season than *An. gambiae* (see Bradley, 1991).

2.3. Environmental Determinants of Disease Transmission

2.3.1. *Temperature*

Malaria is a disease of tropical and temperate countries between the latitudinal limits of 64° North and 57° South (Gill, 1921) with prevalence increasing towards the equator. As the parasites require time to develop into infective stages, female anopheles are not immediately infective after feeding. This extrinsic incubation period (T) is temperature dependent and optimum conditions have been defined between 25 and 30°C, with development ceasing below 16°C and above 40°C (Russell *et al.*, 1946; Gilles, 1993). Intermittent low temperatures have also been found to delay sporogony, with the period immediately after the infective bite being most

Table 1 The effect of temperature on sporogonic duration (Macdonald, 1957; Detinova, 1962), daily vector survival (Martens, 1997), percentage cohort survival against sporogonic duration and larval development (Jepson *et al.*, 1947). Note that thermal death of anophelines occurs below 5°C (Leeson, 1931; de Mellion, 1934) and above 40–45°C (Buxton, 1931; Haddow, 1943; Jepson *et al.*, 1947).

T (°C)	Duration of sporogony (days)	Daily vector survival (%)	Vector survival after period required for sporogony (%)	Larval development (days)
16	∞	89.3	0	47
17	111	89.7	0.001	37
18	56	90.0	0.28	31
20	28	90.3	5.9	23
22	19	90.4	15	18
30	7.9	88.1	37	10
35	5.8	80.8	29	7.9
39	4.8	38.9	1.1	6.7
40	4.6	0	0	6.5

sensitive to temperature decreases. The effects of temperature on various aspects of anophelene biology have been investigated under laboratory conditions (Jepson *et al.*, 1947; Macdonald, 1957; Detinova, 1962; Martens, 1997) and are summarized in Table 1. High temperatures are also associated with increased frequency of feeding, since blood meals are more rapidly digested and the whole gonotrophic cycle accelerated (Boyd, 1930).

2.3.2. *Rainfall*

Rainfall provides surface water in which female anopheles can lay eggs. In arid areas where temperatures are usually suitable, malaria transmission occurs only when rainfall provides temporary breeding habitat for vectors. These areas are often classified as 'malarious near water' since transmission outside the rainy seasons typically occurs only along riverbeds, oases and other man-made surface water sites. Studies have demonstrated an association between abundance of *An. gambiae s.l.* and rainfall (Christie, 1959; White *et al.*, 1972; Molineaux and Gramiccia, 1980; Charlwood *et al.*, 1995). Rainfall effects are often most apparent during epidemics when the rise in malaria cases is often proportional to the amount of precipitation, among other factors (Christophers, 1911; Covell, 1957; Wernsdorfer and McGregor, 1988; Malakooti *et al.*, 1998; Kilian *et al.*, 1999). Classic epidemiological studies in Punjab compared the hospital-based monthly prevalence of fever deaths with contemporaneous total monthly rainfall data (Gill, 1920, 1921; Boyd, 1930) and found periods of increased mortality correlated maximally with rainfall lagged by one month. The relationship is obviously best observed where temperature is not confounding. Excessive rainfall can reduce transmission by flushing larvae out of small pools (Covell, 1957; Molineaux, 1988; Gilles, 1993). The relationship between precipitation and the development of breeding sites also depends on slope of the land, run-off and soil type leading some authors to look at the feasibility of hydrological modelling to improve rainfall–malaria associations (Patz et al., 1998).

2.3.3. Climate Seasonality

The interaction between temperature and rainfall is largely responsible for the seasonal characteristic of malaria transmission. Seasonal variation of infection risk is a common feature of malaria in SSA and is reflected in intra-annual changes in vector densities, entomological inoculation rates and malaria admissions (Christie, 1959; Julvez et al., 1992; Aniedu, 1997; Hay et al., 1998b, 2000). The map showing the heterogeneity in malaria seasonality across Uganda and Kenya (see Plate 6), based on an RS proxy of seasonality (Hay et al., 1998a) is discussed in more detail in Section 3.2.

2.3.4. Atmospheric Moisture

Early documented reports of human malaria describe its association with humid swamps and marshes (Gill, 1921) and several authors have attempted to define optimum conditions of relative humidity (RH) based on such observations (Wernsdorfer and McGregor, 1988; Gilles, 1993). Gill (1921) was the first formally to investigate the effect of changes in RH on Culex fatigans and the transmission of avian malaria. Higher relative humidities were associated with increased vector longevity and greater frequency of feeding. RH also determined the timing and duration of daytime resting behaviour (Boyd, 1930; Russell et al., 1946; Molineaux, 1988).

2.3.5. Altitude

Altitude has long been a subject of interest among malariologists (Schwetz, 1942; Garnham, 1948; Heisch and Harper, 1949; Covell, 1957; Roberts, 1964; Malakooti et al., 1998). Altitude and temperature are explicitly linked, with every 100 m increase in height corresponding to an approximately 0.5°C decline in temperature. The use of altitude can be confusing, however, with the limit for malaria transmission variously reported above 2000 m in Ethiopia (Covell, 1957), 1800 m in the Congo (Schwetz, 1942) and 1950 m in

Kenya (Garnham, 1948). Use of the phrase 'highland malaria' continues, but it is more clearly thought of as temperature-limited unstable transmission.

2.3.6. *Other Environmental Factors*

Many reviews have described the effects of anthropogenic environmental change on malaria transmission (Bradley, 1992; Mouchet *et al.*, 1998) including specifically the effects of agricultural practices (Coosemans, 1985; Robert *et al.*, 1985; Githeko *et al.*, 1993; Dossou-Yoyo *et al.*, 1994; Hay *et al.*, 2000), water resource development (El Gaddal, 1991; Hunter *et al.*, 1993), urbanization (Trape and Zoulani, 1987; Coene, 1993), population movements (Nega, 1991), malaria control (Draper and Smith, 1957, 1960; Draper *et al.*, 1972) and climate change (Lindsay and Birley, 1996; Lindsay and Martens, 1998). There is no scope to elaborate on these here.

3. REMOTE SENSING AND MOSQUITO-BORNE DISEASE

3.1. The Rationale for Remote Sensing

The principal goal of RS in malaria epidemiology is to help map the disease distribution so that control efforts in endemic situations and intervention strategies in epidemic situations may be most directed efficiently. Section 2 highlighted the long understood relationships between mosquito numbers and climate, so that shorter term meteorological variation could be used in predicting the onset and severity of malaria epidemics (Swaroop, 1949). Decades later, our understanding of how factors such as temperature, humidity and rainfall influence mosquito population dynamics has improved, along with the sophistication with which weather variables can be used to predict malaria transmission rates (Onori and Grab, 1980). Furthermore, many mosquito species oviposit in specific aquatic habitats which support characteristic plant communities and which are hence easily identified (Rioux *et al.*, 1968). The use of RS techniques to investigate mosquito and malaria ecology stems from the understanding that aerial and space-borne sensors could provide relevant surrogate information relating to the spatial variation in these meteorological and vegetation variables (Cline, 1970).

Previous reviews of RS applied to the problems of mosquito and malaria control (Roberts and Rodriguez, 1994; Washino and Wood, 1994; Thomson *et al.*, 1997; Dale *et al.*, 1998; Hay *et al.*, 1998a; Connor, 1999) and to the study and control of a range of arthropod-borne diseases (Hay *et al.*, 1997), have explored the relationships between remotely derived estimates of ecological conditions and parameters of malaria distribution and abundance. Related themes, such as the advantages of RS in terms of the rate, geographical extent and spatial and spectral resolution of data collection have also been well documented (Hugh-Jones, 1989; Hay, 1997), as have some of the disadvantages regarding the interpretation and quality of such RS data (Hay *et al.*, 1996; Hay and Lennon, 1999). The aim of this section is not to reiterate these works, but to outline the increased epidemiological understanding and control operation efficiency that can be gained by adopting RS techniques. The extent of current research does not allow a specific focus on *P. falciparum* malaria in the continent of Africa. Instead we look at RS applications to all mosquito-borne diseases, from around the world, with an emphasis on the techniques applied. This background will enable us to anticipate, identify and plan for the opportunities made available by forthcoming advances in satellite-sensor technology.

Details of RS terminology, acronyms and satellite-sensor system specifications are outlined by Hay (this volume). In brief however, low altitude, high spatial resolution sensors are those that can resolve objects of less than 1×1 km at nadir and high altitude, low spatial resolution sensors those that cannot. Passive sensors derive their energy from the reflected and emitted energy of the Sun, and actively sensors generate energy themselves, usually at radar frequencies.

3.2. Low Altitude Aerial Sensors

It is the dependence of mosquitoes on fresh and brackish water habitats in the early stages of their life-cycle that first allowed RS techniques to be exploited. The following sections describes how early studies used RS techniques for the purpose of high spatial resolution habitat mapping.

3.2.1. *Passive Photography*

In 1971, National Aeronautics and Space Administration (NASA) scientists, in combination with personnel from the New Orleans Mosquito Control District (NOMCD), were the first to investigate the use of colour and colour–infrared (CIR) aerial photography in mapping vegetation assemblages associated with the larval habitat of *Aedes sollicitans*, a nuisance saltmarsh mosquito, that in addition is a competent vector for the equine encephalitis flavivirus (Anonymous, 1973). Previous work by NOMCD had shown that female mosquitoes always oviposited in areas of the saltmarsh intermittently flooded by freshwater. The floral assemblage dominated by spikerush

(*Spartina patens*) and wiregrass (*Juncus roemerianus*) was adapted to this same hydrological regime and hence was a reliable indicator of *Ae. sollicitans* larval habitat (Bidlingmayer and Klock, 1955). The report documents that such vegetation assemblages were extremely accurately identified at an 80 ha test site near New Orleans, although accuracy statistics were not provided.

The first operational use of CIR aerial photography was for delineating forested and open wetlands, marshes and residential areas in a new mosquito control area in the Saginaw and Bay counties of Michigan (Wagner *et al.*, 1979). As well as nuisance floodwater *Aedes* mosquitoes, *Culex* spp. that oviposited in tree-hole pools posed a significant public-health problem causing a local epidemic of St Louis encephalitis (SLE) in 1975. The known flight range of each mosquito species was used in combination with information on the distance between residential areas and mosquito habitat to identify control priorities for the two counties. The authors stressed the short time in which an environmental inventory was gathered and management priorities identified. Furthermore, the streamlining of subsequent control efforts led to a campaign of relatively low economic and environmental cost, since the area designated for insecticide treatment was considerably reduced in comparison with the alternative approach of broadcast aerial spraying. In contrast, Hopkins *et al.* (1975) reported that after an outbreak of SLE in Dallas in 1966 the entire county of Texas was sprayed aerially with an organophosphate insecticide (malathion) mist. Moreover, in response to the highly publicized outbreak of a West Nile-like flavivirus in New York (Lanciotti *et al.*, 1999), large areas of the city were sprayed using aerial insecticide. Should these flaviviruses become newly established in urban centres of North America (Anderson *et al.*, 1999), aerial spraying guided by RS would save time, money and unnecessary environmental impact.

The purpose of these studies was to identify larval habitats of mosquito species that constituted a significant nuisance or public health risk, so as to direct larval control more efficiently. Although simple and empirical, species–habitat correlations from CIR aerial photography have been, and continue to be, cited as more cost-effective than conventional ground survey techniques for obtaining information on the distribution of potential larval habitats; for example, for *Psorphora columbiae* in both Louisiana (Fleetwood *et al.*, 1981) and Texas rice fields (Welch *et al.*, 1989a, b) and for *Culex annulirostris* in urban areas of Queensland (Dale and Morris, 1996).

Perhaps unsurprisingly, it has also been observed from a study in Queensland, Australia that *Ae. aegypti* breeding sites (usually water accumulated in discarded containers) were not easily resolved using CIR photography (Moloney *et al.*, 1998), which was hence not thought to be of sufficient resolution to assist in mapping dengue vector habitat to assist control.

3.2.2. *Passive Optical Radiometers*

Investigators were quick to exploit these early successes and move from aerial photography to aerial observations with Multispectral Scanners (MSS). Improvements upon CIR aerial photography were demonstrated by Cibula (1976), and Barnes and Cibula (1979) using an airborne MSS that had a spatial resolution of 2.5 × 2.5 m when flown at an altitude of 1200 m. The greater number of wavelengths over which data were recorded (22 channels from 0.3 to 13 μm) enabled more accurate spectral identification of the *Spartina–Juncus* associations characteristic of favourable egg-laying habitats for *Ae. sollicitans* than by aerial infrared photographs.

Under the aegis of the NASA Global Monitoring and Disease Prediction Program (GMDPP), various authors investigated the relationship between mosquito population dynamics and environmental factors that could be observed remotely (Wood *et al.*, 1994). In phase I of this project, populations of *An. freeborni* in the rice fields of northern and central California were studied (Pitcairn *et al.*, 1988). This mosquito species represented no health risk to the local population, but provided an accessible model for the study of anopheline vectors in irrigated rice habitat (Wood *et al.*, 1991a,b, 1992).

Larval mosquito populations were sampled fortnightly throughout the period of rice crop development in 1985 (Wood *et al.*, 1991a). An airborne MSS collected data simultaneously with a spatial and spectral resolution designed to simulate the Landsat Thematic Mapper (TM). A Normalized Difference Vegetation Index (NDVI) was then calculated for individual fields on each of the sampling dates and the differences in spectral signal between rice fields producing high and low numbers of anopheline larvae were followed throughout the growing season. The NDVI was considered an important variable since Rejmankova *et al.* (1988) had previously demonstrated percentage rice crop cover to be positively correlated with mosquito larval production. The results showed that higher values of NDVI in the early growing season (June) were associated with high mosquito-producing rice fields, but that the spectral separation (between high and low mosquito-producing fields) diminished to a minimum in mid-July, when percentage rice crop cover was at the maximum. A discriminant analysis which incorporated information from Landsat-TM equivalent channels 1, 2, 3, 4, and 7 was able to distinguish between high and low mosquito-producing fields with an overall accuracy of 75%. The uneven distribution of high mosquito-producing fields was also shown to be related to patterns of surrounding land use, with 70% of the high mosquito-producing fields being within 1.5 km of livestock pasture (i.e. potential hosts). A more detailed survey in 1987 repeated the above work and included data on the distance to livestock in a GIS (Wood *et al.*, 1991b). An identical discriminant analysis combined with cattle distance data resolved high and low producing fields with an accuracy of 90%, with errors of omission (not identifying high producing fields) and commission (identifying as a high producing field one which is not) of 10% and 40% respectively (Wood *et al.*, 1992).

3.2.3. *Active Radar Radiometers*

A recent approach has demonstrated that land-based radar may be capable of predicting habitat flooding status (Ritchie, 1993). Radar was used in this study to estimate rainfall in Collier county, Florida. This information, in combination with data from tide tables, was used to guide surveys for recently inundated saltmarsh habitat suitable for oviposition by *Ae. taeniorhynchus*. Rain, tides and rain plus tide events triggered 48%, 26% and 26%, respectively, of proposed inspections and this information system located seven of the eight habitats of mosquitoes found during the study and control period.

3.3. High Altitude Satellite Sensors

The increasing number of successful applications of RS techniques to mosquito habitat mapping in North America provided sufficient impetus for researchers to transfer these approaches to satellite-borne sensor mapping of tropical areas, predominantly in meso-America, where both the logistical and public health problems were significantly greater. This move from analogue photographic to digital MSS techniques was also important, as it facilitated the move to quantitative analyses which, in turn, enabled the investigation of more subtle variation in environmental variables and thus mosquito habitat and ultimately disease suitability.

3.3.1. *Passive Optical Radiometers*

(a) *High Spatial, Low Temporal Resolution Sensors*. Landsat-1 and -2 MSS data were then used for the first time to identify freshwater plant communities associated with the larval habitats of *Ae. vexans* and *Culex tarsalis* in riparian habitat bordering the Niobara river between South Dakota and Nebraska (Hayes *et al.*, 1985). A supervised classification using multiple-date Landsat-MSS scenes identified these aquatic habitats with 95% accuracy. This study demonstrated that such species–habitat correlations could be successfully 'scaled-up' to regional control programmes using relatively high spatial resolution satellite sensor data.

The second phase of the previously cited NASA-GMDPP project investigated the population dynamics of *An. albimanus* and *An. pseudopunctipennis*, malaria vectors in the tropical wetlands of Chiapas,

Mexico (Roberts *et al.*, 1991). Pope *et al.* (1994a) used two Landsat-TM scenes of the area, one from the dry and one from the wet season, to provide an unsupervised classification (Curran *et al.*, this volume) of the region. The resulting clusters were assigned to land-cover types on the basis of CIR aerial photographs and field inspection of 30 test sites. These sites were independently sampled for mosquito density and information was collected on environmental variables affecting water and vegetation characteristics (Rejmankova *et al.*, 1991). The sites were then grouped into 16 habitat types using cluster analysis and correlations were performed between the habitat types and land-cover units (Rejmankova *et al.*, 1992). The land-cover units were subsequently ranked as having high, medium or low mosquito production potential on the basis of these correlations. Incorporating this information into a GIS, sites of high mosquito production around the towns of La Victoria and Efrain Gutierrez were found to occupy only 9% of the designated control area, allowing the potential for substantial streamlining of control campaign effort and resources.

An independent study also demonstrated that Landsat-TM data could be used to identify villages at high risk of malaria transmission within the Tapachula region of Chiapas, Mexico (Beck *et al.*, 1994, 1995). Dry and wet season Landsat-TM scenes were again subject to an unsupervised classification. A stepwise discriminant analysis and linear regression were then used to establish the relationship between vector abundance and land-cover. Land-cover in a 1 km buffer area surrounding the perimeter of the village was analysed since this corresponded to the known flight range of *An. albimanus*. Transitional swamp and unmanaged pasture were found to be the most important land-covers for *An. albimanus*, and their combined area in the buffer zone was sufficient to predict high and low vector abundance in 40 villages to an overall accuracy of 90%. Furthermore, these relationships were sufficiently robust to predict the abundance of adult *An. albimanus* in a further 40 randomly selected villages from the neighbouring Huixtla region to an overall accuracy of 70% using more multi-date Landsat-TM scenes (Beck *et al.*, 1997).

Rejmankova *et al.* (1995) have also shown that the density of adult *An. albimanus* mosquitoes around villages in Belize could be reliably predicted using multispectral Satellite pour l'Observation de la Terre (SPOT) High Resolution Visible (HRV) data. Productive larval habitats were first identified as marshes containing relatively few emergent aquatic plants and a high coverage of cyanobacterial mats. An unsupervised Bayesian maximum likelihood classification was then applied to a single SPOT scene covering a test site occupying northern Belize. The classes generated were subsequently assigned to individual 'landscape elements' based on field observations. Human settlements were identified with ancillary map data and located more precisely on subsequent field visits with a global positioning system

(GPS). These settlements were divided into two groups according to their distance to the larval habitat class. Group 1 was composed of settlements closer than 500 m and group 2 of settlements further than 1 500 m. Based on independent measurements, a landing rate greater than 0.5 mosquitoes per human per minute (during the hours of maximum mosquito activity from 18:30 to 20:00) was used as a threshold for high adult mosquito density (Rodriguez *et al.*, 1996). Group 1 was predicted as having adult mosquito densities higher than this threshold and group 2 lower. These predictions were tested by collecting mosquitoes landing on humans during the hours of peak activity within each of the settlements. The resulting predictions were 100% accurate for group 2 and 89% accurate for group 1.

Roberts *et al.* (1996) investigated the utility of multispectral SPOT-HRV data in predicting the distribution of the malaria vector *An. pseudopunctipennis* in central Belize. Previous investigations had shown that altitude and the presence of filamentous algae in sun-exposed pools were critical determinants for the presence of *An. pseudopunctipennis* larvae (Rejmankova *et al.*, 1993). Using the SPOT and cartographic data, 49 sites were chosen and predicted to have a high or low probability for *An. pseudopunctipennis* presence. The criteria used in site selection were the distance of the houses from waterways and their altitude above such waterways, as well as the amount of forest between the houses and the waterway. The SPOT data provided more contemporary information than the cartographic data and hence allowed map errors to be corrected. They also gave information on important aspects of environmental suitability which could not be attained from the maps, such as the size of a waterway or the degree to which it was open to sunlight. Collections of mosquitoes were then made in a sample of these sites to test the predictions. Four of the eight sites that were predicted as high probability locations for presence of *An. pseudopunctipennis* were positive and all 12 low probability sites were negative. The absence of *An. pseudopunctipennis* at four high probability locations was thought to be due to field population densities of the species being below the threshold of the sampling effort.

Multi-date Indian RS (IRS) satellite Linear Imaging and Self Scanning (LISS) II sensor data have been used to identify 'mosquitogenic conditions' for maps to guide local mosquito control activities in Delhi, northern India (Sharma *et al.*, 1996). Water bodies and marsh areas associated with housing were identified in a supervised classification and then changes in the area of these habitats between images correlated with adult and larval mosquito numbers. Statistics were not provided, making more detailed comment difficult.

Thomas and Lindsay (1999) conducted a retrospective spatial analysis of malaria cases in children from a series of villages in the centre of The Gambia. SPOT HR imagery and a GPS-based supervised classification were used to map exposed alluvial beds saturated with fresh water, that larval-dip surveys had shown to be the dominant mosquito larval producing habitat. A significant, spatially corrected correlation between the area of RS derived breeding habitat within a 2 km radius of a village and entomological innoculation rate (EIR) ($r = 0.85$, $n = 10$, $P = 0.002$), allowed EIR to be extrapolated over a wider area, including hundreds of villages where clinical surveillance was being conducted. From this larger sample it was seen that both asymptomatic and clinically manifested malaria cases were negatively correlated with risk of exposure to malaria. This had been found before in The Gambia and attributed to bed net use (Thomson *et al.*, 1996a), but this could not explain the relationship in this study. The authors concluded that higher malaria incidence in areas of lower exposure was due to a lack of challenge while young, resulting in low population immunity (Snow and Marsh, 1995; Snow *et al.*, 1997).

(b) *Low Spatial, High Temporal Resolution Sensors*. Multi-temporal NDVI data at 8×8 km spatial resolution from the National Oceanic and Atmospheric Administration's (NOAA) Advanced Very High Resolution Radiometer (AVHRR) have been applied to the problem of Rift Valley fever (RVF) epidemics (Linthicum *et al.*, 1987, 1990). The work showed that high NDVI values in Kenya were good indicators of seasonally flooded linear depressions, known as dambos. These habitats were highly suitable for *Aedes* spp. mosquito breeding and were hence closely associated with RVF epidemics. The work progressed to incorporate higher spatial resolution Landsat-TM and multispectral SPOT-HRV imagery to locate individual areas of high RVF risk determined by the NDVI (Linthicum *et al.*, 1991). Operational application of the technique, however, was hindered because investigators were not able to discriminate flooded from dry dambos in such images (see Section 3.3.2). Further refinement of this work has shown that a combination of NOAA-AVHRR NDVI anomaly data and Indian Ocean sea surface temperature data could predict RVF epidemics in Kenya 2–5 months before outbreaks, allowing sufficient time for public health service intervention (Linthicum *et al.*, 1999).

Linthicum *et al.* (1994) also investigated an unusually severe and extensive outbreak of RVF in the West African Senegal river basin in 1987. The RVF outbreak was particularly unusual as it occurred during a period of only moderate rainfall. Analysis of multispectral SPOT-HRV scenes for the period of the epidemic revealed extensive flooding in Mauritania to have peaked in October 1987, as a result of the construction of the Diama and Manatelli dams on the river Senegal. This coincided exactly with the period of maximum RVF disease activity in the area. Furthermore, maximum values of local area coverage (LAC) NDVIs were associated with increased rice production (and hence productive mosquito larval habitats) around the newly flooded regions in Daro and Rosso, the foci of RVF outbreaks. Unfortunately, no statistics were presented in the study.

NDVI data derived from the NOAA-AVHRR have also been used in the Mar Chiquita lake region of central Argentina to predict the abundance of *Ae. albifasciatus* an important nuisance to livestock and vector of western equine encephalitis in the region (Gleiser *et al.*, 1997). The authors argued that such relationships could be used to monitor and predict future surges in the population of this floodwater mosquito.

Thomson *et al.* (1996b, 1997) investigated the potential of coarse spatial resolution GAC and LAC NOAA-AVHRR data, as well as cold cloud duration (CCD) data from the Food and Agriculture Organization's (FAO) African Real Time Environmental Monitoring Using Imaging Satellites (ARTEMIS) programme, to predict malaria risk in The Gambia. They concluded, however, that although there were clear relationships between satellite sensor data and environmental variables associated with malaria transmission, it was difficult to predict how these would affect adult mosquito abundance and behaviour. For instance, along the river Gambia a decrease in rainfall may at times increase available anopheline breeding sites by increasing the number of suitable pools in the alluvial soils at the river margin. They also noted that relationships between malaria incidence and environmental variables were complicated by sociological factors, because in areas where anopheline abundance was greatest, and hence biting most frequent, people were more likely to protect themselves with insecticide-impregnated bed nets (Thomson *et al.*, 1994). Further work, however, has demonstrated more clearly that malaria infection rates in Gambian children could be related to a parameter representing NDVI seasonality after behavioural considerations had been factored out (Thomson *et al.*, 1999). The authors went on to show how these relationships could be used to predict the changes in malaria prevalence resulting from treated and untreated bed net use, as well as giving some important warnings regarding the confidence intervals surrounding relationships between spatially autocorrelated data (see also Curran *et al.*, this volume; Robinson, this volume).

Predictions of malaria seasonality (the combination of disease risk in space and time) in Kenya have been achieved by establishing relationships between childhood malaria cases, with disease data collected during on-going surveillance of severe malaria morbidity in five communities in Africa (Snow *et al.*, 1997), and contemporary imagery from the NOAA-AVHRR and Meteosat-HRR (Hay *et al.*, 1998b). The remotely sensed data were first processed to provide surrogate information on land surface temperature (LST), rainfall (expressed as CCD or the number of hours for which a given pixel was covered by cloud below a threshold determined as rain-bearing), reflectance in the middle infrared (MIR) wavelengths and the NDVI. These variables were then subject to temporal Fourier processing (Rogers *et al.*, 1996) and compared with the mean percentage of total annual malaria admissions recorded each month at three sites in Kenya. The NDVI lagged by

1 month was found to be the most significant variable consistently correlated with malaria admissions across the sites (mean adjusted $r^2 = 0.71$, range 0.61–0.79). Interestingly, Patz *et al.* (1998) showed that NOAA-AVHRR NDVI also explained a large amount of temporal variation in human biting rate in a 2 years study in Kisian, western Kenya. Subsequent regression analyses showed that an NDVI threshold of 0.35–0.40 is required for more than 5% of the annual malaria cases to be presented in a given month (Hay *et al.*, 1988b). Spatial extrapolation of these thresholds allowed the number of months for which malaria admissions could be expected to be defined across Kenya. The resulting 'malaria season' predictions were compared with a map of malaria transmission periods compiled from expert opinion (Butler, 1959). The correspondence is remarkable when the deficiencies inherent in compiling the original map are acknowledged. The authors stressed that the maps produced did not constitute a definitive picture of malaria transmission in Kenya but were a demonstration of methodology. It was considered that much more validation of the relationship between malaria admissions and NDVI was required to check its robustness in space and time. This map for Uganda and Kenya is reproduced at 1×1 km spatial resolution (see Plate 6).

3.3.2. *Active Radar Radiometers*

Imagery derived from C and L bands of an airborne Synthetic Aperture Radar (SAR) (Hay, this volume) has been used to detect dambo flooding status in the search for predictive mechanisms for RVF epidemics in Kenya (Pope *et al.*, 1992). A significant advantage of using SAR was that data collection was independent of cloud coverage, especially important during the East African rains. Pope *et al.* (1992) concluded that the spatial resolution of current satellite-borne SARs was not sufficient to reveal many of the smaller dambos in the region. In a similar study, Pope *et al.* (1994b) demonstrated that airborne SAR data collected over the wetlands of Belize were capable of mapping marshes dominated by *Eleocharis*, an important breeding habitat of *An. albimanus* (Rejmankova *et al.*, 1993).

4. GEOGRAPHIC INFORMATION SYSTEMS AND MOSQUITO-BORNE DISEASE

4.1. The History of Malaria Mapping

Before microcomputers and the advent of GIS, few examples existed of maps defining the global distribution of malaria (Lysenko and Semashko, 1968).

More maps are found at a regional and country scale, reflecting the more local concerns and experience of research and control personnel. Common to all historical maps is their representation of expert opinion, often aided by simple climatic and/or geographical iso-lines and, more rarely, limited empirical evidence. This static information failed to reflect the great spatial and temporal heterogeneity in transmission. GIS-based mapping allows methodologies to be standardized and repeated, and data to be updated and, since it is digital, easily interrogated. Thus GIS moves the mapping of malaria from a largely subjective science to one with quantitative foundations.

4.2. Continental Scale GIS

4.2.1. *Modelling Malaria Burden*

A major advocate of GIS in the study of malaria at a continental level has been the Mapping Malaria Risk in Africa/Atlas du Risque de la Malaria en Afrique (MARA/ARMA) project (le Sueur *et al.*, 1997). The project was conceived to provide comprehensive, empirical, standardized maps of malaria distribution and endemicity to help guide malaria control operations in Africa (Snow *et al.*, 1996). The aim was to collect and geo-reference all available malaria data in the formal and 'grey' literature. Data collection focused initially on the parasite prevalence survey, a proxy of malaria endemicity (Metselaar and Van Theil, 1959) since it is the most commonly measured parameter. This continental database of parasite prevalence was then used to produce evidence-based maps of malaria transmission. Recent additional attempts have been initiated to collect, map and interpolate the more direct but less standardized proxy of endemicity, the EIR (Hay *et al.*, 2000).

The second focus of the MARA/ARMA project involved defining the distribution of endemic malaria and the duration and timing of the transmission seasons. The first completed model (Craig *et al.*, 1999) was based on the assumption that malaria transmission at the continental scale was limited primarily by climate, and could hence be inferred from the distribution of temperature and rainfall at this scale. Continuous climate surfaces of mean monthly temperature and rainfall (1920–1980) were developed from interpolated weather station data at a spatial resolution of about 5×5 km ($0.03 \times 0.03°$) for the analysis (Hutchinson *et al.*, 1995). The model (Craig *et al.*, 1999) used a fuzzy logic concept to express 'degrees of climate suitability' in contrast to rigid Boolean limits that ignored natural gradients and uncertainty. Mean monthly temperatures of 18°C were considered too cold for endemic *P. falciparum* transmission, but they became increasingly suitable to a maximum at 22°C and above (Detinova, 1962; Macdonald, 1957). Zero rainfall in any month was considered unsuitable, and again it increased to maximum suitability for stable transmission at 80 mm and above per month. Monthly temperature and rainfall surfaces were converted to fuzzy suitability maps. Suitable conditions were required to persist for five consecutive months (three in North Africa). Low temperatures affect vector distribution: *An. gambiae* was found to be limited to frost-free regions (de Mellion, 1934) or to areas where absolute minimum temperatures in winter remained above 5°C (Leeson, 1931). Thus areas with minimum winter temperatures <3°C were classified as unsuitable. These empirically derived climate limits were tested by extracting rainfall and temperature profiles in known perennial, seasonal, epidemic and non-malarial areas. The resulting model describes climate suitability for endemic malaria, in which a fuzzy value of one predicts endemic malaria and a fuzzy value of zero predicts highly unstable or no transmission (see Plate 4). The model compared well with available country-level malaria risk maps of Kenya (Nelson, 1959), Tanzania (Wilson, 1956), South Africa before control (Anonymous, 1938), Botswana (Chayabejara *et al.*, 1975; Diseko *et al.*, 1997) and Namibia (de Meillon, 1951).

The malaria distribution model was subsequently used to estimate the number of people at risk of malaria in SSA (Snow *et al.*, 1999b). The fuzzy distribution model was reclassified so that values of 0.5 and above were regarded as malarious. By overlaying this Boolean malaria distribution map with a 1990 population distribution model (Deichmann, 1996), the total population and the number of children aged below 5 years living in predicted malaria-endemic areas were estimated to be 360 million and 66 million, respectively. A more detailed study classified malaria risk differently for southern Africa (fuzzy values <0.5 = no malaria, 0.5 = stable malaria) than for the rest of Africa (where 0 = no malaria, >0 and <0.2 = epidemic risk, 0.2 = stable malaria) (Snow *et al.*, 1999a) (Figure 1). In southern Africa, mortality (0.104 per 1000) and morbidity (11 per 1000) were estimated from surveillance data. For other endemic countries annual mortality estimates were based on 76 independent childhood mortality studies. Since very little is known about the mortality pattern in older children and adults, severe malaria admission rates in children under 10 years old, by year of age, were used to derive a model describing the decline in mortality risk with age, and from this to predict mortality risk for ages above 10 years by extrapolation. Thus the estimated mortality figures by age category were 9.4 per 1000 for 0–4 years, 2.17 for 5–9 years, 0.8 for 10–14 years and 0.13 per 1000 for 15 years and older. Morbidity estimates were based on 51 studies of malarial fever events in children: 976 cases per 1000 for all age groups in epidemic areas, and 999, 239 and 400 per 1000 for 0–4, 5–14 and 15-year-olds, respectively, in stable areas. Finally, the rate of severe malaria (cerebral,

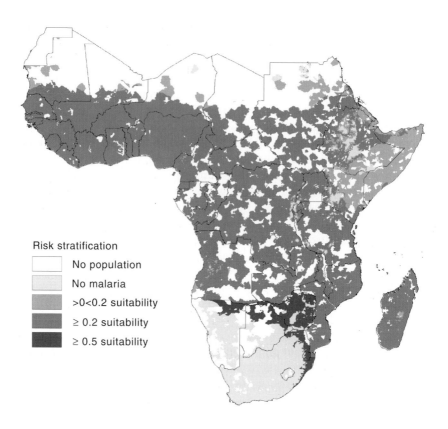

Figure 1 Zones defining different levels of risk of malaria according to the malaria distribution model (Craig *et al.*, 1999): endemic areas in southern Africa (> 0.5), stable transmission areas in the rest of sub-Saharan Africa (>0.2), epidemic-prone areas (>0, <0.2), and negligible malaria risk (<0.5 in southern Africa, 0.0 elsewhere).

severe anaemia), the proportion of people surviving severe malaria, and the risk of longer-term effects (neurological sequelae and human immuno-deficiency virus (HIV) infection following blood transfusion) were estimated from further studies. The overall estimates indicate that there may have been over 200 million clinical cases of malaria in Africa in 1995 and close to 1 million deaths. Furthermore, each year 3000 children may be left with life-long disability (e.g. spasticity or epilepsy) and 19 000 may have contracted HIV after receiving contaminated blood for severe anaemia. These studies illustrate the rigorous way in which GIS techniques can integrate data from various origins. Whilst the final estimates do not differ significantly from previous expert guesses (Greenwood, 1990; Murray and Lopez, 1997), such estimates are achieved using a rational and transparent methodology and are easily updated as new population and disease risk data become available.

4.2.2. *Mapping Vector Distribution*

Following the mapping of African *Anopheles* species collections by Gillies and de Meillon (1968), the records for the *An. gambiae* complex were updated and re-mapped by Davidson and Lane (1981). Coetzee *et al.* (2000) transferred these data to GIS and further updated the collection to contain 2538 records of six of the seven presently recognized species of the *An. gambiae* complex, including surveys from 1944 to the present (see Plate 5).

Lindsay *et al.* (1998) showed that the ranges of *An. gambiae s.s.* and *An. arabiensis* published by Davidson and Lane (1981) were described well by an array of climatic variables (Corbett and Kruska, 1994). Temperature ranges favoured by the two species were similar, but *An. gambiae* occurred in areas with higher annual precipitation (330–3224 mm) than *An. arabiensis* (237–415 mm), which corroborates previous observations; see Section 2.3.2. Lindsay *et al.* (1998) further analysed the relative abundance and found the distribution to be best described by the ratio of precipitation to potential evapotranspiration (ppn/pet) with *An. gambiae* dominating more humid, and *An. arabiensis* dominating more arid, areas respectively. The regression equations were fed back into the climatic data to produce a continent-wide model of proportional abundance of the two species. The expected values correlated well with the available data ($r^2 = 0.745, n = 14, P = 0.002$).

4.2.3. *Climate Change*

There have been several attempts to model changing malaria risk in relation to various scenarios of climate change. Inherent in these studies is the determination of baseline risk against which to measure change and some degree of spatial sophistication in data analysis.

Martin and Lefebvre (1995) used simple Boolean thresholds for temperature and moisture conditions suitable for transmission, with the requirement that these should be temporally coincident for *P. malariae, P. ovale, P. vivax* and *P. falciparum*, and iterated these monthly using $0.5 \times 0.5°$ climate data to map areas of seasonal and perennial transmission globally. Temperatures from 19 to 32°C were considered suitable for *P. falciparum* and between 16 and 32°C for the other three *Plasmodium* species. A moisture index (ppn/pet; see above) of 0.7 defined the lower limit of suitable conditions for all four species. Malaria was considered seasonal if the suitable transmission period lasted 1–7 months and perennial if greater. They then investigated how the outputs from five global circulation models affected their index of annual potential malaria transmission. With some notable exceptions, present day prediction agreed with the maps published previously (Lysenko and Semashko, 1968), although no accuracy statistics were provided. The future scenarios all predicted change in malaria distribution and seasonality, but did not concur on the nature and extent of that change. The five future models predicted 7–28% expansion of total malarious areas, and an increase in seasonal transmission areas of 12–55%, while all except one predicted a reduction in perennial zones by infringement of seasonal zones.

A more biological approach was adopted by Martens *et al.* (1995a,b), Martens (1997, 1999), and Lindsay and Martens (1998). The critical threshold density for malaria transmission (the number of infective mosquito bites per human host) were derived by re-arranging the equation for the basic reproductive number (R_0) for malaria (Anderson and May, 1991a). The reciprocal of this figure was then defined as the 'epidemic potential'. Importantly, due to the complexity of estimating the change in vector abundance resulting from changing meteorological conditions, this variable was held constant (i.e. = 1). Risk for the various *Plasmodium* species were then derived using a $5 \times 7.5°$ (laitude × longitude) global climate surface. The changes in epidemic potential in combination with a world population map were then used to make educated guesses about the changes in the total population exposed under various climate change scenarios. There was again no attempt at statistical validation of predictions of the present against current conditions. These studies contributed to the climate change debate by illustrating the sensitivity of malaria transmission to climate perturbations but were inadequate to define the current risk accurately.

The potential effect of predicted climate change on the distribution of *An. gambiae s.l.* has also been investigated for areas south of the equator (Rogers, 1996). The range published by Gillies and de Meillon (1968) was analysed against annual summary climate variables for 1961–1990. *An. gambiae s.l.* was specifically considered present in grid cells where it had been recorded, and absent in grid cells where other *Anopheles* species, but not *An. gambiae*, had been collected. Present distribution was predicted and compared with future global climate change scenarios. Interestingly, habitat suitability for the *An. gambiae* complex showed a net increase under all change scenarios.

A more extensive and global analysis of current and potential future *P. falciparum* malaria distribution by Rogers and Randolph (2000) resulted in similar conclusions. The recorded present day distribution of *P. falciparum* malaria was used to establish the current multivariate climatic constraints, using maximum likelihood methods based on the mean, max and minimum values of interpolated long-term climate records of temperature, rainfall and saturation vapour pressure. The prediction of the present day distribution was significantly better than previous attempts (Martin and Lefebvre, 1995;

Martens *et al.*, 1995a,b; Martens, 1997, 1999; Lindsay and Martens, 1998) and this was attributed to the incomplete biological modelling approaches adopted in these studies. These results were applied to future climate scenarios (HadCM2 'medium-high' and HadCM2 'high') to predict future distributions. The potential future distributions showed remarkably few changes, even under the most extreme scenario, due to co-varying rainfall and moisture variables limiting any temperature driven expansion.

4.3. Regional Scale GIS

4.3.1. *Surveillance*

In South Africa malaria has been notifiable at the district level since 1956. Simple mapping of these malaria case data to administrative boundaries (see Plate 7) highlighted the lowlands adjacent to Mozambique as consistently high-risk districts (Sharp and le Sueur, 1996) and indicated expansion of malaria in the northern districts. le Sueur *et al.* (1995) and Sharp and le Sueur (1996) divided the northern high-risk districts of Kwazulu–Natal province into 20 control areas, and each area was subdivided into 10 sectors of approximately 30×30 km to investigate these trends. Annual malaria incidence varied widely from more than 300 per 1000 on the Mozambique border to less than 2 per 1000 in coastal regions from which *An. funestus* had been eradicated. An irrigation system around the inland town of Jozini was also implicated in the local rise of cases allowing control operations to be focused accordingly. The surveillance was further expanded to include the high-risk districts of Mpumalanga province (Booman *et al.*, 2000). Malaria case data were entered within one week of collection, and the malaria control programme managers were able to generate near real-time reports and malaria distribution maps. Guided by the GIS surveillance system, indoor insecticide spraying is now limited to villages with incidence exceeding 8 cases per 1000. This is especially important as major redistribution in the health budget towards the management of the HIV/acquired immune deficiency syndrome (AIDS) epidemic in South Africa has increased pressure for efficient use of limited resources for malaria control.

A community health care system has also been established in Tigray, Ethiopia (Ghebreyesus *et al.*, 1996) with the purpose of reducing malaria morbidity and mortality through early diagnosis and treatment, vector control and preventing infection during pregnancy with prophylaxis. The programme has combined data from the Tigray malaria control programme with spatial base maps to display and survey present health service infrastructure, clinic-based reporting of malaria, health service accessibility, community-based treatment of malaria cases, and coverage of early diagnosis and treatment at the periphery. This supports planning of activities and monitoring of success of the programme in different parts of the region (Ghebreyesus *et al.*, 1999).

In Israel, where malaria was eradicated in 1950, a national GIS surveillance system has been used to keep track of imported malaria cases and mosquito breeding sites (Kitron *et al.*, 1994). Depending on the proximity of the water bodies to towns where malaria cases have been reported, and depending on the known flight ranges of the vectors, breeding sites are targeted for intensive vector control, through larviciding and/or drainage and channelling of run-off water, and increased surveillance.

A similar system in Trinidad resulted in all local and imported malaria cases referred to the Insect Vector Control Division being carefully recorded since 1965. Follow-up blood surveys and notification of surrounding health facilities ensured that practically no case was missed. All cases were geo-referenced and analysed retrospectively for space–time clustering using a k-nearest neighbour analysis (Chadee and Kitron, 1999). Of a total of 213 cases over a 30 year period, 164 were imported and were concentrated around the two large port cities on the west coast. All cases of *P. falciparum* were imported, and no clustering was found. Some *P. malariae* cases were concentrated in the island interior, however, with significant clustering in time of the second ($J = 30$, $n = 57$, $P < 0.01$) and seventh ($J = 41$, $n = 57$, $P < 0.02$) nearest neighbours, indicating local transmission. The vectors were bromeliad breeding *An. bellator* and *An. homunculus*. Most *P. vivax* cases were imported, but nine cases were clustered closely in time in a small port town; these were associated with local transmission, probably by *An. aquasalis* and *An. albitarsis* which breed in surrounding marsh and rice fields, following the importation of a single case. This study thus provided the confidence for the government to restrict detailed and costly surveys to rural cases of *P. malariae* where bromeliad malaria can occur, and in coastal areas where people live in proximity to marshes.

Su (1994) established a GIS to monitor dengue and its vectors *Aedes* spp. in Kaohsiung city, Taiwan. The system allowed monitoring of vector breakouts, dengue cases and distribution of breeding sites, which gave a better overview of the situation. This supported the local government in their investigations and decision making and thus contributed towards more effective control of the disease.

4.3.2. *Patterns of Disease Transmission*

In Kenya, malaria survey data have been aggregated and displayed at the district administrative unit level to illustrate parasite distribution and prevalence (Omumbo *et al.*, 1998). The continental malaria distribution model described in Section 4.2.1 was used to mask areas where malaria was predicted as unstable or absent, to avoid the misuse of surveys possibly carried out during epidemics. The survey results were then analysed against underlying environmental factors (Snow *et al.*, 1998a). A total of 124 community-based, cross-sectional malaria surveys, representing at least 100 children 0–10 years old and conducted since 1960, was included, and prevalence ratios were classified into high (70%), intermediate (20–69%) and low (<20%) transmission categories. The best explanatory climate variables were identified in a step-wise discriminant analysis, and these were then used to allocate each sample location to one of the categories. Agreement between allocated and observed values was high: of 124 measurements, 96 were allocated to their correct category. Interestingly, environmental conditions in April (also May) contributed most strongly towards malaria prevalence. The total number of suitable months with 60 mm of rainfall and temperatures above 19°C were also correlated with prevalence (univariate $r = 0.62, n = 124, P < 0.001$). The entire country was then categorized into high, intermediate and low risk areas, based on the regression equation. Populations exposed to different levels of risk were extracted, and mortality rates estimated: 26 000 children under 5 years old were estimated to die each year in endemic areas, and a further 3300–14 500 in epidemic areas. Additionally, 145 000 children were estimated to require intensive clinical management in an average year.

A clear example of application is provided by a situation analysis of community-based health care programmes in Kenya (Snow *et al.*, 1998b) which indicated that, apart from the Ministry of Health, 23 different non-governmental organizations were active in health care. The analysis showed that not only were several populations covered by more than one programme, which illustrates the lack of inter-sectoral collaboration due to lack of appropriate information, but malaria intervention programmes were also not ideally positioned in relation to the distribution of malaria. By comparing the map of malaria endemicity with the distribution of target populations for insecticide-treated bed net implementation, a discrepancy emerged between the people exposed to malaria and those with access to treated nets (Shretta *et al.*, 1998). More targeted and informed intervention was made possible simply by knowing how many people are exposed to malaria and where.

The statistical method used to analyse the Kenyan data treated each survey as an independent measurement and disregarded possible spatial autocorrelation. A subsequent analysis carried out on malaria survey data collected in Mali went a step further by investigating spatial trends (Kleinschmidt *et al.*, 1999). Using a logistic regression method, the authors modelled larger-scale risk of malaria in Mali on relationships between prevalence and underlying environmental factors, namely various measurements of climate including the number of months with more than

60 mm of rain, NDVI, and distance to water. Spatial dependence in the model was investigated through a variogram, which suggested spatial dependence over distances of 15–20 km analysis (Robinson, this volume). Local variation in malaria risk, not accounted for by the original regression model, was then modelled by kriging the residuals at every survey point (Robinson, this volume). The accuracy of the model was then improved by adjusting the predicted values to resemble the observed values more closely, based on the kriged residuals. Distance to water (categorical), NDVI during the wet season, number of months with >60 mm rain, and average maximum temperature at the end of the dry season significantly explained malaria prevalence in Mali. One initially counter-intuitive outcome was that prevalence was higher at intermediate distances from water than at very short and very long distances. The likely explanation for this was that people living close to water make frequent use of bed nets to protect themselves from the great nuisance of mosquitoes, and are thus inadvertently protected from malaria (see Section 3.3.1).

Hightower *et al.* (1998) gave a detailed account of their experience in mapping 1169 houses using GPS in 15 villages along the shore of Lake Victoria, as well as health facilities, vector breeding sites, roads, rivers and the lake shoreline. Weekly mosquito trap collections and malaria prevalence in children less than 5 years old, were analysed against the distance of houses from the nearest major larval breeding site. Mosquito numbers were significantly correlated (r^2 not given, $n = 362$, $P = 0.0039$) with distance to breeding sites during the dry, but not during the wet, months (r^2 not given, $n = 395$, $P = 0.1530$). Parasite prevalence was weakly correlated with distance to breeding site in the wet months (r^2 not given, $n = 445$, $P = 0.3437$), but not during the dry months. The authors reported large variation in vector abundance, between and within villages, for both *An. gambiae s.s.* and *An. funestus* and even between the species.

Smith *et al.* (1995) analysed this local scale variation in abundance of *An. funestus* and *An. gambiae s.l.* in Namawala village in Tanzania. The houses were geo-referenced using a GPS and mosquitoes sampled by light trap every 2 weeks in 43 houses, and every night in one sentinel house. Vector abundance was highly variable, with between-house variation as pronounced as total seasonal variation in the village as a whole. Four different models (fixed effect, random effect, mixed, and conditional auto-regressive) were used to describe catches and extrapolate temporal and spatial variation. Many factors, including effect of windows and adjoining animal sheds, position of the house (central or peripheral, proximity to rice fields), seasonality, sampling frequency and total numbers of mosquitoes all affected the outcome of the models. This study illustrated the importance of site selection when extrapolations are made from highly focal measurements, such as a few sentinel houses. Mosquito density was highest in the valleys where breeding sites were more permanent. Strong association of mosquito density with distance from breeding sites has been demonstrated again and again (Trape *et al.*, 1992; Ribeiro *et al.*, 1996); see also below.

A densely-populated suburb of Maputo, Mozambique was mapped by digitizing 1:10 000 maps and geo-referencing individual homes via GPS (Thompson *et al.*, 1997). A detailed longitudinal study of malaria prevalence, clinical incidence and vector abundance was then carried out over 2.5 years from December 1992 to June 1995. Initial results showed clearly that prevalence of *P. falciparum* decreased sharply from between 40–60% adjacent to swampy vector breeding sites to 5–11% only 500 m away. Incidence of clinical malaria also decreased markedly. In fact, the relative risk of clinical malaria, assessed by multivariate analysis using a Cox model, was 6.2 times higher for individuals living less than 200 m from the breeding sites than for individuals living 500 m or more away (confidence interval 3.5–11.1; $n = 857$; $P = 0.0001$). EIR was estimated to be 20 infectious bites per person per year in the high-risk areas, but was too low to measure in areas where malaria prevalence was below 11%. The high variability in prevalence, EIR and incidence over very short distances was attributed partly to very high population density: mosquitoes need not fly far to find their first host, and vector dispersion, hence transmission, is thus concentrated within the immediate vicinity of the breeding sites.

Some examples outside Africa further illustrate the use of GIS in malaria and mosquito research and control. In Kataragama, a highly endemic region in Sri Lanka, aerial photographs were used to set up a GIS including the position 423 houses representing 1875 inhabitants (Gunawardena *et al.*, 1995). People in houses made of mud, or with incomplete walls and thatch or palm leaf roofs, were at 2.5 times greater risk of being infected with malaria than people in houses with complete and plastered brick walls and tiled or corrugated iron roofs. Distance to water increased risk of malaria only in the case of poorly constructed houses ($r^2 = -0.31, n = 182, P = 0.0001$), but not in modern style houses ($r^2 = 0.14$, $n = 161$, $P = 0.0676$). The authors recommended that the government should replace all mud-and-thatch houses within 200 m of a water body (which included 76% of the people in low-quality housing) with high-quality houses. This would reduce malaria incidence by 36% and would save the government money spent on malaria control. The money savings would off-set the initial expenditure after only 7.2 years, thereafter resulting in a net return of investments (Gunawardena *et al.*, 1998).

In a study of 100 villages in Nadiad *taluka* (i.e. district), Gujarat, India (Sharma and Srivastava, 1997) layers of soil type, water table, water quality, relief and hydro-geomorphology were reclassified into zones of high, medium and low risk of waterlogging (which provide breeding sites). Risk layers were overlaid, yielding a mosaic of zones each with a certain combination of conditions. Each zone was then given the risk level of the dominant factor in that zone. The resulting risk map was correlated with annual malaria incidence per zone in non-irrigated, but not in irrigated, areas. Each factor on its own could not adequately explain incidence. Malaria was also more prevalent where the water table was high, and where surface water and irrigation canals were present, than in villages along natural rivers, indicating the preferred breeding habitats of the local vector (Malhotra and Srivastava, 1995).

4.3.3. *Disease Clustering*

Patterns of malaria morbidity have been investigated by Snow *et al.* (1993) in a study involving 50 000 people living in 5000 scattered homesteads throughout Kilifi district of Kenya. Hand-drawn maps were used to identify and geo-reference with GPS the homes of all children presenting with severe *P. falciparum* malaria at Kilifi district hospital from May 1989 to April 1992. The number of pairs of cases found to live within 1, 1.5, 2, 2.5 and 3 km from one another, and presenting within 3, 5, 7, 10 and 14 days from one another, were compared with the number of cases that could be expected to occur close in time and space by chance. A significant number of cases was found to occur within 5 days and 2 km of one another. The clustering was unexplained in this study, but microclimatic differences causing local changes in the vector populations, or local importation of different strains of the parasite, were both postulated as possible causes.

In a further study of severe childhood malaria in the same population, from May 1991 to April 1993 all homesteads in the Kilifi district were geo-referenced with, and captured in, a GIS (Armstrong Schellenberg *et al.*, 1998). Using a simple 'equal population area' algorithm, households were grouped into four different sizes of artificial areas, each containing about 1000, 500, 200 and 100 children respectively. In the second study year, cases of severe malaria were distributed more or less homogeneously across these areas, but in the first year there was significant heterogeneity at every level, with more admissions from people with easy access to the hospital. Using the Monte Carlo simulation method to check for overall clustering of cases, the childhood population surrounding a circular area from which 2, 3 or 4 cases of severe malaria had been admitted within 3, 7, 14 and 31 days of one another was compared against similar figures taken from simulations that assumed a random distribution. There was evidence of clustering in both time and space, particularly in the first study year. Lack of similar clustering in all-cause mortality data suggested that the clusters were real and not due to limitations in the numerator or denominator data.

Spatial cluster analysis of cases of La Crosse encephalitis (transmitted by *Aedes* mosquitoes) in Illinois, USA was used to identify 'hot spots' of the disease in the state and the extent of clustering. Analysis revealed that most clustering occurred at a distance 10 km at state-level, and at a distance of 3 km at city-level (Kitron *et al.*, 1997). Local clusters of disease were concentrated in houses near breeding sites such as undeveloped wood lots, large piles of tyres, or other discarded containers, which could be targeted for clean-up and vector control.

5. CONCLUSIONS AND THE FUTURE

Mosquito and malaria control are problems that operate at a range of spatial scale, from the catchment population of rural health clinics concerned with individual patients, to the regional and continental interests of international donor agencies concerned with the continental distribution of the disease and the quantification of total malaria incidence. The variety of RS platforms and sensors provides data relevant to each of these spatial scales and GIS provides an appropriate platform for integrating and interrogating such information. We have seen a wide range of RS imagery utilized with respect to mosquito-borne diseases and the work has progressed significantly from habitat characterizations at high spatial resolution to work that has inferred habitat quality (and hence disease risk) for both larval and adult forms from a village to a continental scale. International agencies, in cooperation with developing countries, are now able to develop programmes to validate real-time predictions of malaria incidence and prevalence across continental areas. Research relating remotely sensed variables to mosquito life history parameters is an essential basis for future 'disease forecasting' for a malaria early warning system (MEWS). This could operate in much the same way that drought and famine conditions are now routinely monitored using satellite sensor data (Hutchinson, 1991), and this is explored further by Myers *et al* (this volume).

Data capture and mapping, associated with even the most rudimentary GIS analysis, has made evidence-based response by malaria control programmes possible, and many examples have been cited. Maps provide a powerful lobbying tool and contribute towards planning and monitoring malaria control operations through community health care, targeted insecticide spraying and drug distribution. Country maps of continental models of malaria distribution, duration and timing of transmission season, population distribution and distribution of surveys have served to highlight discrepancies between supply and need of resources, and have stimulated discussion about other possibilities and requirements. Equity of access to bed nets at country level, bed net coverage, services (availability of drugs, etc.); market analysis, trade tariffs, distribution network, 'money season' versus 'malaria season' for involvement of the private sector, poverty index, and migration maps including refugee camps are all important issues that play a role in national malaria control, which would benefit from spatial representation and analysis. Finally, once certain generic data sets have been generated or gathered (e.g. boundaries, population, topography, cadastral) a GIS system can easily be adapted for further topics in malaria, such as the distribution of drug and insecticide resistance. It can be adapted equally well for other diseases and even for other sectors, such as agriculture or infrastructure planning and development.

The perception that RS and GIS are not appropriate in technologically developing regions persists and is manifest in the form of frequent objections to the cost of image processing equipment, lack of access to imagery, expertise and ground truth and the novelty of the techniques (Bos, 1990; Arambaulo and Astudillo, 1991; Barinaga, 1993; Kleiner, 1995; Oppenshaw, 1996). These problems are real concerns, but are diminishing as (i) computer processing and data storage facilities become relatively cheaper, (ii) image-processing and GIS software has become widely accessible and can automate many tasks, (iii) recent changes in the philosophy of digital image dissemination have resulted in data becoming more widely and freely available (Mulcahy and Clarke, 1994; Justice *et al.*, 1995), and (iv) the history of successful application of RS and GIS techniques in epidemiology expands. An excellent review of many of these issues has been given by Kithsir Perera and Tateishi (1995).

The future will see a further increase in the range of data available as well as its speed of dissemination. This will present new challenges and opportunities for synergism among the epidemiological, RS and GIS research communities. The degree of sophistication with which studies are utilizing high RS data is set to advance (Goetz *et al.*, this volume) including the way in which a variety of images is used to obtain refined information for a given area (Pohl and van Genderen, 1998), the way we deal with scale (Quattrochi and Goodchild, 1997; Walsh *et al.*, 1999) and the more widespread use of SAR (Henderson and Lewis, 1998). We think these efforts would be best focused on addressing the following: (i) defining seasonal variation in disease risk to optimize timing of health information, insecticide treatment of bed nets, intermittent chemotherapy or chemoprophylaxis during pregnancy; (ii) identification of endemic- and epidemic-prone areas to provide information on geographic location of areas amenable to interventions aimed at eradicating infection risk (e.g. residual housespraying or mass drug administration); (iii) epidemic early warning systems to provide timely information on conditions suitable for epidemics (prior to their occurrence); and (iv) population distribution with respect to health service provision for determining optimum health service delivery for new clinical management or bed net re-treatment centres.

ACKNOWLEDGEMENTS

We are grateful to the David Rogers and Sarah Randolph for comments on this manuscript. SIH is an Advanced Training Fellow with the Wellcome Trust (No. 056642) and RWS is a Wellcome Trust Senior Research Fellow in Basic Biomedical Sciences (No. 033340).

REFERENCES

Anderson, J.F., Andreadis, T.G., Vossbrinck, C.R. *et al.* (1999). Isolation of West Nile virus from mosquitoes, crows, and a Cooper's hawk in Connecticut. *Science* **286**, 2331–2333.

Anderson, R.M. and May, R.M. (1991a). A framework for discussing the population biology of infectious diseases. In: *Infectious Diseases of Humans: Dynamics and Control*, pp. 13–23. Oxford: Oxford University Press.

Anderson, R.M. and May, R.M. (1991b). Indirectly transmitted microparasites. In: *Infectious Diseases of Humans: Dynamics and Control*, pp. 374–429. Oxford: Oxford University Press.

Aniedu, I. (1997). Dynamics of malaria transmission near two permanent breeding sites in Baringo District. *Indian Journal of Medical Research* **105**, 206–211.

Anonymous (1938). *Malaria Areas in the Union of South Africa*. Pretoria: Union of South Africa, Department of Public Health.

Anonymous (1973). *The Use of Remote Sensing in Mosquito Control*. Houston, TX: Health Applications Office, NASA.

Arambaulo, P.V.,III and Astudillo, V. (1991). Perspectives on the application of remote sensing and geographic information system to disease control and health management. *Preventive Veterinary Medicine* **11**, 345–452.

Armstrong-Schellenberg, J., Newell, J.N., Snow, R.W. *et al.* (1998). An analysis of the geographical distribution of severe malaria in children in Kilifi District, Kenya. *International Journal of Epidemiology* **27**, 323–329.

Barinaga, M. (1993). Satellite data rocket disease-control efforts into orbit. *Science* **261**, 31–32.

Barnes, C.M. and Cibula, W.G. (1979). Some implications of remote sensing technology in insect control programs including mosquitoes. *Mosquito News* **39**, 271–282.

Beck, L.R., Rodriguez, M.H., Dister, S.W. *et al.* (1994). Remote sensing as a landscape epidemiologic tool to identify villages at high risk for malaria transmission. *American Journal of Tropical Medicine and Hygiene* **51**, 271–280.

Beck, L.R., Wood, B.L. and Dister, S.W. (1995). Remote sensing and GIS–new tools for mapping human health. *Geo Info Systems* **5**, 32–37.

Beck, L.R., Rodriguez, M.H., Dister, S.W. *et al.* (1997). Assessment of a remote sensing-based model for predicting malaria transmission risk in villages of Chiapas, Mexico. *American Journal of Tropical Medicine and Hygiene* **56**, 99–106.

Beier, J.C. (1998). Malaria parasite development in mosquitoes. *Annual Review of Entomology* **43**, 519–543.

Bidlingmayer, W.L. and Klock, J.W. (1955). Notes on the influence of salt-marsh topography on tidal action. *Mosquito News* **15**, 231–235.

Bloom, D.E. and Sachs, J.D. (1998). Geography, demography, and economic growth in Africa. *Brookings Papers on Economic Activity*, 207–295.

Booman, M., Durrheim, D.N., La Grange, J.J.P. *et al.* (2000). Planning malaria control using a geographic information system in Mpumalanga province, South Africa. *Bulletin of the World Health Organization*, in press.

Bos, R. (1990). Application of remote-sensing. *Parasitology Today* **6**, 39.

Boyd, M.F. (1930). *An Introduction to Malariology*. Cambridge, MA: Harvard University Press.

Bradley, D.J. (1991). Malaria. In: *Disease and Mortality in Sub-Saharan Africa* (R.G. Feacham and D.T. Jamison, eds), pp. 190–202. Oxford: Oxford University Press.

Bradley, D.J. (1992). Malaria: old infections, changing epidemiology. *Health Transition Review* **2**, 137–183.

Butler, R.J. (1959). *Atlas of Kenya: A Comprehensive Series of New and Authenticated Maps Prepared from the National Survey and other Governmental Sources with Gazetteer and Notes on Pronunciation and Spelling*. Nairobi: The Survey of Kenya.

Buxton, P.A. (1931). The measurement and control of atmospheric humidity in relation to entomological problems. *Bulletin of Entomological Research* **22**, 431–447.

Chadee, D.D. and Kitron, U. (1999). Spatial and temporal patterns of imported malaria cases and local transmission in Trinidad. *American Journal of Tropical Medicine and Hygiene* **61**, 513–517.

Charlwood, J.D., Kihonda, J., Sama, S. *et al.* (1995). The rise and fall of *Anopheles arabiensis* (Diptera: Culicidae) in a Tanzanian village. *Bulletin of Entomological Research* **85**, 37–44.

Chayabejara, S., Sobti, S.K., Payne, D. and Braga, F. (1975). *Malaria Situation in Botswana*. Brazaville, Congo: Regional Office for Africa, World Health Organization.

Christie, M. (1959). A critical review of the role of the immature stages of anopheline mosquitoes in the regulation of adult numbers, with particular reference to *Anopheles gambiae*. *Tropical Diseases Bulletin* **56**, 385–399.

Christophers, S.R. (1911). *Paludism: Epidemic Malaria of the Punjab, with a Note on a Method of Predicting Epidemic Years*. Simla: Government of India.

Cibula, W.G. (1976). *Applications of Remotely Sensed Multidpectral Data to Automated Analysis of Marshland Vegetation*. NASA Technical Note TN D-8139. Houston, TX: Lyndon B. Johnson Space Center, NASA.

Cline, B.L. (1970). New eyes for epidemiologists: aerial photography and other remote sensing techniques. *American Journal of Epidemiology* **92**, 85–89.

Coene, J. (1993). Malaria in urban and rural Kinshasa: the entomological input. *Medical and Veterinary Entomology* **7**, 127–137.

Coetzee, M., Craig, M.H. and le Sueur, D. (2000). Mapping the distribution of members of the *Anopheles gambiae* complex in Africa and adjacent islands. *Parasitology Today* **16**, 74–77.

Connor, S.J. (1999). Malaria in Africa: the view from space. *Biologist* **46**, 22–25.

Coosemans, M.H. (1985). Comparaison de l'endémie malarienne dans une zone de riziculture et dans une zone de culture de coton dans la Plaine de la Rusizi, Burundi. *Annales de la Société Belge de Médecine Tropicale* **65**, 187–200.

Corbett, J.D. and Kruska, R.L. (1994). *African Monthly Climate Surfaces, v. 1.0. Three Arc-min Resolution. Based on Climate Coefficients from Centre for Resource and Environmental Studies, Canberra, Australia*. Nairobi, Kenya: ICRAF/ILRAD.

Covell, G. (1957). Malaria in Ethiopia. *Journal of Tropical Medicine and Hygiene* **60**, 7–16.

Craig, M.H., Snow, R.W. and le Sueur, D. (1999). A climate-based distribution model of malaria transmission in sub-Saharan Africa. *Parasitology Today* **15**, 105–111.

Dale, P.E.R. and Morris, C.D. (1996). *Culex annulirostris* breeding sites in urban areas—using remote sensing and digital image analysis to develop a rapid predictor of potential breeding areas. *Journal of the American Mosquito Control Association* **12**, 316–320.

Dale, P.E.R., Ritchie, S.A., Territo, B.M., Morris, C.D., Muhar, A. and Kay, B.H. (1998). An overview of remote sensing and GIS for surveillance of mosquito vector habitats and risk assessment. *Journal of Vector Ecology* **23**, 54–61.

Davidson, G. and Lane, J. (1981). *Distribution maps for the Anopheles gambiae complex*. London: London School of Hygiene and Tropical Medicine.

de Meillon, B. (1951). Malaria survey of south-west Africa. *Bulletin of the World Health Organization* **4**, 333–417.

de Mellion, B. (1934). Observations on *Anopheles funestus* and *Anopheles gambiae* in the Transvaal. *Publications of the South African Institute of Medical Research* **6**, 195.

Deichmann, U. (1996). *Africa Population Database*. Santa Barbara, CA: National Center for Geographic Information and Analysis, United Nations Environment Programme (http://grid2.cr.usgs.gov/globalpop/africa).

Desowitz, R.S. (1999). Milestones and millstones in the history of malaria. In: *Malaria: Molecular and Clinical Aspects*, (M. Wahlgren and P. Perlmann, eds), pp. 1–16. Amsterdam: Harwood Academic.

Detinova, T.S. (1962). *Age Grouping Methods in Diptera of Medical Importance, with Special Reference to Some Vectors of Malaria*. Geneva: World Health Organization.

Diseko, R., Rumisha, D.W. and Pilatwe, T.R. (1997). Study to evaluate community perspective on the use of impregnated linen for malaria control in Botswana: the Botswana malaria control program in a rural district. *Malaria and Infectious Disease in Africa* **7**, 29–52.

Dossou-Yoyo, J., Doannio, J., Rivière, F. and Duval, J. (1994). Rice cultivation and malaria transmission in Bouaké city (Côte d'Ivoire). *Acta Tropica* **57**, 91–94.

Draper, C.C. and Smith, A. (1957). Malaria in the Pare area of N.E. Tanganyika. Part I: epidemiology. *Transactions of the Royal Society of Tropical Medicine and Hygiene* **51**, 137–151.

Draper, C.C. and Smith, A. (1960). Malaria in the Pare area of N.E. Tanganyika. Part II: effects of three years' spraying of huts with Dieldrin. *Transactions of the Royal Society of Tropical Medicine and Hygiene* **54**, 342–357.

Draper, C.C., Lelijveld, J.L.M., Matola, Y.G. and White, G.B. (1972). Malaria in the Pare area of Tanzania. IV: Malaria in the human population 11 years after the suspension of residual insecticide spraying, with special reference to the serological findings. *Transactions of the Royal Society of Tropical Medicine and Hygiene* **66**, 905–912.

El Gaddal, A.A. (1991). The experience of the Blue Nile Health Project in the control of malaria and other water associated diseases. In: *Malaria and Development in Africa: A Cross-sectional Approach*. Washington, DC: American Association for the Advancement of Science, Sub-Saharan African programme.

Fleetwood, S.C., Chambers, M.D. and Terracina, L. (1981). An effective and economical mapping system for the monitoring of *Psorphora columbiae* in rice and fallow fields in south western Louisiana. *Mosquito News* **41**, 174–177.

Fujioka, H. and Aikawa, M. (1999). The malaria parasite and life-cycle. In: *Malaria: Molecular and Clinical Aspects*. (M. Wahlgren and P. Perlmann, eds), pp. 19–55. Amsterdam: Harwood Academic.

Garnham, P.C.C. (1948). The incidence of malaria at high altitude. *Journal of the National Malaria Society, USA* **7**, 275–283.

Ghebreyesus, T.A., Alemayehu, T., Bosman, A., Witten, K.H. and Teklehaimanot, A. (1996). Community participation in malaria control in Tigray region Ethiopia. *Acta Tropica* **61**, 145–156.

Ghebreyesus, T.A., Witten, K.H., Getachew, A., O'Neill, K., Bosman, A. and Teklehaimanot, A. (1999). Community-based malaria control in Tigray, Northern Ethiopia. *Parassitologia* **41**, 367–371.

Gill, C.A. (1920). The relationship of malaria and rainfall. *Indian Journal of Medical Research* **7**, 618–632.

Gill, C.A. (1921). The role of meteorology in malaria. *Indian Journal of Medical Research* **8**, 633–693.

Gilles, H.M. (1993). Epidemiology of malaria. In: *Bruce-Chwatt's Essential Malariology* (H.M. Gilles and D.A. Warrell, eds), 3rd edn, pp. 124–163. London: Edward Arnold.

Gillies, M.T. and Coetzee, M. (1987). *A Supplement to the Anophelinae of Africa South of the Sahara*. Johannesburg: South African Institute for Medical Research.

Gillies, M.T. and de Meillon, B. (1968). *The Anophelinae of Africa south of the Sahara*. Johannesburg: South African Institute for Medical Research.

Githeko, A.K., Service, M.W., Mbogo, C.M., Atieli, F.K. and Juma, F.O. (1993). *Plasmodium falciparum* sporozoite and entomological inoculation rates at the Ahero rice irrigation scheme and the Miwani sugar-belt in western Kenya. *Annals of Tropical Medicine and Parasitology* **87**, 379–391.

Gleiser, R.M., Gorla, D.E. and Ludueña Almeida, F.F. (1997). Monitoring the abundance of *Aedes* (*Ochlerotatus*) *albifasciatus* (Macquart 1838) (Diptera: Culicidae) to the south of Mar Chiquita Lake, central Argentina, with the aid of remote sensing. *Annals of Tropical Medicine and Parasitology* **91**, 917–926.

Greenwood, B.M. (1990). Populations at risk. *Parasitology Today* **6**, 188.

Gunawardena, D.M., Muthuwattac, L., Weerasingha, S. *et al.* (1995). *Spatial Analysis of Malaria Risk in an Endemic Region of Sri Lanka*. Ottawa: International Development Research Centre.

Gunawardena, D.M., Wickremasinghe, A.R., Muthuwatta, L. *et al.* (1998). Malaria risk factors in an endemic region of Sri Lanka, and the impact and cost implications of risk factor-based interventions. *American Journal of Tropical Medicine and Hygiene* **58**, 533–542.

Haddow, A.J. (1943). Measurements of temperature and light in artificial pools with reference to the larval habitat of *Anopheles (Myzomyia) gambiae* Giles and *A. (M.) funestus* Giles. *Bulletin of Entomological Research* **34**, 89–93.

Hay, S.I. (1997). Remote sensing and disease control: past, present and future. *Transactions of the Royal Society of Tropical Medicine & Hygiene* **91**, 105–106.

Hay, S.I. and Lennon, J.J. (1999). Deriving meteorological variables across Africa for the study and control of vector-borne disease: a comparison of remote sensing and spatial interpolation of climate. *Tropical Medicine and International Health* **4**, 58–71.

Hay, S.I., Tucker, C.J., Rogers, D.J. and Packer, M.J. (1996). Remotely sensed surrogates of meteorological data for the study of the distribution and abundance of arthropod vectors of disease. *Annals of Tropical Medicine and Parasitology* **90**, 1–19.

Hay, S.I., Packer, M.J. and Rogers, D.J. (1997). The impact of remote sensing on the study and control of invertebrate intermediate host and vectors for disease. *International Journal of Remote Sensing* **18**, 2899–2930.

Hay, S.I., Snow, R.W. and Rogers, D.J. (1998a). From predicting mosquito habitat to malaria seasons using remotely sensed data: practice, problems and perspectives. *Parasitology Today* **14**, 306–313.

Hay, S.I., Snow, R.W. and Rogers, D.J. (1998b). Prediction of malaria seasons in Kenya using multitemporal meteorological satellite sensor data. *Transactions of the Royal Society of Tropical Medicine and Hygiene* **92**, 12–20.

Hay, S.I., Rogers, D.J., Toomer, J.F. and Snow, R.W. (2000). Annual *Plasmodium falciparum* entomological inoculation rates (EIR) across Africa. I. Literature survey, internet access and review. *Transactions of the Royal Society of Tropical Medicine and Hygiene* **94**, 113–127.

Hayes, R.O., Maxwell, E.L., Mitchell, C.J. and Woodzick, T.L. (1985). Detection, identification, and classification of mosquito larval habitats using remote sensing scanners in earth-orbiting satellites. *Bulletin of the World Health Organization* **63**, 361–374.

Heisch, R.B. and Harper, J.O. (1949). An epidemic of malaria in the Kenya highlands transmitted by *Anopheles funestus*. *Journal of Tropical Medicine and Hygiene* **52**, 187–190.

Henderson, F.M. and Lewis, A.J. (1998). *Principles and Applications of Imaging Radar*. New York: Wiley & Sons.

Hightower, A.W., Ombok, M., Otieno, R. *et al.* (1998). A geographic information system applied to a malaria field study in western Kenya. *American Journal of Tropical Medicine and Hygiene* **58**, 266–272.

Hopkins, C.C., Hollinger, F.B., Johnson, R.F., Dewlett, H.J., Newhouse, V.F. and Chamberlin, R.W. (1975). The epidemiology of St Louis encephalitis in Dallas, Texas, 1966. *American Journal of Epidemiology* **102**, 1–15.

Hugh-Jones, M. (1989). Applications of remote sensing to the identification of the habitats of parasites and disease vectors. *Parasitology Today* **5**, 244–251.

Hunter, J.M., Rey, L., Chu, K.Y., Adekolu-John, E.O. and Mott, K.E. (1993). *Parasitic Diseases in Water Resources Development. The Need for Intersectoral Negotiation*. Geneva: World Health Organization.

Hutchinson, C.F. (1991). Use of satellite data for famine early warning in sub-Saharan Africa. *International Journal of Remote Sensing* **12**, 1405–1421.

Hutchinson, M.F., Nix, H.A., McMahan, J.P. and Ord, K.D. (1995). *Africa—A Topographic and Climatic Database*. Canberra, Australia: Center for Resource and Environmental Studies, Australian National University.

Jepson, W.F., Moutia, A. and Courtois, C. (1947). The malaria problem in Mauritius: the bionomics of Mauritian anophelines. *Bulletin of Entomological Research* **38**, 177–208.

Julvez, J., Develoux, M., Mounkaila, A. and Mouchet, J. (1992). Diversité du paludisme en zone sahelo-Saharienne. Une revue à propos de la situation au Niger, Afrique de l'Ouest. *Annales de la Société Belge de Médecine Tropicale* **72**, 163–177.

Justice, C.O., Bailey, G.B., Maiden, M.E., Rasool, S.I., Strebel, D.E. and Tarpley, J.D. (1995). Recent data and information system initiatives for remotely sensed measurements of the land surface. *Remote Sensing of Environment* **51**, 235–244.

Kilian, A.H.D., Langi, P., Talisuna, A. and Kabagambe, G. (1999). Rainfall pattern, El Niño and malaria in Uganda. *Transactions of the Royal Society of Tropical Medicine and Hygiene* **93**, 22–23.

Kithsir Perera, L. and Tateishi, R. (1995). Do remote sensing and GIS have a practical applicability in developing countries (including some Sri Lankan experiences). *International Journal of Remote Sensing* **16**, 35–51.

Kitron, U., Pener, H., Costin, C., Orshan, L., Greenberg, Z. and Shalom, U. (1994). Geographic information system in malaria surveillance: mosquito breeding and imported cases in Israel, 1992. *American Journal of Tropical Medicine and Hygiene* **50**, 550–556.

Kitron, U., Michael, J., Swanson, J. and Haramis, L. (1997). Spatial analysis of the distribution of Lacrosse encephalitis in Illinois, using a geographic information system and local and global spatial statistics. *American Journal of Tropical Medicine and Hygiene* **57**, 469–475.

Kleiner, K. (1995). Satellites wage war on disease. *New Scientist* **148**, 9.

Kleinschmidt, I., Bagayoko, M., Clarke, G.P.Y., Craig, M. and Le Sueur, D. (2000). A spatial statistical approach to mapping malaria. *International Journal of Epidemiology* **29**, 355–361.

Lanciotti, R.S., Roehrig, J.T., Deubel, V. *et al.* (1999). Origin of the West Nile virus responsible for an outbreak of encephalitis in the northeastern United States. *Science* **286**, 2333–2337.

le Sueur, D., Ngxongo, S., Stuttaford, M. *et al.* (1995). *Towards a Rural Information System*. Ottawa: International Development Research Centre.

le Sueur, D., Binka, F., Lengeler, C. *et al.* (1997). An atlas of malaria in Africa. *Africa Health* **19**, 23–24.

Leeson, H.S. (1931). *Anopheline Mosquitoes in Southern Rhodesia*. London: London School of Hygiene and Tropical Medicine.

Lindsay, S.W. and Birley, M.H. (1996). Climate change and malaria transmission. *Annals of Tropical Medicine and Parasitology* **90**, 573–588.

Lindsay, S.W. and Martens, W.J.M. (1998). Malaria in the African highlands: past, present and future. *Bulletin of the World Health Organization* **76**, 33–45.

Lindsay, S.W., Parson, L. and Thomas, C.J. (1998). Mapping the ranges and relative abundance of the two principal African malaria vectors, *Anopheles gambiae sensu stricto* and *An. arabiensis*, using climate data. *Proceedings of the Royal Society of London, Series B* **265**, 847–854.

Linthicum, K.J., Bailey, C.L., Glyn Davies, F. and Tucker, C.J. (1987). Detection of Rift Valley fever viral activity in Kenya by satellite remote sensing imagery. *Science* **235**, 1656–1659.

Linthicum, K.J., Bailey, C.L., Tucker, C.J. *et al.* (1990). Application of polar-orbiting, meteorological satellite data to detect flooding in Rift Valley Fever virus vector mosquito habitats in Kenya. *Medical and Veterinary Entomology* **4**, 433–438.

Linthicum, K.J., Bailey, C.L., Tucker, C.J. *et al.* (1991). Towards real-time prediction of Rift Valley fever epidemics in Africa. *Preventive Veterinary Medicine* **11**, 325–334.

Linthicum, K.J., Bailey, C.L., Tucker, C.J. *et al.* (1994). Man-made ecological alterations of Senegal river basin on Rift Valley fever transmission. *Sistema Terra* **3**, 44–47.

Linthicum, K.J., Anyamba, A., Tucker, C.J., Kelley, P.W., Myers, M.F. and Peters, C.J. (1999). Climate and satellite indicators to forecast rift valley fever epidemics in Kenya. *Science* **285**, 397–400.

Lysenko, A.Y. and Semashko, I.N. (1968). Geography of malaria. In: *Medical Geography* (A.W. Lebedew, ed.). Moscow: Academy of Sciences, USSR.

Macdonald, G. (1957). *The Epidemiology and Control of Malaria*. Oxford: Oxford University Press.

Malakooti, M.A., Biomndo, K. and Shanks, D. (1998). Re-emergence of epidemic malaria in the highlands of western Kenya. *Emerging Infectious Diseases* **4**, 671–676.

Malhotra, M.S. and Srivastava, A. (1995). Diagnostic features of malaria transmission in Nadiad using remote sensing and GIS. In: *GIS for Health and the Environment* (D. de Savigny and P. Wijeyaratne, eds), pp. 109–114. Ottawa: International Development Research Centre.

Martens, W.J.M. (1997). *Health Impacts of Climate Change and Ozone Depletion: An Eco-Epidemiological Modelling Approach*. Bilthoven: National Institute of Public Health and the Environment.

Martens, P. (1999). How will climate change affect human health? *American Scientist* **87**, 534–541.

Martens, W. J. M., Niessen, L.W., Rotmans, J., Jetten, T.H. and McMichael, A.J. (1995a). Potential impact of global climate change on malaria risk. *Environmental Health Perspectives* **103**, 458–464.

Martens, W.J.M., Jetten, T.H., Rotmans, J. and Niessen, L.W. (1995b). Climate change and vector-borne diseases—a global modelling perspective. *Global Environmental Change, Human and Policy Dimensions* **5**, 195–209.

Martin, P.H. and Lefebvre, M.G. (1995). Malaria and climate—sensitivity of malaria potential transmission to climate. *Ambio* **24**, 200–207.

Metselaar, D. and Van Theil, P.M. (1959). Classification of malaria. *Tropical and Geographical Medicine* **11**, 157–161.

Molineaux, L. (1988). Malaria: the epidemiology of human malaria as an explanation of its distribution, including some implications for control. In: *Principles and Practice of Malariology* (W.H. Wernsdorfer and I. McGregor, eds), pp. 913–998. New York: Churchill Livingstone.

Molineaux, L. and Gramiccia, G. (1980). *The Garki Project: Research on the Epidemiology and Control of Malaria in the Sudan Savanna of West Africa*: Geneva, World Health Organization.

Moloney, J.M., Skelly, C., Weinstein, P., Maguire, M. and Ritchie, S. (1998). Domestic *Aedes aegypti* breeding site surveillance: limitations of remote sensing as a predictive surveillance tool. *American Journal of Tropical Medicine and Hygiene* **59**, 261–264.

Mouchet, J., Manguin, S., Sircoulon, J. *et al.* (1998). Evolution of malaria in Africa for the past 40 years: impact of climatic and human factors. *Journal of the American Mosquito Control Association* **14**, 121–130.

Mulcahy, K.A. and Clarke, K.C. (1994). Government digital cartographic data policy and environmental research needs. *Computers Environment and Urban Systems* **18**, 95–101.

Murray, C.J. and Lopez, A.D. (1997). Mortality by cause for eight regions of the world: Global Burden of Disease Study. *Lancet* **349**, 1269–1276.

Nabarro, D.N. and Tayler, E.M. (1998). The 'roll back malaria' campaign. *Science* **280**, 2067–2068.

Nega, A. (1991). Population migration and malaria transmission in Ethiopia. In: *Malaria and Development in Africa: A Cross-sectional Approach*. Washington, DC: American Association for the Advancement of Science.

Nelson, G.S. (1959). *Atlas of Kenya*. Nairobi: Surveys of Kenya, Crown Printers.

Omumbo, J., Ouma, J., Rapouda, B., Craig, M.H., Le Sueur, D. and Snow, R.W. (1998). Mapping malaria transmission intensity using geographical information systems (GIS): an example from Kenya. *Annals of Tropical Medicine and Parasitology* **92**, 7–21.

Onori, E. and Grab, B. (1980). Indicators for the forecasting of malaria epidemics. *Bulletin of the World Health Organization* **58**, 91–98.

Oppenshaw, S. (1996). Geographic information systems and tropical diseases. *Transactions of the Royal Society of Tropical Medicine and Hygiene* **90**, 337–339.

Patz, J.A., Strzepek, K., Lele, S. *et al.* (1998). Predicting key malaria transmission factors, biting and entomological inoculation rate, using modelled soil moisture in Kenya. *Tropical Medicine & International Health* **3**, 818–827.

Pitcairn, M.J., Rejmankova, E., Wood, B.L. and Washino, R.K. (1988). Progress report on the use of remote sensing data to survey mosquito larval abundance in California rice fields. *Proceedings of the Californian Mosquito and Vector Control Association* **56**, 158–159.

Pohl, C. and van Genderen, J.L. (1998). Multisensor image fusion in remote sensing: concepts, methods and applications. *International Journal of Remote Sensing* **19**, 823–854.

Pope, K.O., Sheffner, E.J., Linthicum, K.L. *et al.* (1992). Identification of central Kenyan Rift Valley fever virus vector habitats with Landsat TM and evaluation of their flooding status with airborne imaging radar. *Remote Sensing of Environment* **40**, 185–196.

Pope, K.O., Rejmankova, E., Savage, H.M., Arredondo-Jimenez, J.I., Rodriguez, M.H. and Roberts, D.R. (1994a). Remote sensing of tropical wetlands for malaria control in Chiapas, Mexico. *Ecological Applications* **4**, 81–90.

Pope, K.O., Rey-Benayas, J.M. and Paris, J.F. (1994b). Radar remote sensing of forested and wetland ecosystems in the Central American tropics. *Remote Sensing of Environment* **48**, 205–219.

Quattrochi, D.A. and Goodchild, M.F. (1997). *Scale in Remote Sensing and GIS.* Boca Raton, FL: CRC Press.

Rejmankova, E., Rejmanek, M., Pitcairn, M.J. and Washino, R.K. (1988). Aquatic vegetation in rice fields as a habitat for *Culex tarsalis* and *Anopheles freeborni. Proceedings and Papers of the Annual Conference of the California Mosquito and Vector Control Association* **15**, 160–163.

Rejmankova, E., Savage, H.M., Rejmanek, M., Roberts, D.R. and Arredondo-Jimenez, J.I. (1991). Multivariate analysis of relationships between habitats, environmental factors and occurrence of anophelene mosquito larvae (*Anopheles albimanus* and *A. pseudopunctipennis*) in southern Chiapas, Mexico. *Journal of Applied Ecology* **28**, 827–841.

Rejmankova, E., Savage, H.M., Rodriguez, M.H., Roberts, D.R. and Rejmanek, M. (1992). Aquatic vegetation as a basis for classification of *Anopheles albimanus* Wiedmann (Diptera: Culicidae) larval habitats. *Environmental Entomology* **21**, 598–603.

Rejmankova, E., Roberts, D.R., Harbach, R.E. *et al.* (1993). Environmental and regional determinants of *Anopheles* (Diptera: Culicidae) larval distribution in Belize, Central America. *Environmental Entomology* **22**, 978–992.

Rejmankova, E., Roberts, D.R., Pawley, A., Manguin, S. and Polanco, J. (1995). Predictions of adult *Anopheles albimanus* densities in villages based on distances to remotely sensed larval habitats. *American Journal of Tropical Medicine and Hygiene* **53**, 482–488.

Ribeiro, J.M., Seulu, F., Abose, T., Kidane, G. and Teklehaimanot, A. (1996). Temporal and spatial distribution of anopheline mosquitoes in an Ethiopian village: implications for malaria control strategies. *Bulletin of the World Health Organization* **74**, 299–305.

Rioux, J.A., Croset, H., Corre, J.-J., Simonneau, P. and Gras, G. (1968). Phyto-ecological basis of mosquito control: cartography of larval biotypes. *Mosquito News* **28**, 572–582.

Ritchie, S.A. (1993). Application of radar rainfall estimates for surveillance of *Aedes taeniorhynchus* larvae. *Journal of the American Mosquito Control Association* **9**, 228–231.

Robert, V., Gazin, P., Boudin, C., Molez, J.F., Ouédraogo, V. and Carnevale, P. (1985). La transmission du paludisme en zone de savane arborée et en zone rizicole des environs de Bobo-Dioulasso (Burkina Faso). *Annales de la Société Belge de Médecine Tropicale* **65**, 201–214.

Roberts, D.R. and Rodriguez, M.H. (1994). The environment, remote sensing, and malaria control. *Annals of the New York Academy of Sciences* **740**, 396–402.

Roberts, D., Rodriguez, M., Rejmankova, E. *et al.* (1991). Overview of field studies for the application of remote sensing to the study of malaria transmission in Tapachula, Mexico. *Preventive Veterinary Medicine* **11**, 269–275.

Roberts, D.R., Paris, J.F., Manguin, S. *et al.* (1996). Predictions of malaria vector distribution in Belize based on multispectral satellite data. *American Journal of Tropical Medicine and Hygiene* **54**, 304–308.

Roberts, J.M.D. (1964). The control of epidemic malaria in the highlands of Western Kenya. Part II. The campaign. *Journal of Tropical Medicine and Hygiene* **67**, 191–199.

Rodriguez, A.D., Rodriguez, M.H., Hernandez, J.E. *et al.* (1996). Landscape surrounding human settlements and *Anopheles albimanus* (Diptera: Culicidae) abundance in southern Chiapas, Mexico. *Journal of Medical Entomology* **33**, 39–48.

Rogers, D.J. (1996). Changes in disease vector distributions. In: *Climate Change and Southern Africa: An Exploration of Some Potential Impacts and Implications in the SADC Region* (M. Hulme, ed.), pp. 49–55. Norwich: Climate Change Research Unit, University of East Anglia.

Rogers, D.J., Hay, S.I. and Packer, M.J. (1996). Predicting the distribution of tsetse-flies in West-Africa using temporal Fourier processed meteorological satellite data. *Annals of Tropical Medicine and Parasitology* **90**, 225–241.

Rogers, D.J. and Randolph, S.E. (2000). The global spread of malaria in a future, warmer world. *Science*, in press.

Russell, P.F., West, L.S. and Manwell, R.D. (1946). *Practical Malariology*. Philadelphia: W.B. Saunders.

Sachs, J.D. and Warner, A.M. (1997). Sources of slow growth in African economies. *Journal of African Economies* **6**, 335–376.

Schwetz, J. (1942). Recherches sur la limite altimetrique du paludisme dans le Congo Orientale et sur la cause de cette limite. *Annales de la Société Belge de Médecine Tropicale* **22**, 183–209.

Service, M.W. (1985). *Anopheles gambiae*—Africa's principal malaria vector, 1902–1984. *Bulletin of the Entomological Society of America* **31**, 8–12.

Service, M.W. (1993). The *Anopheles* vector. In: *Bruce-Chwatt's Essential Malariology* (H.M. Gilles and D.A. Warrell, eds), 3 edn, pp. 96–123. London: Edward Arnold.

Sharma, V.P. and Srivastava, A. (1997). Role of geographic information systems in malaria control. *Indian Journal of Medical Research* **106**, 198–204.

Sharma, V.P., Dhiman, R.C., Ansari, M.A. *et al.* (1996). Study on the feasibility of delineating mosquitogenic conditions in and around Delhi using Indian remote sensing satellite data. *Indian Journal of Malariology* **33**, 107–125.

Sharp, B.L. and le Sueur, D. (1996). Malaria in South Africa—the past, the present and selected implications for the future. *South African Medical Journal* **86**, 83–89.

Shretta, R., Omumbo, J.A. and Snow, R.W. (1998). *Community Based Health Care Activity and its Relationship to the Delivery of Insecticide-treated Bed Nets in Kenya*. Nairobi: Ministry of Health, Government of Kenya.

Smith, T., Charlwood, J.D., Takken, W., Tanner, M. and Spiegelhalter, D.J. (1995). Mapping the densities of malaria vectors within a single village. *Acta Tropica* **59**, 1–18.

Snow, R.W. and Marsh, K. (1995). Will reducing *Plasmodium falciparum* transmission alter malaria mortality among African children. *Parasitology Today* **11**, 188–190.

Snow, R.W. and Marsh, K. (1998). New insights into the epidemiology of malaria relevant for disease control. *Tropical Medicine: Achievements and Prospects, British Medical Bulletin* **54**, 293–309.

Snow, R.W., Armstrong-Schellenberg, J.R.M., Peshu, N. *et al.* (1993). Periodicity and space-time clustering of severe childhood malaria on the coast of Kenya. *Transactions of the Royal Society of Tropical Medicine and Hygiene* **87**, 386–390.

Snow, R.W., Marsh, K. and Le Sueur, D. (1996). The need for maps of transmission intensity to guide malaria control in Africa. *Parasitology Today* **12**, 455–457.

Snow, R.W., Omumbo, J.A., Lowe, B. *et al.* (1997). Relation between severe malaria morbidity in children and level of *Plasmodium falciparum* transmission in Africa. *Lancet* **349**, 1650–1654.

Snow, R.W., Gouws, E., Omumbo, J. *et al.* (1998a). Models to predict the intensity of *Plasmodium falciparum* transmission: applications to the burden of disease in Kenya. *Transactions of the Royal Society of Tropical Medicine and Hygiene* **92**, 601–606.

Snow, R.W., Mwenesi, H. and Rapuoda, B. (1998b). *Malaria: A Situation Analysis for Kenya*. Nairobi: Ministry of Health.

Snow, R.W., Craig, M., Deichmann, U. and Marsh, K. (1999a). Estimating mortality, morbidity and disability due to malaria among Africa's non-pregnant population. *Bulletin of the World Health Organization* **77**, 624–640.

Snow, R.W., Craig, M.H., Deichmann, U. and le Sueur, D. (1999b). A preliminary continental risk map for malaria mortality among African children. *Parasitology Today* **15**, 99–104.

Su, M.-D. (1994). Framework for application of geographic information system to the monitoring of dengue vectors. *Kaohsiung Journal of Medical Science* **10**, S94–S101.

Swaroop, S. (1949). Forecasting of epidemic malaria in the Punjab, India. *American Journal of Tropical Medicine* **29**, 1–17.

Thomas, C.J. and Lindsay, S.W. (1999). Local-scale variation in malaria infection amongst rural Gambian children estimated by satellite remote sensing. *Transactions of the Royal Society of Tropical Medicine and Hygiene* **94**, 159–163.

Thompson, R., Begtrup, K., Cuamba, N. *et al.* (1997). The Matola malaria project: a temporal and spatial study of malaria transmission and disease in a suburban area of Maputo, Mozambique. *American Journal of Tropical Medicine and Hygiene* **57**, 550–559.

Thomson, M.C., D'Alessandro, U., Bennett, S. *et al.* (1994). Malaria prevalence is inversely related to vector density in The Gambia, West Africa. *Transactions of the Royal Society of Tropical Medicine and Hygiene* **88**, 638–643.

Thomson, M., Connor, S., Bennett, S. *et al.* (1996a). Geographical perspectives on bednet use and malaria transmission in The Gambia, West Africa. *Social Science and Medicine* **43**, 101–112.

Thomson, M.C., Connor, S.J., Milligan, P.J.M. and Flasse, S.P. (1996b). The ecology of malaria – as seen from Earth observation satellites. *Annals of Tropical Medicine and Parasitology* **90**, 243–264.

Thomson, M., Connor, S., Milligan, P. and Flasse, S. (1997). Mapping malaria risk in Africa: what can satellite data contribute? *Parasitology Today* **13**, 313–318.

Thomson, M.C., Connor, S.J., D'Alessandro, U., Rowlinson, B., Diggle, P., Cresswell, M. and Greenwood, B. (1999). Predicting malaria infection in Gambian children from satellite data and bed net use surveys: the importance of spatial correlation in the interpretation of results. *American Journal of Tropical Medicine and Hygiene* **61**, 2–8.

Trape, J.F. and Zoulani, A. (1987). Malaria and urbanization in Central Africa: the example of Brazzaville. Part II. Results of entomological surveys and epidemiological analysis. *Transactions of the Royal Society of Tropical Medicine and Hygiene* **81**, 10–18.

Trape, J.F., Lefebvre-Zante, E., Legros, F. *et al.* (1992). Vector density gradients and the epidemiology of urban malaria in Dakar, Senegal. *American Journal of Tropical Medicine and Hygiene* **47**, 181–189.

Wagner, V.E., Hill-Rowley, R., Narlock, S.A. and Newson, H.D. (1979). Remote sensing: a rapid and accurate method of data acquisition for a newly formed mosquito control district. *Mosquito News* **39**, 283–287.

Walsh, S.J., Evans, T.P., Welsh, W.F., Entwisle, B. and Rindfuss, R.R. (1999). Scale-dependent relationships between population and environment in northeastern Thailand. *Photogrammetric Engineering & Remote Sensing* **65**, 97–105.

Washino, R.K. and Wood, B.L. (1994). Application of remote sensing to arthropod vector surveillance and control. *American Journal of Tropical Medicine and Hygiene* **50**, 134–144.

Welch, J.B., Olson, J.K., Hart, W.G., Ingle, S.G. and Davis, M.R. (1989a). Use of aerial color IR as a survey technique for *Psorophora columbiae* oviposition habitats in Texas (USA) ricelands. *Journal of the American Mosquito Control Association* **5**, 147–160.

Welch, J.B., Olson, J.K., Yates, M.M., Benton, A.R., jr and Baker, R.D. (1989b). Conceptual model for the use of aerial color infrared photography by mosquito control districts as a survey technique for *Psorophora columbiae* oviposition habitats in Texas ricelands. *Journal of the American Mosquito Control Association* **5**, 369–373.

Wernsdorfer, W.H. and McGregor, I. (1988). *Malaria: Principles and Practice of Malariology*. Oxford: Oxford University Press.

White, G.B. (1974). *Anopheles gambiae* complex and disease transmission in Africa. *Transactions of the Royal Society of Tropical Medicine and Hygiene* **68**, 278–295.

White, G.B., Magayuka, S.A. and Boreham, P.F.L. (1972). Comparative studies on sibling species of the *Anopheles gambiae* Giles complex (Dipt., Culicidae): bionomics and vectorial activity of Species A and Species B at Segera, Tanzania. *Bulletin of Entomological Research* **62**, 295–317.

WHO (1998). *Roll Back Malaria*. Geneva: World Health Organization.

Wilson, D.B. (1956). *Atlas of Tanzania*. Dar es Salaam: Survey Division, Department of Lands and Surveys, Government Printers.

Wood, B.L., Beck, L.R., Washino, R.K., Palchick, S.M. and Sebesta, P.D. (1991a). Spectral and spatial characterization of rice field mosquito habitat. *International Journal of Remote Sensing* **12**, 621–626.

Wood, B., Washino, R., Beck, L. *et al.* (1991b). Distinguishing high and low anopheline-producing rice fields using remote sensing and GIS technologies. *Preventive Veterinary Medicine* **11**, 277–288.

Wood, B.L., Beck, L.R., Washino, R.K., Hibbard, K.A. and Salute, J.S. (1992). Estimating high mosquito-producing rice fields using spectral and spatial data. *International Journal of Remote Sensing* **13**, 2813–2826.

Wood, B.L., Beck, L.R., Dister, S.W. and Spanner, M.A. (1994). Global monitoring and disease prediction program. *Sistema Terra* **3**, 38–39.

Young, M.D. (1976). *Tropical Medicine*. Philadelphia: W.B. Saunders.

Ticks and Tick-borne Disease Systems in Space and from Space

S.E. Randolph

Department of Zoology, South Parks Road, Oxford OX1 3PS, UK

ABSTRACT

Analyses within geographical information systems (GISs) indicate that small- and large-scale ranges of hard tick species (Ixodidae) are determined more by climate and vegetation than by host-related factors. Spatial distributions of ticks may therefore be analysed by statistical methods that seek correlations between known tick presence/absence and ground- or

ADVANCES IN PARASITOLOGY VOL 47
0065-308-X $30.00

remotely-sensed (RS) environmental factors. In this way, local habitats of *Amblyomma variegatum* in the Caribbean and *Ixodes ricinus* in Europe have been mapped using Landsat RS imagery, while regional and continental distributions of African and temperate tick species have been predicted using multi-temporal information from the National Oceanic and Atmospheric Administration–Advanced Very High Resolution Radiometer (NOAA-AVHRR) imagery. These studies illustrate ways of maximizing statistical accuracy, whose interpretation is then discussed in a biological framework. Methods such as discriminant analysis are biologically transparent and interpretable, while others, such as logistic regression and tree-based classifications, are less so. Furthermore, the most consistently significant variable for predicting tick distributions, the RS Normalized Difference Vegetation Index (NDVI), has a sound biological basis in that it is related to moisture availability to free-living ticks and correlated with tick mortality rates. The development of biological process-based models for predicting the spatial dynamics of ticks is a top priority, especially as the risk of tick-borne infections is commonly related not simply to the vector's density, but to its seasonal population dynamics. Nevertheless, using statistical pattern-matching, the combination of RS temperature indices and NDVI successfully predicts certain temporal features essential for the transmission of tick-borne encephalitis virus, which translate into a spatial pattern of disease foci on a continental scale.

1. TICKS AS PESTS—PARASITES AND VECTORS

Throughout the world, in tropical, temperate and even tundra regions, ticks are serious veterinary and medical pests. With 80% of the world population of 1200 million cattle exposed to ticks, global annual losses in the 1970s were estimated at US $7 billion (McCosker, 1979). More recently, production losses and control costs incurred by tick-borne *Theileria parva* alone in 11 countries in eastern, central and southern Africa were estimated at US $168 million per year (Norval *et al.*, 1992).

Ticks pose a dual threat to their livestock hosts. As ectoparasites, they cause direct harm by the significant debilitating effect they have on the host's physique through blood loss and disturbance to foraging, resulting in reduced live-weight gain (Pegram *et al.*, 1989) and milk production (Norval *et al.*, 1997). With their sawing rather than piercing mouthparts, ticks also damage the hide and so breach the animal's outer defences to secondary infections, allowing invasion by bacteria, fungi and macroparasites such as screw-worm larvae.

As vectors, ticks surpass all other arthropods in the variety of microparasites that they transmit to humans as well as to livestock (Sonenshine, 1991). These include fungi, viruses, bacteria, rickettsia and protozoal parasites (all of which will hereafter be referred to as 'pathogens', the tick-borne agents of disease).

Curiously, it is in their role as parasites that ticks most resemble insect vectors in that the damage they do is more or less directly related to the ticks' distribution and abundance. As vectors, on the other hand, ticks are much more complex than insects, relying as they do on a few large blood meals during each life cycle, as they progress through larval, nymphal and adult stages. Soft ticks of the family Argasidae are more similar to typical haematophagous insects as the nymphal and adult instars each take relatively small blood meals (up to five times their body weight) at intervals of a few days, so that pathogens acquired at one meal may be transmitted soon afterwards. Despite the significance of Argasid ticks as vectors, particularly of arboviruses such as West Nile fever virus amongst birds, of African swine fever virus amongst domestic swine, and of relapsing fever spirochaetes amongst humans, in whose dwellings these soft ticks live, they have been the subject of virtually no quantitative epidemiology.

Consequently, this chapter will focus on the hard ticks of the family Ixodidae, which take very large meals (10–100 times their body weight) only once per life stage. Acquisition and transmission of pathogens are therefore separated by long periods of free-living existence during which the tick develops from one stage to the next. This takes place in sheltered microhabitats on the ground, where the tick is subjected to a buffered range of the prevailing diurnal and seasonal fluctuations of climate. As with most terrestrial arthropods, ticks' development rates are temperature-dependent, while their survival is limited by water loss under dry conditions. Geographically and seasonally variable rates of development and mortality impose major constraints on both tick population dynamics and pathogen transmission dynamics, determining the pace of transmission and the basic reproduction number (R_0) of the pathogen (Randolph, 1998). Furthermore, within the triangle of host–tick–pathogen interactions, factors such as the host's immune response moderate the expression of disease and the potential for pathogen maintenance. Commonly, therefore, the presence of ticks is not sufficient on its own to pose a significant risk of associated disease, although it is, of course, a prerequisite.

In this chapter, I shall emphasize the different spatial scales at which information about the environment, including remotely sensed (RS) data, has been used within geographical information systems (GIS) to predict the distribution of ticks. Most useful of the statistical methods that seek correlations between extrinsic factors and tick presence are those that remain biologically transparent and appropriate. We are not yet in a position to adopt the ideal versatile, but more demanding, biological process-based approach to create predictive maps for tick-borne diseases on continental scales. A feedback loop, however, between determining statistical associations and understanding biological processes may together describe and explain spatial patterns. The conclusion is that temporal and spatial patterns are intimately linked in tick-borne disease systems, and both may be detected remotely from satellites.

2. PREDICTING THE DISTRIBUTION OF TICKS

2.1. Determinants of Species Distributions

Given the intimate, prolonged association between tick and host during each blood meal, but the even longer non-parasitic phases of each life cycle, it is not intuitively obvious whether tick species ranges are limited primarily by host-related biotic or environmental abiotic factors. Clearly ticks cannot survive in the absence of suitable hosts, but the less specific the tick–host relationship, the less likely it is that hosts will be a limiting factor; vertebrates upon which ticks can feed are more or less ubiquitous. This is certainly true of the important temperate tick species, *Ixodes ricinus* in Europe and *I. scapularis* in USA, which both feed on virtually any vertebrate with which they come into contact (Milne, 1949; Wilson, 1998).

A GIS may be used to establish the relative importance of hosts and environment in determining tick distributions. Kitron *et al.* (1992) showed that *I. dammini* (≡ *I. scapularis*) is less widespread than its principal host, the white-tailed deer, and so is probably not limited by it. Within a county in Illinois, USA, the distances from a known tick focus at which deer were shot were used to generate a relational digitized database, overlaid on computerized maps of the county. When the spatial distributions of tick-infested and non-infested deer were compared, infestations were shown to be significantly clustered around a single important source of ticks in the locality, despite the much wider distribution of deer. Similarly, in another county in the same state, a single focus of tick infestations was associated with sandy soil and deciduous forests (Kitron *et al.* 1991). Although analyses on these scales could have been performed laboriously with a pencil and ruler, computerized GISs speed up the process, encouraging analyses that produce such elegantly convincing conclusions as a basis for larger scale studies.

In an extensive study of African tick species, Cumming (1998, 1999a) established that the overwhelming majority are not host-specific and show no evidence of being host-limited; for example, of the 55 species collected frequently from cattle, 48 were far less widely distributed than cattle; a further 22 species are absent from a large part of the range of their particular non-cattle hosts. Thus it appears that ticks occupy only a sub-set of their hosts' ranges, with the broad-scale boundaries to their ranges being set by abiotic factors such as vegetation and climate that determine tick development and survival rates on the ground.

Conversely, at local scales the absence or scarcity of hosts in any parts of a region with a tick-permissive environment will cause patchy distributions, so that, for example, ticks may be absent in fields or residential areas but present in adjoining woodlands inhabited by appropriate vertebrate species. This habitat heterogeneity itself will show clear correlations with vegetation indices and microclimatic factors.

Tick distributions, therefore, may be predicted by relatively straightforward statistical methods that seek correlations between environmental factors and tick presence. The irony of this approach is that without good descriptive maps of a species' distribution, based on ground observations, we do not have the essential starting point to generate predictive maps based on statistical pattern-matching. Yet if we already have such good descriptive maps, we hardly need predictive maps. The fact is, of course, our descriptive maps are never complete, and, more to the point, they can only ever be static without repeated and expensive field work. The objective of deriving predictive maps is to understand the processes underlying the patterns of distribution so that we may fill in the gaps in our primary observations and also update the distributions as they shift with changing environmental conditions. The challenge, therefore, is to identify which combination of environmental factors are the best predictors, and to relate them to tick biology.

2.2. Identifying Local Tick Habitats Using High Resolution Satellite Imagery

Over limited areas, habitats may be classified according to their observed levels of tick infestation. If these habitats can be discriminated by key environmental indicators, they and similar habitats may be mapped beyond the area of the original field observation, building up a predictive map of the local, often patchy distribution of ticks. This habitat-matching approach was used in the first attempts to apply satellite data to modelling tick distributions, using high-resolution Landsat derived imagery.

Pioneers in this approach were Hugh-Jones and his colleagues, who identified habitats of the bont tick *Amblyomma variegatum*, in the Caribbean islands of St Lucia and Guadeloupe, whither it had spread from its African origins probably through the movement of infested cattle (Walker, 1987). On

St Lucia, the major infestation was in the dry, rough grazing characterized by the shrub *Prosopis juliflora*. By comparing the historic Landsat Multispectral Scanner (MSS) imagery with published island vegetation maps, Hugh-Jones and O'Neil (1986) were able to show that this tick habitat was being expanded by the islanders through land and grazing mismanagement. Next, in Guadaloupe, contemporary Landsat Thematic Mapper (TM) imagery (bands 1, 2, 3, 4, 5 and 7) was used to identify *A. variegatum* habitats on known infested farms. Discriminant analysis of the plant composition, grazing cover, soil character and depth, slope and rainfall indicated a number of habitats with different degrees of tick infestation—lightly infested 'dry meadows', moderately infested 'fond' (karst) and 'foothills', and heavily infested 'dry scrub' and 'rocky grasslands' (Hugh-Jones *et al.*, 1992). Cooler temperatures and high annual rainfall (>4000 mm) ensured that the high altitude grazing above 500 m was essentially tick free. When the various grazing sites on each farm visible in the original unclassified image were clustered according to the variances of the band values and by vegetation biomass (estimated as the Normalized Difference Vegetation Index, NDVI; Hay, this volume) and moisture, high tick infestation levels were associated with large variances in the band values, i.e. where the vegetation was very heterogeneous. Within these heterogeneous sites, those with high vegetation (NDVI) and moisture indices had more ticks than those with low values. This result offers some testable insights into the preferred habitats of *A. variegatum* elsewhere within its range: is it also true in its ancestral range in Africa that this tick is more likely to occur in humid places with mixed grazing and bushes than in dry open grassland?

In Europe, the single most important vector species is the tick *I. ricinus*, transmitter of *Babesia* spp. (protozoans), louping ill virus and *Ehrlichia* spp. (rickettsia) to livestock, and *Borrelia burgdorferi s.l.* (bacteria) and tick-borne encephalitis virus to humans. Long-term interest in the latter disease in central Europe has resulted in an impressive 20 year (1953–1972) record of the fine-scale distribution of the tick amongst different types of vegetation within forested areas southwest of Prague, in the Czech Republic (Cerny, 1988). Ticks are most abundant in deciduous and mixed forests and least abundant in coniferous forests and non-wooded areas. This data-set was related to Landsat-MSS imagery to establish that habitats characterized by different vegetation, and therefore associated with different tick infestation levels, could be identified remotely (Daniel and Kolár, 1990). Satellite imagery can therefore be used to predict areas of high tick populations. Where these exist alongside habitats indicative of human occupation (housing developments, agricultural areas, recreational facilities), also visible on Landsat-MSS images, forecasts of high epidemiological risk can direct preventive control measures.

The major advantage of this approach to modelling tick distributions is the high spatial resolution offered by Landsat-MSS imagery, which can match the natural landscape heterogeneity and mosaic incidence of ticks. Information on fine-scale tick distributions, when made available to farmers in the Caribbean or the inhabitants of Prague, will permit avoidance action to minimize contact between ticks and livestock or humans. The validation of any predictions demand equally fine-scale ground data, but these (if they exist at all) together with affordable satellite images are available over very limited areas. The islands of Guadeloupe measure c. 65×48 km, while the Czech study site measured c. 41 km^2 within an image area of c.185×170 km.

Furthermore, because of the cost and limited temporal availability of good cloud-free images (Hay, this volume), analysis tends to be based on an image for a single day (23 April 1986 for Guadeloupe, 3 August 1984 for the Czech study). Such a static picture for a single point in time offers little insight into the dynamic processes underlying the observed pattern. In the case of the well studied tick *I. ricinus*, it is not difficult to relate the results of Daniel and Kolár's (1990) study to the tick's dependence on high (>85%) relative humidity within its microhabitat. This is typically found in moist leaf litter on the floor of deciduous woodlands, but rarely beneath conifers. Thus, prior knowledge of tick biology can help explain the new results, rather than the new results adding to our understanding of tick biology. On the other hand, even as a static mapping tool, high spatial resolution satellite sensor images, with their multispectral signals and the much greater ease with which they can be geo-registered, are a significant advance over aerial photography in displaying the environmental features of tick habitats.

2.3. Wide-scale Distributions of Ticks—from Climate to Satellite Imagery

A general knowledge of microclimatic requirements for tick survival, principally in terms of moisture and temperature conditions, gained from intensive biological studies in the laboratory or field, may be extrapolated to regional and even continental areas if those requirements can be identified extensively. Before the routine availability of satellite imagery, early attempts to predict wide-scale distributions of ticks relied on direct or interpolated climate databases. Best known of the climate-matching approaches is the programme CLIMEX, which calculates the climatic suitability of geographical regions for arthropod species using a temperature-dependent growth index, moderated by four stress indices (hot, cold, dry, wet) (Sutherst and Maywald, 1985). Following its relative success in capturing the distribution of the one-host cattle tick *Boophilus microplus* in northern Australia, but its marked over-prediction of suitable regions for the same tick species in Africa (Sutherst and Maywald, 1985), CLIMEX was applied to a variety of other tick species in Africa. Maps of the distribution of ecoclimatic suitability index (EI) for *R. appendiculatus* were compared visually, but not statistically, with the recorded distribution of the tick (Lessard *et al.*, 1990; Perry *et al.*, 1990), from which it appears that, on a scale from 0 to 100, an annual EI value of anything over 1 may coincide with the presence of the tick in eastern to southern Africa (Perry *et al.*, 1991), whilst a value of less than 1 generally indicated absence. Very high EI values, however, in many parts of western Africa do not coincide with the presence of *R. appendiculatus*. The same predictions of climatic suitability were used to warn of the possible spread of *R. appendiculatus* northwards into western and central Ethiopia (Perry *et al.*, 1990; Norval *et al.*, 1991a).

Unwarranted faith was put in CLIMEX, even when its results contradicted ground observations. When studying the contiguous, mostly non-overlapping distributions of two *Amblyomma* species in Africa, Norval *et al.* (1994) concluded 'The distribution of predicted climatic suitability is almost the opposite of the actual distribution of *A. hebraeum* and *A. variegatum* in Zimbabwe. ... As the CLIMEX predictions of the distributions of *A. variegatum* and *A. hebraeum* for Africa as a whole are generally excellent (Sutherst and Maywald, 1985; Norval *et al.*, 1991b, 1992), it can be concluded that the problem does not lie with the model's predictions. The reason for the anomaly must therefore be the role played by non-climatic factors.' In fact the CLIMEX predictions for *A. variegatum* are nearly as poor for the rest of Africa as they are for Zimbabwe (Norval *et al.*, 1991b, 1992).

A second climate-based approach to modelling tick distributions relied on the model BIOCLIM (Nix, 1986; Norval *et al.*, 1992). For each of a selection of geographic points throughout the distribution of the tick species, BIOCLIM generates 24 climatic attributes from which are derived indices that summarize annual and seasonal mean conditions, extreme values and intra-year seasonality. Computer-selected thresholds and limits of each of the indices are matched across a geographical grid to predict the potential distribution. On a pan-African scale, this model gave a better match between the predicted and known distributions of *R. appendiculatus*, although, as Norval *et al.* (1992) point out, on a smaller scale the match was not so good in some areas of highland eastern Africa. At the time, it was suggested that higher resolution (8 km rather than 25 km) interpolated climate databases might have improved the precision of both these climate-based methods. A logistic regression model, however, based on interpolated climate and elevation data for Africa (Hutchinson *et al.*, 1996) reduced to the same resolution (25 km), achieved a highly accurate prediction (Cumming, 1999b).

The new tool of remote sensing, using the advanced very high resolution radiometer (AVHRR) on the National Oceanic and Atmospheric

Administrations's (NOAA's) meteorological satellites, which allowed the direct detection of environmental factors at 8 km resolution over whole continents, rapidly superseded the use of interpolated climate databases in the search for predictors of tick distributions. Indeed, Lessard *et al.* (1990) and Perry *et al.* (1990) presented visual comparisons of the recorded distribution of *R. appendiculatus* and the satellite-derived maximum mean monthly NDVI values in the same publication in which they compared the CLIMEX-derived ecoclimatic index with this tick's distribution. They showed that *R. appendiculatus* may occur anywhere with a mean maximum NDVI of 0.15 or greater (on a scale of 0 to >0.4), but at that stage no statistical correlations had been found between NDVI and *R. appendiculatus* distribution (Perry *et al.*, 1991), and the view was that the dynamic nature of NDVI values would make their use in predicting tick habitats very complicated (Kruska and Perry, 1991). We now realise that it is this very dynamism that makes NDVI, and other seasonally variable environmental factors, so useful and appropriate for discriminating between areas where tick populations can and cannot survive. Tick habitats, like all habitats, are characterized by specific seasonal profiles of these factors.

2.4. Biological Basis of NOAA-AVHRR Satellite Data as Predictors of Tick Spatial Patterns

The widespread use of NDVI as a single, catch-all satellite-derived index of environmental suitability is in danger of obscuring, rather than illuminating, the biological basis for its repeatedly proven usefulness. To rescue its credibility, we must explore and quantify the geographically variable relationships between NDVI and whatever environmental factor is critical to an organism's distribution and abundance. The most robust ecological explanations are always founded on the demographic processes, birth and death, whose rates determine population performance. For ticks, recruitment to populations is determined by temperature-dependent development rates, complicated by diapause in temperate and semi-temperate regions, while loss from populations is determined by mortality rates, whose measurement is complicated by variable feeding success rates.

For natural tick populations, these processes have as yet been teased apart and their rates quantified only for the relatively simple situation of continuous overlapping generations of *R. appendiculatus* in tropical (east) and semi-temperate (southern) Africa (Randolph, 1994, 1997). Here, the NDVI is indeed correlated with the single critical environmental factor most responsible for the tick's seasonally variable mortality rates, and, as a consequence, is correlated directly with those mortality rates, according to the appropriate time delay (Figure 1). Thus in Burundi, Uganda and

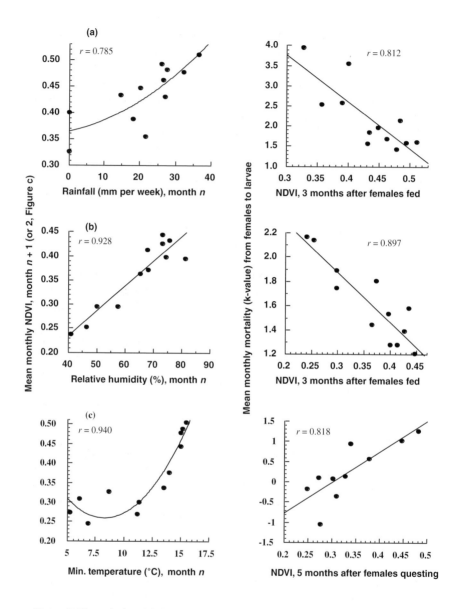

Figure 1 The relationships between (on the left) NDVI and mean monthly values of the critical climatic factor most strongly correlated with the index of mortality between females and larvae of the tick *Rhipicephalus appendiculatus*; and (on the right) that mortality index and mean monthly NDVI at (a) Kirundo, Burundi; (b) Mwanza, Tanzania; and (c) Lake McIlwaine, Zimbabwe. The values are annual synoptic means, because NDVI records are not contemporaneous with climate and tick records. (Redrawn from Randolph, 1994.)

Tanzania, where moisture availability to eggs and larval ticks is critical, the NDVI is positively correlated with rainfall or atmospheric relative humidity one or two months previously. Accordingly, mortality between the female and larval stages is *negatively* correlated with the NDVI 3 months after the females have fed. In Zimbabwe, by contrast, female-to-larval mortality is most strongly positively correlated with minimum temperatures 3 months after the females were questing. Low temperatures are associated with low NDVI values 2 months later, so here mortality from females to larvae is *positively* correlated with the NDVI 5 months after the female ticks have quested. The opposite signs of these NDVI-mortality correlations are therefore informative in themselves as long as they are recognized.

This satellite sensor-derived vegetation index, therefore, does seem to be a reliable marker for tick population performance. This provides a sound biological justification for using this variable in a purely statistical GIS framework to define the environmental characteristics of sites where ticks do occur and others where they do not.

The first attempt to do this on a broad scale for vectors focused on tsetse and ticks in Kenya, Tanzania and Zimbabwe (Rogers and Randolph, 1993). Tick distribution data are more difficult to analyse than tsetse data because published maps often record only the points at which ticks have been collected (Norval *et al.*, 1992). Areas in between the points may well be suitable for ticks but are assumed by the analysis to be unsuitable. The fit between predicted and observed distributions is thereby compromised. Nevertheless, linear discriminant analysis of elevation, five temperature variables and four NDVI variables (i.e. various permutations of minimum, maximum and mean values) yielded 72–73% correct predictions of *R. appendiculatus* in these African countries, with consistently more false positives (false predictions of presence—*c.* 25%) than false negatives (false predictions of absence—*c.* 1%) (see Plate 8). It seems that temperature and NDVI together can correctly define the major environmental constraints debarring *R. appendiculatus* from certain areas, although the false positives suggest there are other potentially inhabited areas where this tick is as yet unrecorded.

In both these regions of Africa, the maximum of the monthly maximum temperature was the most significant predictor, with ticks present in cooler rather than warmer places. As maximum temperature here is strongly negatively correlated with rainfall (which was not available for inclusion in the discriminant analysis) on a seasonal basis, this result implies that places (generally at high elevations) with higher rainfall are more favourable, in accordance with the negative correlation between seasonal mortality and rainfall (Randolph, 1994). The second most significant predictor was NDVI. Whereas in Kenya and Tanzania monthly *minimum* NDVI was limiting, consistent with the negative correlation between seasonal mortality and

NDVI (see above), in Zimbabwe monthly *maximum* NDVI was limiting, consistent with the positive equivalent correlation there. In this southern country, however, nearer to the edge of the range of *R. appendiculatus*, the second variable, NDVI, did not significantly improve the description of the tick's distribution, while well within the distributional limits more than one variable (temperature, NDVI and elevation) all make major contributions to the observed distribution. Not surprisingly, the edge of an organism's range may be defined by a single limiting factor, whereas in the more generally permissive conditions well within the range this single factor may only be limiting in combination with other factors.

Thus the purely statistical discriminant analysis yields information about the determinants of this organism's distribution that are not only biologically transparent and interpretable, but also entirely consistent with the tick's biology as revealed by the complementary process-based analysis. In comparison, some analytical methods achieve a high descriptive accuracy at the expense of biological insight. Tree-based classifications (also called automatic interaction detection methods), for example, are recommended for use as a preliminary to more robust statistical analysis by other methods (Green, 1978). One such model, applied to *I. ricinus* in the Trentino region of the Italian Alps (Merler *et al.*, 1996), indicated that altitude and geological substratum are the principal predictors of tick distributions there, prevailing over small-scale variables such as vegetation type and cover density.

2.5. Taking Account of Regional Heterogeneities

Heterogeneities occur at all spatial scales. The above results for Kenya, Tanzania and Zimbabwe exemplify regional heterogeneities in eco-climatic conditions and the tick's response to them (which may or may not have a genetic basis—DoRosario and Coluzzi, 1997). In scaling up predictions from regional to continental distributions, we must take account of such heterogeneities, otherwise they will blur the very distinctions we are seeking. A single average set of abiotic conditions across disparate regions of presence and absence sites will not capture the true defining features in any one place.

In principle, there is no reason why disparate regions should follow a rigid geographical pattern, although in practice spatial autocorrelation of abiotic factors (Robinson, this volume) makes a degree of geographical coherence likely. Environmental conditions, as detected by satellite sensors, may be used to divide sites from large continental areas into more homogeneous groups or 'clusters' before the analysis begins. It is a matter of choice whether clusters of environmental similarity are identified before or after dividing the

Table 1 The statistical accuracy of the predicted pan-African distribution of the tick *A. variegatum* in the absence of clustering, and under two of the best clustering regimes.

	Zero clustering	4 Absence clusters 3 Presence clusters	5 Absence clusters 2 Presence clusters
Correct (%)	71	81	83
Sensitivity (ability to detect presence)	0.960	0.885	0.835
Specificity(ability to detect absence)	0.662	0.795	0.883
False negatives (%)	1	2	8
False positives (%)	28	17	10

data set according to the presence or absence of the species of interest; there are pros and cons to each method (Rogers, this volume).

Results for the African tick *A. variegatum* illustrate the value of clustering. Toomer (1996) used a total of 28 variables (means, minima, maxima, amplitudes, ratios, etc.) from two AVHRR derived signals—NDVI and Ch4, a thermal infrared band; from a Meteosat High Resolution Radiometer (HRR)—Cold Cloud Duration, a surrogate of rainfall; and a digital elevation model (DEM). Having divided the digitized distribution of *A. variegatum* (from Walker, 1987) into sites of presence and absence, the effects of clustering each set (up to five clusters) were explored. For each degree of clustering, only the 10 most significant of the 28 predictor variables, as ranked by linear discriminant analysis, were used to model the tick's distribution. Clustering resulted in a significant increase in predictive accuracy, both statistically (Table 1) and visually (see Plate 9).

As well as confirming empirically the intuitive benefits of clustering, these results also illustrate that the statistics do not always give an unequivocal indication of which map represents the best prediction of risk. In going from four absence and three presence clusters to five absence and two presence clusters, the overall prediction improves a little but the map becomes more conservative; more absence sites are identified correctly (increased specificity, fewer false positives), and a lot of the remaining false positives (across the Democratic Republic of Congo [Zaire] into northern Angola and down the east side of South Africa) show a reduced probability of habitat suitability for the tick, which may imply a lower abundance if the tick is actually present. On the other hand, not all the sites with recorded presence are detected (decreased sensitivity and markedly more false negatives). If the objective of the risk map is to warn of potential threat, to alert control services and to direct attention to hitherto uncharted localities where the tick may be lurking, it may be better to err, within limits, on the side of false alarms than of false complacency.

2.6. Statistical Cautions and Caveats

With the flood of ever better RS environmental data, more versatile analytical software packages and faster computers, risk maps for arthropod vectors based on habitat similarity models may now be produced with seductively good empirical results. The statistical accuracy of the model, however, depends on decisions about, for example, the modelling method to be used, the area over which estimates are to be made and the minimum number of observations necessary for a meaningful hypothesis of species distribution. Taking the distribution of the southern African tick *A. hebraeum* as an example, Cumming (1999b) explored the objective basis for these decisions, using Threshold Receiver Operating Characteristics (ROC) plots (Fielding and Bell, 1997) to compare the fit of different models with varying inputs, driven by 36 climate variables (average monthly minimum and maximum temperature, and rainfall) (Hutchinson *et al.*, 1996). ROC plots (with sensitivity on the *y*-axis and 1-specificity on the *x*-axis) describe the compromise that a model attains between including known presences (a pressure for a greater geographical area) and excluding absences (a pressure for a smaller area). The area under the ROC curve (AUC) provides an objective measure of accuracy (Fielding and Bell, 1997), varying from 0.5 for randomness to 1.0 for a perfect fit.

First Cumming established that, when using the standard procedures available on SPSS version 7.5 (SPSS Inc., Chicago), logistic regression was fractionally more accurate (AUC = 0.932) than linear discriminant analysis (AUC = 0.915), although the latter is less powerful than customized non-linear discriminant analysis methods (Rogers *et al.*, 1996; Rogers, this volume). Unlike discriminant analysis, however, logistic regression models are in many ways 'black boxes' (Cumming 1999b), as the coefficient values cannot be translated into statements of biological meaning even with a prior understanding of the biology of the species concerned. Furthermore, at the same threshold probability, the logistic regression predicted a very much smaller area of tick presence than the discriminant analysis. This highlights the difference between conclusions based on statistical or visual comparisons, and introduces yet another decision, what threshold probability is to be taken to indicate tick presence.

Second, Cumming investigated the effects of the relative numbers of positive and negative records on the apparent accuracy of the logistic regression model. If more negative records were included by increasing the total area considered, in this case by using climatic data for the whole of

Africa rather than just for south of 8°S (the northern limit of all bar one positive record), the AUC value increased from 0.845 to 0.963. As Cumming (1999b) points out, the prediction that a species cannot occur in a given area, even *A. hebraeum* in the Sahara desert, is as valid as the prediction that it can occur elsewhere; neither the effect of adding more positives nor of adding more negatives is necessarily a problem, provided that it is recognized as statistical rather than biological and so does not lead to the inaccurate acceptance or rejection of hypotheses.

It is important, however, to include in the training set positive and negative records from the full geographical range of a species if statistics are not to give a misleading impression of the reliability of a predictive map. Predicted distributions of *I. scapularis* in North America and *I. ricinus* in Europe, with <6.5% false positives or negatives in each case, have been generated by the geostatistical technique of cokriging (Estrada-Peña, 1998, 1999). Kriging provides a means of interpolating values for unrecorded locations and calculating a measure of variance around the estimated values, while cokriging allows predictions of a sparsely sampled variable (here, the presence of ticks) on the basis of its cross-correlation with a more frequently sampled variable (here, NDVI and ground temperature derived from NOAA-AVHRR satellite data). The predictive map for North America (showing habitat suitability beyond 60°N, west of the Great Slave Lake in Canada) extended far beyond the area of the training set of presence/absence records, so that accuracy statistics based on that training set do not reflect the reliability of the whole predicted map. Furthermore, the data were not clustered on either continent, on the explicit assumption that the cross-covariance matrix between the sparsely sampled and frequently sampled data (i.e. the ticks' response to the main climatic features) remains the same over the entire area of the predictions. This assumption needs to be investigated urgently.

3. PREDICTING THE RISK OF TICK-BORNE INFECTIONS

As Kitron (1998) wrote:

> The epidemiology of vector-borne zoonoses should be viewed as a matrix, which includes dynamic interactions of populations of vectors, humans and/or domestic animals, reservoir hosts, pathogens in any of these vertebrate populations, and environmental determinants of transmission.

While such complexity may indeed be disentangled and its component parts displayed by powerful spatial statistics within a GIS, computer power must be balanced by clear insight into the biology of these interactions. With the identification of the essential key factor that unlocks a full explanation, the complexity becomes more transparent. Then one can take the explanations drawn from the biology and the environmental surfaces, and apply them to future scenarios of environmental change. The two most significant vector-borne infections of the temperate northern hemisphere, Lyme disease and tick-borne encephalitis (TBE), illustrate the non-uniform nature of the epidemiological puzzles to be solved.

3.1. Lyme Disease

To date, almost all landscape epidemiological studies of tick-borne pathogens have concerned Lyme disease in USA, mostly over areas of a few thousand square kilometres (Glass et al., 1995, in Baltimore, Maryland; Nicholson and Mather, 1996, for Rhode Island; and Maupin et al., 1991; Dister et al., 1997, for Westchester county, New York), although Kitron and Kazmierczak (1997) covered the whole of Wisconsin state (c.250 000 km²). Glass et al., (1995) and Dister et al., (1997) included data derived from Landsat-TM images in their GISs, alongside other, conventionally mapped landscape features, while Kitron and Kazmierczak (1997) used AVHRR-NDVI as the sole environmental variable.

The spirochaetes *Borrelia burgdorferi s.l.* that cause Lyme disease have a very high transmission potential, largely because of the long duration of infectivity, several months at least, in the reservoir hosts (Randolph et al., 1996). Consequently, more or less wherever there are competent vector ticks, enzootic cycles of Lyme spirochaetes persist and pose a certain risk of infection to humans. All the above studies concluded that the incidence of Lyme disease cases was, not surprisingly, high in areas of high recorded tick densities (or where environmental features correspond to those known to favour tick survival—Glass et al., 1995). What the GIS analyses established was that many suitable tick habitats now coincide with human habitation, especially in the northeastern states: residential properties close to broad-leaf woodlands ensure high contact rates, and consequent spirochaete transmission, between ticks and humans. The necessary wildlife hosts are abundant in the same habitats (Dister et al., 1997): deer to support the tick population (Wilson et al., 1985), and mice and birds as competent reservoirs (Fish, 1995) for the single genotype of pathogenic Lyme spirochaete, *B. burgdorferi s.s.*, found in USA (Postic et al., 1994). In this robust and relatively simple zoonotic system, therefore, prediction of the primary risk factor reverts to the prediction of tick distributions. At lower latitudes in USA, however, where the seasonal course of the tick's life cycle and host choice both change, simple tick distribution is no longer a sufficient predictor of Lyme disease (U. Kitron, *pers. comm.*). It should also be noted that none of

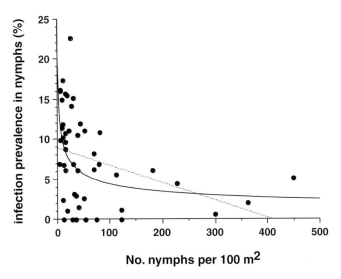

Figure 2 The relationship between the infection prevalence of *Borrelia burgdorferi* *s.l.* in nymphal *Ixodes ricinus* ticks and the density of unfed nymphal ticks sampled from the vegetation at 48 sites in 16 European countries. (Reproduced from Gray *et al.*, 1998, including the original linear relationship (dotted line), $R^2 = 0.12$, $P = 0.0139$. The data are better described by $Y = 22.936X^{-0.355}$, $R^2 = 0.24$, $P < 0.001$ (solid line).)

the statistical methods yet gives estimates of tick abundance, although relative levels may be implied by the varying probabilities with which ticks are predicted to be present in any one place: the greater the probability, the greater the suitability of the habitat and therefore the higher the tick abundance is likely to be.

In the temperate Old World, much greater biotic diversity within Lyme disease systems results in a spatially variable risk of infection: infection prevalence in unfed nymphal ticks varies from zero to *c.* 23% largely independently of tick density (Gray *et al.*, 1998) (Figure 2). The two major elements of biotic diversity are the genetic diversity of *B. burgdorferi s.l.* (Postic *et al.*, 1994) and the tick's wide range of hosts (Milne, 1949). Many of these hosts contribute to transmission of spirochaetes, but do so in different ways because they are differentially competent to transmit the different genospecies of *B. burgdorferi s.l.* to ticks (Kurtenbach *et al.*, 1998a,b), and feed different fractions of the tick population. Thus, for example, in some woodlands in southern Britain, pheasants feed such a large proportion of the nymphal tick population that they apparently inhibit the circulation of the mammal-specific *B. afzelii* (associated with dermatological disorders), and permit a significant prevalence only of *B. garinii* (associated with neurological symptoms) and the apathogenic *B. valaisiana*. As very few larvae feed on pheasants, infection prevalence in unfed nymphs in these woods is very low (2.6%). Amplification of infection occurs principally in nymphs feeding on pheasants, giving a high infection prevalence (16%) only in unfed adult ticks, which are less abundant than nymphs and therefore pose less of a risk to humans. Elsewhere, high infection prevalences of *B. afzelii* may be found in larvae and nymphs that have fed on rodents (Humair *et al.*, 1993; Humair and Gern, 1998) and therefore in unfed nymphs (5–34%) that may bite humans. Until we have identified environmental markers for the key elements of this biotic diversity, we are unlikely to produce reliable predictive risk maps for this complex zoonosis.

3.2. Tick-borne Encephalitis

Although relying on the same vector ticks as do Lyme spirochaetes, and some of the same ubiquitous hosts (mice of the genus *Apodemus*), TBE virus shows a very different geographical pattern. It is limited to discrete foci within central Europe, the Baltic States and the Russian Federation (Immuno, 1997), and is absent in many regions where both ticks and hosts abound (Labuda and Randolph, 1999). Only where infected nymphs feed on rodents alongside large numbers of infectible larvae (Randolph *et al.*, 1996, 1999) can persistent circulation of the TBE virus be achieved by the transmission of non-systemic infections between co-feeding ticks (Labuda *et al.*, 1993). This requires a particular synchronization of the two life stages that only occurs in some areas of the tick's range (Randolph *et al.*, 2000a).

The mapped European foci (Immuno, 1997) were analysed by logistic regression (Randolph *et al.*, 2000b) using DEM and layers of Fourier-processed (Rogers, this volume) 8×8 km resolution NOAA-AVHRR satellite imagery from the NASA Pathfinder programme (James and Kalluri, 1994) as predictor variables. In general, monthly satellite images were produced by maximum value composition (MVC) of daily data (Hay *et al.*, this volume). In addition, the daily land surface temperature (LST) data were filtered using the cloud mask information of the Pathfinder data, and averaged to give a mean monthly LST image as well as the maximum LST from the MVC exercise. To accommodate the wide latitudinal range and considerable landscape heterogeneity within the area of interest (outlined in Figure 5), the analysis was performed on the central European and Baltic regions separately, within which five (central Europe) or four (Baltic) clusters were distinguished, largely on the basis of elevation. Within each eco-climatic zone, half of all TBE-present and TBE-absent pixels were used as a training set, with stepwise inclusion of significant variables. The remaining 50% of the pixels were used to test the prediction of TBE distribution. The threshold posterior probabilities used to assign each pixel to a presence or absence class were decided objectively for each eco-climatic zone to maximize the goodness of fit as measured by the kappa coefficient of agreement (Robinson, this volume). In this case, the overall kappa coefficients for the training and test sets in each region were not significantly different (central Europe: training set $\kappa = 0.45$, test set $\kappa = 0.43$; Baltic: training set $\kappa = 0.79$, test set $\kappa = 0.78$).

The resultant predictive map of TBE risk corresponds closely to the recorded foci both visually (Figure 5) and statistically (82 and 90% correct in central Europe and the Baltic region, respectively). It captures the overall boundaries of the limited extent of TBE virus within the two regions, and much of the fine detail of the patchy distribution within central Europe: the crescentic focus round the eastern end of Austria, with a gap north and east of Vienna; the two major foci in the Czech Republic, in the south-west (Bohemia) and south-east (Moravia), with absences in a corridor along the German border and through the north of the country; the string of foci along the Slovak/Hungarian border, extending south-westwards through western Hungary into northern Croatia and Slovenia; the major foci in southern Germany. Many of the false positives coincide with regions where infection has been recorded but not fully mapped, e.g. along the Slovak/Hungarian border and in the far south-east of the Czech Republic (M. Labuda, *pers. comm.*), in the Carpathian foothills in Romania (G. Nicolescu, *pers. comm.*) and in Germany between and north of the major foci (Immuno, 1997). Of the relatively few false negatives, many reflect the fact that the mapped polygons, drawn round administrative districts from which TBE virus infection has been reported, do not signify the presence of enzootic cycles in all encompassed pixels. Within the major Bohemian focus in the Czech Republic, for example, there are micro-foci that correspond with the vegetation types suitable for good tick survival, detectable by high spatial resolution (30×30 m) Landsat-TM imagery (Daniel *et al.*, 1998).

The five predictor satellite variables that contribute most significantly to the continental map of TBE risk shed light on the biological basis for the spatial patterns. They are the annual amplitude of NDVI, and various factors (e.g. amplitudes, phases) of a land surface temperature (LST) index (Price, 1984) and middle infrared radiation from AVHRR channel 3 which together characterize the seasonal profile of ground temperature (Randolph *et al.*, 2000b). For these images, annual NDVI amplitude is strongly correlated with maximum NDVI, which is the single most important variable indicating general suitability of the habitat for *I. ricinus* populations (Estrada-Peña, 1999). This reflects the tick's requirement of high humidity at ground level for long-term survival and active host-seeking activity, which is typically found in deciduous woods (Daniel *et al.*, 1998) characterized by high annual NDVI amplitude (Defries and Townshend,

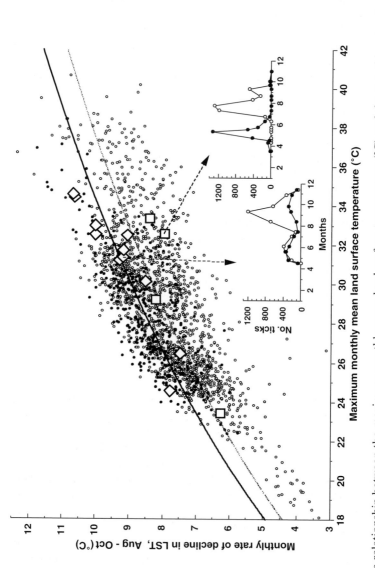

Figure 3 The relationship between the maximum monthly mean land surface temperature (°C) and the monthly rate of autumnal cooling from August to October for 418 TBE-present (●) and 1 574 TBE-absent (○) pixels from a grid across Europe and the Baltic States. TBE present, $Y = 18.152 \log X - 17.878, r = 0.777$; TBE-absent, $Y = 17.565 \log X - 17.647, r = 0.777$. Superimposed are TBE-present (◇) and TBE-absent (□) sites at which field data reveal contrasting patterns of seasonal dynamics of the tick *Ixodes ricinus*, as illustrated in the insets—larvae (○) and nymphs (●). (Redrawn from Randoph *et al.*, 2000a.)

1994). Meanwhile, the particular seasonal profile of LST characteristic of TBE foci determines tick development rates and therefore the pattern of seasonal population dynamics. TBE virus occurs predominantly where the rate of autumnal cooling is above the average relative to peak summer temperatures (Figure 3). This form of seasonal LST profile is associated with the high degree of synchrony in the feeding activity of larval and nymphal ticks, that is seen consistently within, but not outside, TBE foci (Randolph et al., 2000a). It is proposed that rapid autumnal cooling inhibits host-seeking activity of larvae that emerge from eggs laid during the summer, sending them into behavioural diapause (Belozerov, 1982) until reactivated together with the nymphs by rising temperatures in the spring. Under other temperature regimes, either larvae could complete their feeding in the autumn, or eggs would pass the winter in developmental diapause (Belozerov, 1982). This would result in a delay in larval activity (Figure 3) while the eggs hatch the following year, throwing the larvae and nymphs out of synchrony.

As the risk of this disease is not simply related to density of ticks, but is intimately linked with their seasonal population dynamics, both vegetation and thermal indices are necessary to predict the specific conditions required for virus circulation. The clear links between the critical satellite signals and the biological processes underlying the spatial pattern of TBE virus distribution provide a firm causal basis for the predictive risk map.

4. CONCLUSIONS AND FUTURE DIRECTIONS

Compared with the array of studies already in the literature for insect-borne disease systems, especially for pathogens vectored by mosquitoes (Hay et al., this volume), the use of GISs and RS information in the study of tick-borne pathogens has yet to blossom following its early start (Hugh-Jones and O'Neil, 1986). The principal, however, is now well established. Studies to date have shown that both local tick habitats and broad-scale species ranges may be identified and mapped successfully within GISs, using a combination of ground-based and RS information. The increasingly sophisticated statistical packages used to match areas of tick presence with environmental factors tend to give seductively good fits between predictions and observations, which are in danger of outstripping our understanding of biological processes on the ground. Nevertheless, the single most consistent predictor variable shown to be important in these exercises does make sense biologically. Types of vegetation and the related index of photosynthetic activity (NDVI), identifiable from satellite images at spatial resolutions from a few tens of metres (e.g. Landsat TM imagery) to a few kilometres (e.g.

NOAA-AVHRR imagery) are associated with essential moisture availability to ticks. Like all terrestrial arthropods, ticks are susceptible to moisture stress during the free-living phases of their life cycles, and NDVI has been directly correlated with tick mortality rates in Africa. Hitherto only a small number of tick species has been subjected to these sorts of analyses. It would be interesting to investigate some of the more dry-adapted species, such as those of the genus *Hyalomma*. NDVI may prove to be an equally robust predictor for such species, but perhaps with oppositely signed correlations, as with contrasting tsetse species (Rogers and Randolph, 1991).

As academic exercises, these studies show considerable promise in developing our understanding. As practical tools, however, we must ask, what use are the predictions to be gained from GIS exercises, with or without RS data? The answer depends on the spatial scale at which those predictions are made. Maps of local risk may be used on the ground to facilitate avoidance action and control operations. This is especially true for pests such as ticks, with their limited mobility and therefore highly focal distribution within well-defined eco-zones (Daniel *et al.*, 1998). Such maps, however, give no indication of the full geographical extent of the threat posed by each tick species or tick-borne pathogen. Predictions over regional or continental areas are too coarse-scale to direct tactical control, but are essential for strategic economic planning and the targeting of national resources within veterinary and public health services.

To generate the ideal integration between fine- and coarse-scale predictions over local and continental areas, both for the present and for the future, biological process-based models of both tick populations and pathogen transmission dynamics are essential. They will provide the necessary link, as the same fundamental biological processes operate across all scales and, we imagine, into the future. Quantitative explanations and predictions generated from intensive studies, preferably of natural populations, may be extrapolated to extensive patterns. Recognizing the overriding importance of climatic factors in driving the natural dynamics of tick-borne disease systems is a crucial step forward towards this goal. In this process, the danger of applying inappropriate quantitative relationships between ticks and abiotic factors derived from one study site, to the same species in distant places where interactions may differ quantitatively and even qualitatively, cannot be ignored. The generic population model for *R. appendiculatus* in Africa, however, gives accurate predictions of the tick's seasonal population dynamics from the north of its range in Uganda to the south in the Cape of South Africa, with only minor adjustments for locally varying determinants (Randolph and Rogers, 1997).

With increasingly accurate use of RS data in place of relevant seasonal climatic information (Green and Hay, 2000; Goetz, this volume) to drive these biological models, we shall be less constrained by the availability of appropriately recorded input variables. Arthropods are particularly affected by near-surface air temperature and saturation deficit. Clever manipulations of an expanding array of satellite signals may capture the subtle environmental limitations on tick population performance better than can standard meteorological records. At the same time, greater understanding of the impact of purely climatic variables must be developed for future predictions based on climate change scenarios. Satellite imagery cannot go beyond the present, but it is ideal for real-time monitoring of any changes up to the present, and for checking climate scenarios.

When risk of tick-borne infections is simply related to the density of ticks, as appears to be the case for Lyme disease in northern USA, epidemiological risk may be mapped using the same techniques as for ticks themselves. For all tick-borne pathogens, however, the impact of climatic seasonality on the variable abundance of each tick stage is a critical determinant of the quantitative transmission potential (Randolph, 1998), which will be captured more accurately by biological models than statistical pattern-matching. In some cases, the impact of seasonality is qualitative, depending on the existence or not of enzootic cycles. For TBE, for example, the combination of RS NDVI and temperature indices successfully predicts elements of the temporal dynamics of the system, which translate into a spatial pattern of foci on a continental scale. Given the precise conditions required for the circulation of such pathogens (Section 3.2), reliable predictions of future change in risk depend on a greater knowledge of detailed seasonal climate change than is available at present. It is entirely possible that, with the anticipated hotter and drier summers in Europe, enzootic cycles of TBE virus will be disrupted and their spatial distribution diminished (Rogers and Randolph, 2000).

Throughout this chapter, the varying *spatial* scales of epidemiological patterns have been emphasized, while *temporal* dynamics have been addressed mostly at the ecological scale of seasons. Could the same tools and techniques be applied to processes on the far longer time scale over which evolution occurs? Can we match the new and improving, but hitherto purely descriptive, phylogenies that are emerging from molecular-based analyses, with GISs and RS data to investigate the environmental context of the evolution of vector-borne pathogens? If climate is the major limiting factor on vector-borne pathogen systems, as appears to be the case, climate is likely to have been a significant factor driving and constraining the evolution of these pathogens. Our new ability to characterize global environmental patterns with RS data and to match these to statistical patterns and biological processes within GISs, should allow us to predict, not necessarily the appearance of new pathogens in the future, but renewed appearances and spread of existing pathogens.

ACKNOWLEDGEMENTS

I gratefully acknowledge valuable insights and advice from David Rogers, Robert Green, Simon Hay, Martin Hugh-Jones and Uriel Kitron. Robert Green made a major contribution to the spatial analysis of tick-borne encephalitis presented here. This review was written while the author was a Wellcome Trust Senior Research Fellow in Basic Biomedical Science.

REFERENCES

Belozerov, V.N. (1982). Diapause and biological rhythms in ticks. In: *Physiology of Ticks* (F.D. Obenchain and R. Galun, eds.), pp. 469–500. Oxford: Pergamon Press.

Cerny, V. (1988). *Long Term Change of* Ixodes ricinus *Occurrence Influenced by the Intensive Forest Economic Activities*. Research Report no. VI-2-3/4. Ceské Budejovice: Institute of Parasitology, Czechoslovakian Academy of Sciences.

Cumming, G.S. (1998). Host preference in African ticks (Acari: Ixodidae): a quantitative data set. *Bulletin of Entomological Research* **88**, 379–406.

Cumming, G.S. (1999a). Host distributions do not limit the species ranges of most African ticks (Acari: Ixodidae). *Bulletin of Entomological Research* **89**, 303–327.

Cumming, G.S. (1999b). The evolutionary ecology of African ticks. D.Phil. thesis, Oxford University.

Daniel, M. and Kolár, J. (1990). Using satellite data to forecast the occurrence of the common tick *Ixodes ricinus* (L.). *Journal of Hygiene, Epidemiology, Microbiology and Immunology* **34**, 243–252.

Daniel, M., Kolár, J., Zeman, P., Pavelka, K. and Sádlo, J. (1998). Predictive map of *Ixodes ricinus* high-incidence habitats and a tick-borne encephalitis risk assessment using satellite data. *Experimental and Applied Acarology* **22**, 417–433.

Defries, R.S. and Townshend, J.R.G. (1994). NDVI-derived land cover classification at a global scale. *International Journal of Remote Sensing* **15**, 3567–3586.

Dister, S.W., Fish, D., Bros, S.M., Frank, D.H. and Wood, B.L. (1997). Landscape characterization of peridomestic risk for Lyme disease using satellite imagery. *American Journal of Tropical Medicine and Hygiene* **5**, 687–692.

DoRosario, V.E. and Coluzzi, M. (1997). Advances in the study of Afrotropical vectors. *Annals of Tropical Medicine and Parasitology* **91**, S125–S126.

Estrada-Peña, A. (1998). Geostatistics and remote sensing as predictive tools of tick distribution: a cokriging system to estimate *Ixodes scapularis* (Acari: Ixodidae) habitat suitability in the United States and Canada from advanced very high resolution radiometer satellite imagery. *Journal of Medical Entomology* **35**, 989–995.

Estrada-Peña, A. (1999). Geostatistics as predictive tools to estimate *Ixodes ricinus* (Acari: Ixodidae) habitat suitability in the western Palearctic from AVHRR satellite imagery. *Experimental and Applied Acarology* **23**, 337–349.

Fielding, A.H. and Bell, J.F. (1997). A review of methods for the assessment of prediction errors in conservation presence/absence models. *Environmental Conservation* **24**, 38–49.

Fish, D. (1995). Environmental risks and prevention of Lyme disease. *American Journal of Medicine* **98**, 2S-9S.

Glass, G.E., Schwartz, B.S., Morgan, J.M., Johnson, D.T., Noy, P.M. and Israel, E. (1995). Environmental risk factor for Lyme disease identified with geographic information systems. *American Journal of Public Health* **85**, 944–948.

Gray, J.S., Kahl, O., Robertson, J.N. *et al.* (1998). Lyme borreliosis habitat assessment. *Zentralblatt für Bakteriologie* **287**, 211–228.

Green, P.E. (1978) *Analyzing Multivariate Data*. Hinsdale, IL: Dryden Press.

Green, R.M. and Hay, S.I. (2000). Mapping of climatic variables across tropical Africa and temperate Europe using meteorological satellite data. *Remote Sensing of Environment*, in press.

Hugh-Jones, M.E. and O'Neil, P. (1986). The epidemiological uses of remote sensing and satellites. In: *Proceedings, 4th International Symposium on Veterinary Epidemiology and Economics*, pp. 113–118. Singapore: Singapore Veterinary Association.

Hugh-Jones, M., Barre, N., Nelson, G. *et al.* (1992). Landsat-TM identification of *Amblyomma variegatum* (Acari: Ixodidae) habitats in Guadeloupe. *Remote Sensing and Environment* **40**, 43–55.

Humair, P.-F. and Gern. L. (1998). Relationship between *Borrelia burgdorferi sensu lato* species, red squirrels (*Scuirus vulgaris*) and *Ixodes ricinus* in enzootic areas of Switzerland. *Acta Tropica* **69**, 213–227.

Humair, P.-F., Turrian, N., Aeschlimann, A. and Gern. L. (1993). *Borrelia burgdorferi* in a focus of Lyme borreliosis: epizootiologic contribution of small mammals. *Folia Parasitologica* **40**, 65–70.

Hutchinson, M.F., Nix, H.A., MacMahon, J.P. and Ord, K.D. (1996). *A Topographic and Climatic Database for Africa—Version 1.1*. Canberra: Australian National University.

Immuno Ag (1997). *Tick-borne Encephalitis (TBE) and its Immunoprophylaxis*. Vienna: Immuno Ag.

James, M.E. and Kalluri, S.N.V. (1994). The Pathfinder AVHRR land data set—an improved coarse resolution data set for terrestrial monitoring. *International Journal of Remote Sensing* **15**, 3347–3363.

Kitron, U. (1998). Landscape ecology and epidemiology of vector-borne diseases: tools for spatial analysis. *Journal of Medical Entomology* **35**, 435–445.

Kitron U. and Kazmierczak, J.J. (1997). Spatial analysis of the distribution of Lyme disease in Wisconsin. *American Journal of Epidemiology* **145**, 558–566.

Kitron, U., Bouseman, J.K. and Jones, C.J. (1991). Use of the ARC/INFO GIS to study the distribution of Lyme disease ticks in Illinois. *Preventive Veterinary Medicine* **11**, 243–248.

Kitron, U., Jones, C.J., Bouseman, J.K., Nelson, J.A. and Baumgartner, D.L. (1992). Spatial analysis of the distribution of *Ixodes dammini* (Acari: Ixodidae) on white-tailed deer in Ogle County, Illinois. *Journal of Medical Entomology* **29**, 259–266.

Kruska, R. and Perry, B.D. (1991). Evaluation of grazing lands of Zimbabwe using the AVHRR normalized difference vegetation index. *Preventive Veterinary Medicine* **11**, 363–365.

Kurtenbach, K., Peacey, M.F., Rijpkema, S.G.T., Hoodless, A.N., Nuttall, P.A. and Randolph, S.E. (1998a). Differential transmission of the genospecies of *Borrelia burgdorferi sensu lato* by game birds and small rodents in England. *Applied and Environmental Microbiology* **64**, 1169–1174.

Kurtenbach, K., Sewell, H., Ogden, N.H., Randolph, S.E. and Nuttall, P.A. (1998b). Serum complement sensitivity as a key factor in Lyme disease ecology. *Infection and Immunity* **66**, 1248–1251.

Labuda, M. and Randolph, S.E. (1999). Survival of tick-borne encephalitis virus: cellular basis and environmental determinants. *Zentralblatt für Bakteriologie* **289**, 513–524.

Labuda, M., Nuttall, P.A., Kozuch, O. *et al.* (1993). Non-viraemic transmission of tick-borne encephalitis virus: a mechanism for arbovirus survival in nature. *Experientia* **49**, 802–805.

Lessard, P., L'Eplattenier, R., Norval, R.A.I. *et al.* (1990). Geographical information systems for studying the epidemiology of cattle diseases caused by *Theileria parva*. *The Veterinary Record* **126**, 255–262.

Maupin, G.O., Fish, D., Zultowsky, J., Campos, E.G. and Piesman, J. (1991). Landscape ecology of Lyme diseases in a residential area of Westchester County, New York. *American Journal of Epidemiology* **133**, 1105–1113.

Merler, S., Furlanello, C., Chemini, C. and Nicolini, G. (1996). Classification tree methods for analysis of mesoscale distribution of *Ixodes ricinus* (Acari: Ixodidae) in Trentino, Italian Alps. *Journal of Medical Entomology* **33**, 888–893.

McCosker, P.J. (1979). Global aspects of the management and control of ticks of veterinary importance. *Recent Advances in Acarology* **2**, 45–53.

Milne, A. (1949). The ecology of the sheep tick *Ixodes ricinus* L. Host relationships of the tick. Part 2. Observations on hill and moorland grazings in Northern England. *Parasitology* **39**, 173–194.

Nicholson, M.C. and Mather, T.N. (1996). Methods for evaluating Lyme disease risks using geographic information systems and geospatial analysis. *Journal of Medical Entomology* **33**, 711–720.

Nix, H. (1986). A biogeographic analysis of Australian elapid snakes. In: *Atlas of Elapid Snakes of Australia* (R. Longmore ed.) *Australia Flora and Fauna series*, no. 7, pp. 4–15. Canberra: Australian Government Publishing Service.

Norval, R.A.I., Perry, B.D., Gebreab, F. and Lessard, P. (1991a). East Coast fever: a problem of the future for the horn of Africa? *Preventive Veterinary Medicine* **10**, 163–172.

Norval, R.A.I., Perry, B.D., Kruska, R. and Kundert, K. (1991b) The use of climate data interpolation in estimating the distribution of *Amblyomma variegatum* in Africa. *Preventive Veterinary Medicine* **11**, 365–366.

Norval, R.A.I., Perry, B.D. and Young, A.S. (1992). *The Epidemiology of Theileriosis in Africa*. London: Academic Press.

Norval, R.A.I., Perry, B.D., Meltzer, M.I., Kruska, R.L. and Boothroyd, T.H. (1994). Factors affecting the distributions of the ticks *Amblyomma hebraeum* and *A. variegatum* in Zimbabwe: implications of reduced acaricide usage. *Experimental and Applied Acarology* **18**, 383–407.

Norval, R.A.I., Sutherst, R.W., Kurki, J., Kerr, J.D. and Gibson, J.D. (1997). The effects of the brown-ear tick, *Rhipicephalus appendiculatus* on milk production of Sanga cattle. *Medical and Veterinary Entomology* **11**, 148–154.

Pegram, R.G., Lemche, J., Chizyuka, H.G.B. *et al.* (1989). Effect of tick control on liveweight gain of cattle in central Zambia. *Medical and Veterinary Entomology* **3**, 313–320.

Perry, B.D., Lessard, P., Norval, R.A.I., Kundert, K. and Kruska, R. (1990). Climate, vegetation and the distribution of *Rhipicephalus appendiculatus* in Africa. *Parasitology Today* **6**, 100–104.

Perry, B.D., Kruska, R., Lessard, P., Norval, R.A.I. and Kundert. K. (1991). Estimating the distribution and abundance of *Rhipicephalus appendiculatus* in Africa. *Preventive Veterinary Medicine* **11**, 261–268.

Postic, D., Assous, M., Grimont, P.A.D. and Baranton, G. (1994). Diversity of *Borrelia burgdorferi sensu lato* as evidenced by restriction fragment length polymorphism of rrf(5S)-rrl(23S) intergenic spacer amplicon. *International Journal of Systematic Bacteriology* **44**, 743–752.

Price, J. C. (1984). Land surface temperature measurement for the split window channels of the NOAA 7 advanced very high resolution radiometer. *Journal of Geophysical Research* **89**, 7231–7237

Randolph, S.E. (1994). Population dynamics and density-dependent seasonal mortality indices of the tick *Rhipicephalus appendiculatus* in eastern and southern Africa. *Medical and Veterinary Entomology* **8**, 351–368.

Randolph, S.E. (1997). Abiotic and biotic determinants of the seasonal dynamics of the tick *Rhipicephalus appendiculatus* in South Africa. *Medical and Veterinary Entomology* **11**, 25–37.

Randolph, S.E. (1998). Ticks are not insects: consequences of contrasting vector biology for transmission potential. *Parasitology Today* **14**, 186–192.

Randolph, S.E. and Rogers, D.J. (1997). A generic population model for the African tick *Rhipicephalus appendiculatus*. *Parasitology* **115**, 265–279.

Randolph, S.E. and Rogers, D.J. (2000). Fragile transmission cycles of tick-borne encephalitis virus may be disrupted by predicted climate change. *Proceedings of the Royal Society of London, Series B*, in press.

Randolph, S.E., Gern, L. and Nuttall, P.A. (1996). Co-feeding ticks: epidemiological significance for tick-borne pathogen transmission. *Parasitology Today* **12**, 472–479.

Randolph, S.E., Miklisová, D., Lysy, J., Rogers, D.J. and Labuda, M. (1999). Incidence from coincidence: patterns of tick infestations on rodents facilitate transmission of tick-borne encephalitis virus. *Parasitology* **118**, 177–186.

Randolph S.E., Green, R.M., Peacey, M.F. and Rogers, D.J. (2000a). Seasonal synchrony: the key to tick-borne encephalitis foci identified by satellite data. *Parasitology* **120**, in press.

Randolph S.E., Green, R.M., Labuda M. and Rogers, D.J. (2000b). Satellite imagery predicts and explains tick-borne encephalitis distribution. *Proceedings of the National Academy of Sciences of the USA*, submitted.

Rogers, D.J. and Randolph, S.E. (1991). Mortality rates and population density of tsetse flies correlated with satellite imagery. *Nature* **351**, 739–741.

Rogers, D.J. and Randolph, S.E. (1993). Distribution of tsetse and ticks in Africa: past, present and future. *Parasitology Today* **9**, 266–271.

Rogers, D.J., Hay, S.I. and Packer, M.J. (1996). Predicting the distribution of tsetse flies in West Africa using temporal Fourier processed meteorological satellite data. *Annals of Tropical Medicine and Parasitology* **90**, 225–241.

Sonenshine, D. (1991). *Biology of Ticks*, Vol. 1. Oxford: Oxford University Press.

Sutherst, R.W. and Maywald, G.F. (1985). A computerised system for matching climates in ecology. *Agriculture, Ecosystems and Environment* **13**, 281–299.

Toomer, J. (1996). Predicting pan-African tick distributions using remotely sensed surrogates of meteorological and environmental conditions. Unpublished BA project report, University of Oxford.

Walker, J.B. (1987). The tick vectors of *Cowdria ruminantium* (Ixodoidea, Ixodidae, genus *Amblyomma*) and their distribution. *Onderstepoort Journal of Veterinary Research* **54**, 353–379.

Wilson, M.L. (1998). Distribution and abundance of *Ixodes scapularis* (Acari: Ixodidae) in North America: ecological processes and spatial analysis. *Journal of Medical Entomology* **35**, 446–457.

Wilson, M.L., Adler, G.H. and Speilman, A. (1985). Correlation between abundance of deer and that of the deer tick, *Ixodes dammini* (Acari: Ixodidae). *Annals of the Entomological Society of America* **78**, 172–176.

The Potential of Geographical Information Systems and Remote Sensing in the Epidemiology and Control of Human Helminth Infections

S. Brooker and E. Michael

Wellcome Trust Centre for the Epidemiology of Infectious Disease, Department of Zoology, University of Oxford, South Parks Road, Oxford OX1 3FY, UK

ADVANCES IN PARASITOLOGY VOL 47
0065-308-X $30.00

ABSTRACT

Geographic information systems (GIS) and remote sensing (RS) technologies are being used increasingly to study the spatial and temporal patterns of infectious diseases. For helminth infections, however, such applications have only recently begun despite the recognition that infection distribution patterns in endemic areas may have profound effects on parasite population dynamics and therefore the design and implementation of successful control programmes. Here, we review the early applications of these technologies to the major human helminths (geohelminths, schistosomes and the major lymphatic filarial worms), which demonstrate the potential of these tools to serve as: (1) an effective data capture, mapping and analysis tool for the development of helminth atlases; (2) an environment for modeling the spatial distribution of infection in relation to RS and environmental variables, hence furthering the understanding of the impact of density-independent factors in underlying observed parasite spatial distributions and their effective prediction; and (3) a focal tool in parasite control programming given their abilities to (i) better define endemic areas, (ii) provide more precise estimates of populations-at-risk, (iii) map their distribution in relation to health facilities and (iv) by facilitating the stratification of areas by infection risk probabilities, to aid in the design of optimal drug or health measure delivery systems. These applications suggest a successful role for GIS/RS applications in investigating the spatial epidemiology of the major human helminths. It is evident that further work addressing a range of critical issues include problems of data quality, the need for a better understanding of the population biological impact of environmental factors on critical stages of the parasite life-cycle, the impacts and consequences of spatial scale on these relationships, and the development and use of appropriate spatially-explicit statistical and modeling techniques in data analysis, is required if the true potential of this tool to helminthology is to be fully realized.

1. INTRODUCTION

Estimates suggest that some 1273 million people worldwide are infected with *Ascaris lumbricoides*, 902 million with *Trichuris trichiura*, 1277 million with hookworm (Bundy *et al.*, 2000), 200 million with schistosomiasis (Savioli *et al.*, 1997) and approximately 120 million with lymphatic filariasis (Michael *et al.*, 1996). The impact of these infections on the nutrition, education, development and productivity of individuals (Stephenson, 1987; Watkins and Pollitt, 1997), together with the advent of cheap and effective anthelmintic drugs, has revived global interest in their control (Warren *et al.*, 1993; Ottersen *et al.*, 1997; Savioli *et al.*, 1997). While major control efforts are now being launched, however, the challenge of how to deliver drugs to affected communities in a cost-effective manner continues.

One aspect of effective community-based control, namely the role and impact of helminth population dynamics, has been extensively studied and is remarkably well understood (Anderson and May, 1991). This work has traditionally focused on changes in patterns with time and host age, and has been fundamental to furthering our understanding of the epidemiology of helminth infections and in predicting and evaluating the impact of control. Infection and disease, however, are also distributed in space. It is increasingly recognized that such spatial patterns may have profound effects on parasite population dynamics (Kareiva, 1990) and that an understanding may aid the rational design and implementation of parasite control programmes. In particular, endemic areas may be defined, population at risk may be estimated, areas may be stratified by risk or prevalence, optimal control delivery systems may be designed and their efficacy monitored (Mott *et al.*, 1995; WHO, 1999).

Despite this, very little is known about the spatial distribution of the major helminth infections in endemic regions or of the factors influencing their distributions. This has been due in part to the difficulty of storing, processing and presenting geographical data on helminth infection (Mott *et al.* 1995; Openshaw, 1996; Nuttall *et al.*, 1998). Today, these problems have been resolved by geographical information systems (GIS), which not only allow spatial information to be stored and analysed on a desktop computer, but also, by facilitating the integration of remotely sensed data, allow the investigation of helminth co-distribution with environmental variables at various spatial scales (Openshaw, 1996).

The primary aim of this chapter is to assess the potential of the new technologies of GIS and remote sensing (RS) in the epidemiology and control of helminth infections, drawing upon recent work carried out in this area for geohelminthiases, schistosomiasis and lymphatic filariasis. First, we describe the use of GIS as a tool for capturing existing infection data in the development and analysis of descriptive maps of helminth infections, highlighting both the usefulness and drawbacks of such an approach. In particular, we note the need to determine infection probability in areas for which little or no data exist. In order to demonstrate the potential of using RS data for predicting helminth distribution in areas without survey data, we review the available experimental and field evidence regarding the importance of environmental factors in determining the transmission biology of each of the major helminth species. Particular attention is paid both to the spatial scales at which these factors act on the parasite life-cycle and the complexity of the life-cycle itself, both of which have implications

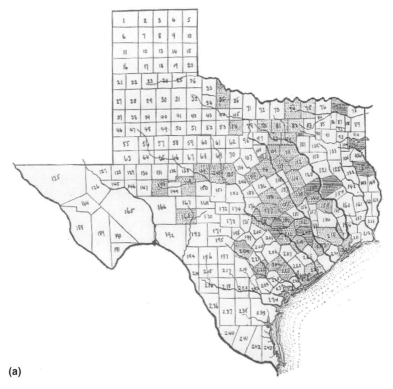

Figure 1 (a) The geographic distribution of hookworm (uncinariasis) in Texas. (Reproduced from Smith, 1903.) (b, opposite) Map of east Texas showing the prevalence of hookworm for the years 1912–1914 (upper numbers) and 1933–1944 (lower numbers). (Reproduced from Scott, 1945.)

for using RS, and other environmental data, for modelling spatial distributions of infection at scales relevant to control programming. We then proceed to consider recent applications of RS to modelling helminth infection distribution. Finally, we focus on the potential public health applications of GIS and RS approaches for the major helminths. The focus is predominantly on work carried out in Africa, as the priority for helminth control is greatest for this continent (Savioli *et al.* 1997).

2. EARLY CARTOGRAPHY OF HELMINTH INFECTIONS

Mapping the spatial distribution of helminth infections has a surprisingly long tradition. In papers published in 1903, Stiles and Smith separately

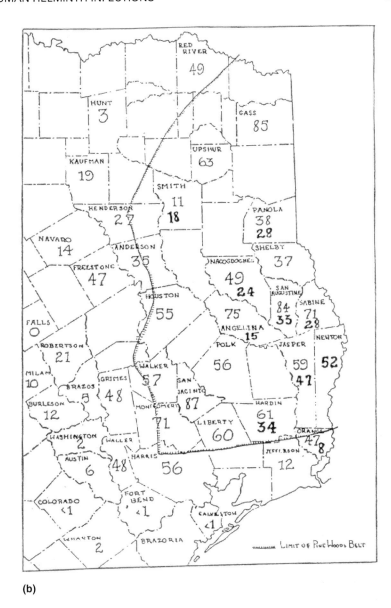

(b)

presented maps of Texas, which displayed the prevalence of hookworm infection (Figure 1a). Their maps demonstrated that infection was typically restricted to the eastern part of the state where the soils were sandiest. A further map of the distribution derived from a hookworm survey of Texas schoolchildren confirmed this pattern, and showed that hookworm was restricted to the pine belt area, where the soil was also sandy (Scott, 1945) (Figure 1b). These maps, and their interpretation, provide some of the earliest examples of deriving inferences about the relationship between the environment and helminth disease. Early impetus for mapping helminth infection arose, as now, from a vital need for reliable information concerning infection distributions for the purposes of enacting parasite control. Not surprisingly, the impact of the 1910–1914 Rockefeller hookworm campaigns was lowest in areas where soil and climate are favourable for the parasites, and most effective in reducing transmission in less favourable environments (Keller *et al.*, 1940).

A feature of these early studies was the use of data from single large-scale surveys covering a wide geographical area. Such dedicated surveys have become less common, especially in Africa, where large-scale surveys have been conducted in only a few countries and data for the majority are normally scattered through the literature. The utility of capturing such data is illustrated by the seminal *Atlas of the Global Distribution of Schistosomiasis* (Doumenge *et al.*, 1987), which maps the occurrence of schistosomiasis in 76 countries worldwide, along with tables of observed prevalence for each study location. These traditional cartographic approaches are especially important for advocacy and for laying the foundations for the epidemiological understanding of helminths, but have the major disadvantage that the derived maps cannot easily be updated, and comparison between different maps is difficult. Modern GIS offer solutions to both of these problems (Burrough and McDonnell, 1998).

3. USING GIS TO MAP THE GEOGRAPHIC DISTRIBUTION OF HELMINTHS

Previous estimates of human helminth infection prevalence have typically been made at the national level, by extrapolating prevalence data from the few available studies to the country as a whole (Crompton and Tulley, 1987; Utroska *et al.*, 1989; Bundy *et al.*, 1991). While such an approach has proved effective in advocacy, as demonstrated by the seminal article 'This wormy world' by Stoll (1947), it is of limited practical relevance to describing and understanding the spatial distribution of infection within and/or between countries or to the targeting of control efforts. The absence of detailed spatial information on helminth infection has been partly addressed by recent GIS based mapping initiatives which aim to describe prevalence patterns and to highlight areas where further information is required (Michael and Bundy, 1997; Brooker *et al.*, 2000a; Bundy *et al.*, 2000).

3.1. Schistosomiasis and Geohelminthiases

A current WHO initiative is collating information from the literature on the prevalence of infection with the two major schistosome species in sub-Saharan Africa (*Schistosoma haematobium* and *S. mansoni)* and the major intestinal nematodes (hookworms, *A. lumbricoides* and *T. trichiura*). Currently, the database incorporates 583 references, which covers 3486 independent cross-sectional surveys conducted since 1970 (see Brooker *et al.*, 2000a for further details). A map of average population density (Deichmann, 1996) masks 336 administrative areas (14.6%) or 948 administrative areas (41.1%) with an average population density less than 5 or less than 20 persons per square km, respectively. Calculated on this basis, the current atlas provides information on schistosomiasis for 33.0% of administrative areas with population density greater than 5 persons per square km, and for geohelminths in 16.5% (see Plate 11). Such maps provide the most detailed data available on the geographical distribution of helminth infection in Africa. When combined with information on the distribution of population in relation to basic health and social infrastructure, the information could provide an operational tool for planning, monitoring and managing public health programmes (see Section 6).

3.2. Lymphatic Filariasis

The global geographical distribution of filariasis prevalences (both infection and chronic disease cases combined), estimated for 96 endemic countries in the 1993 World Bank Global Burden of Disease (GBD) study (Michael *et al.*, 1996), shows different spatial patterns for Bancroftian and Brugian filariasis (Figure 2a,b) (see Section 4.3 for the biology of these parasites). In general, there is a strong regional variation in the prevalence of Bancroftian filariasis, with higher and more variable infection levels in the sub-Saharan and Pacific Islands compared with Asia and South America. The prevalence for Brugian filariasis appears to be relatively more uniform, although there is a slightly higher prevalence in eastern regions. Direct comparisons using crude health maps, however, may be complicated by random variation in the observed rates, especially when the underlying population sizes vary considerably (Bailey and Gatrell, 1995). Poisson probability mapping (Cliff and Hagett, 1988) was therefore used to plot the statistical significance of the difference between risk of Bancroftian filariasis in each study area and the average risk over the entire map. This procedure not only stabilizes the individual prevalence rates for population size variations but also provides a tool for highlighting truly anomalous areas (Bailey and Gatrell, 1995). The resulting global probability map for Bancroftian filariasis (Figure 3) shows reduced

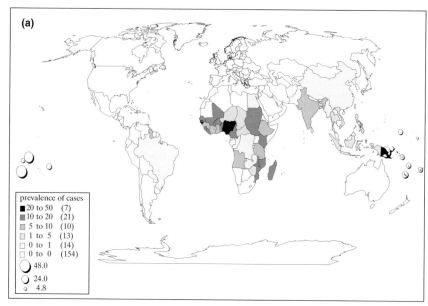

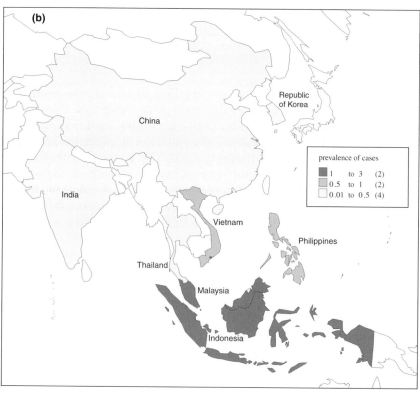

heterogeneity in the 'between-country' distribution of cases, but also confirms the impression from Figure 2 that the underlying case rate for the disease is not constant across the world. This adjusted pattern differs from the traditional view, which considered the Indian sub-continent to be the most important region for filariasis (Sasa, 1976; WHO, 1984), although it is clear that India remains an important endemic area (Figure 3). This result has important implications for planning control strategies as it argues for a geographically targeted strategy for control. This conclusion would be further strengthened by information on infection and disease at regional or district levels. Present work in collaboration with the WHO aims to initiate the first stage of this process by establishing such a GIS platform.

3.3. Limitations of the Empirical Mapping Approach

Despite the important contribution made by current atlases to describing and analysing the spatial distribution of helminth infections, there are a number of limitations to the approach, not least the inherent variability in such survey data. As Stoll (1947) stressed:

> I need scarcely remind you of what some of those hurdles are: so many parasitological surveys of but small numbers of people, frequently by other design than to represent fair samples of an area, done by workers of varying aims and by techniques of even more variable efficiency in relation to the task at hand.

For lymphatic filariasis, for example, differences in survey methods and timings must be corrected to provide truly comparable data to represent infection prevalence precisely in every district (Michael and Bundy, 1998).

A second drawback is the absence of detailed age-stratification in the data to take account of the age dependency in infection levels. Note that, for the schistosomiasis and geohelminth atlases, however, most of the surveys (56.7%) were carried out in pre-school children, and so provide estimates for a moderately well defined age group. This enhances the comparability of the data, and the same results can also be used to predict prevalence among pre-school children and adults, and therefore among the total community (Guyatt et al., 1999).

Figure 2 Geographical distribution of Bancroftian (a) and Brugian (b) filariasis case prevalences based on the crude global burden of disease estimates. Circles denote the corresponding prevalences (%) estimated for various Pacific Islands and vary in size proportionately with the prevalence on each island. The figures in parenthesis indicate the number of countries.

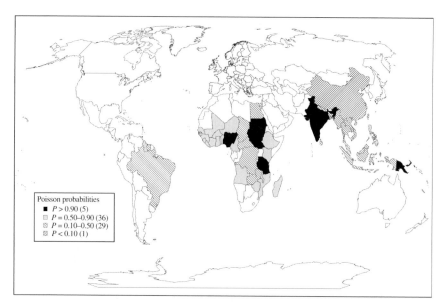

Figure 3 Global Poisson probability map for Bancroftian filariasis case prevalences. The map shows $P_i >$ mean values calculated using the methods described by Michael and Bundy (1997), and may be interpreted by considering that there is a high probability ($P > 0.90$) that the prevalence estimated in each black area is higher than the mean global value (MGV); there is 'equivocal evidence' that the risk of each dark shaded area is higher than the MGV ($P = 0.5–0.9$) and that of each light shaded area is lower than the MGV ($P = 0.1–0.5$); and finally, there is a high probability that the risk of each medium shaded area is lower than the MGV ($P < 0.1$). (Case probabilities for all other endemic Pacific Island countries lay between 0.5 and 0.9.)

Finally, the data are mapped within defined administrative boundaries because of simplicity in data handling, and because control approaches are increasingly implemented at the district level. Such assignment, however, belies the inherent spatial heterogeneity of infection. A more useful unit of analysis might be based on the precise location of communities and the number of people in a given area, or on ecological zones without regard to administrative boundaries. Information on infection per district also varies considerably, ranging from absence of data, one or a few surveys, to an abundance of records (Michael *et al.*, 1996; Brooker *et al.*, 2000a). This is an area where integrated applications of GIS and RS to fill the gaps in empirical data, could be important in deriving more reliable maps of helminth distribution.

4. POTENTIAL OF GIS AND REMOTE SENSING FOR PREDICTING HELMINTH DISTRIBUTION

It has long been known that the transmission of infectious diseases is influenced by many natural and human factors, including geography, climate, human demography and intervention. For helminths, mention has been made of the effect of environment as a density-independent factor affecting parasite transmission, as measured by the basic reproductive number (R_0) (Anderson and May, 1991), and thus observed patterns of infection and disease. The environment may influence the population biology of helminths by altering and affecting intermediate host abundance, infective stage longevity, infectivity and parasite development either in the external habitat or within poikilothermic intermediate hosts such as snails or mosquitoes (Pavlovsky, 1966). These findings suggest that, as for other infectious diseases, predicting the large-scale distribution of helminths based on increasingly available environmental and RS information is justified.

A common approach used to quantify the relationship between the environment and the spatial distribution of infectious diseases is to investigate the statistical correlations that exist between them (Rogers and Randolph, 1991; Rogers and Williams, 1993; Hay et al., 1998). Such analyses have value in identifying significant relationships, which may vary from place to place. For example, Rogers and Randolph (1993) were able to predict the distributions of Glossina morsitans in Zimbabwe, Kenya and Tanzania to an overall accuracy of 82% from a discriminant analysis of several sets of RS-based environmental data, and found that the key variables contributing most to the prediction varied between the study regions. If prediction is to be more general, it is important to develop biological understandings of the observed statistical relationships (Rogers and Williams, 1993; Hay et al., 1997; Rogers, this volume).

When considering the environmental factors that may influence the distribution of helminth infection, it is helpful to consider the impact of parasite life-cycles, as these have a profound influence on the geographic distribution of infection (Shope, 1999). For example, an infectious agent transmitted between people may persist anywhere people travel, and thus have a very wide geographic distribution, affected more by human behaviour than by climate. By contrast, three-factor host–parasite complexes involve transmission through a vector or intermediate host, and are immediately limited geographically by the distribution of the vector or human component of the complex (Shope, 1999). For the human helminths, these considerations would suggest a relatively broader geographical range for the direct life-cycle geohelminths (restricted only by human habitation, behaviour and factors influencing parasite egg survival in the external environment) compared with a more focal spatial distribution for the vector-borne schistosome and filarial parasites (limited by both human and vector populations and climatic effects on the parasite within the vector host), thereby providing some indication of likely factors and patterns of spatial structure to be expected when investigating the geographical distributions of these parasites.

A more important issue central to using RS data to reveal ecological relationships between parasite distributions and the environment relates to the problem of spatial scale (Wiens, 1989; Levin, 1992; Allen and Hoekstra, 1992). The problem is that many biological responses are scale-dependent (Wiens, 1989), such that direct extrapolations from experimental data on the impact of biotic factors on infection processes to explain distribution in nature is realistic only if small scale processes can be scaled up meaningfully to predict spatial patterns on a large spatial scale (Levin and Pacala, 1997).

The issue of scale is also fundamental operationally for filling in gaps in the current GIS atlases. There are two main issues. The first is the level at which prediction is required and the second is the information available for making predictions. In the context of helminth control, the rationale for prediction is to provide information on the spatial patterns of infection and disease at the administrative level at which control resources are likely to be mobilized, usually the district level. Furthermore, although RS data are available at fine spatial scales, the satellite systems most widely used in the RS community are those with a broad spatial scale of 1–8 km (Hay, this volume), which are many orders of magnitude greater than the scales at which most ecological studies are carried out.

This does not, however, remove the need for the generalized biological approach to large scale prediction of spatial patterns. Rather, it necessitates the careful consideration of smaller scale observations in order to identify those factors that are also relevant to large scale spatial patterning (Wiens *et al.*, 1993). In this section, we review those factors that may be readily available at coarse spatial scales from GIS/RS sources for each of the major helminths.

4.1. Schistosomes

In Africa schistosomiasis is caused predominantly by infection with *S. haematobium* or *S. mansoni* which result in urinary and intestinal schistosomiasis, respectively (Jordan and Webbe, 1993). These parasites have an indirect life-cycle involving a molluscan intermediate host. Sexual reproduction of mature worms occurs in the mammalian host and asexual multiplication and differentiation occurs in an intermediate molluscan host. A very short-lived free-swimming stage, the miracidium, hatches from the eggs that pass out of the human host via the faeces or urine. The miracidium penetrates the tissues of the snail host and develops into a first and second stage sporocyst. These sporocysts then give rise, via asexual reproduction, to numerous cercariae, which leave the snail to become free-swimming in the aquatic habitat, and infect the human host by cercarial penetration of the skin.

The molluscan intermediate hosts for schistosomes are freshwater snails; see Sturrock (1993), Rollinson and Simpson (1987) and Brown (1994) for details on their taxonomy and geographical distribution. In Africa, intermediate hosts can be broadly divided into two predominant groups that reflect their geographical distribution and parasite-specificity. *S. haematobium* is transmitted almost exclusively by snails from the genus *Bulinus*, which comprises several species groups: *truncatus, tropicus, forskalii* and *africanus*; and *S. mansoni* is transmitted by the genus *Biomphalaria*, with the main species groups being *pfeifferi, alexandrina, choanomphala*, and *sudanica*.

Generally, these infections have a focal spatial distribution. The generative mechanisms of such heterogeneity are varied, and reflect both human and environmental factors. Heterogeneity in human behaviour associated with exposure to infection, namely water contact patterns, is undoubtedly of great importance in determining transmission dynamics within a defined community and spatial patterns at a micro-level (Woolhouse *et al.*, 1991; Davies *et al.*, 1999). While understanding such heterogeneities is of considerable epidemiological importance (Woolhouse, 1994), the concern here is those environmental factors that may govern schistosomiasis distribution over large spatial scales and can be easily derived from coarse scale RS and GIS information (Hay, this volume). For further details of the numerous other factors that impact on the population or transmission dynamics of the parasite and intermediate host, the reader is referred to Berrie (1970), Appleton (1978), Sturrock (1993), Brown (1994) and Woolhouse (1994).

4.1.1. *Temperature*

Vector-borne pathogens have well defined minimum, optimum, and maximum temperatures for their development or replication in the vector. The optimum is typically the temperature in which development occurs in the shortest time, while the minimum and maximum temperatures represent thermal limits of development. These temperatures are known to affect several population processes of the schistosome life-cycle, including the life-expectancy and average infection rate of miracidia (Anderson *et al.*, 1982), the shedding of human schistosome cercariae (Pitchford and Visser, 1965; Shiff *et al.*, 1975; Sturrock, 1993), and the length of the pre-patent period (Anderson and May, 1979). These processes will certainly influence transmission and be important at small spatial scales (Woolhouse *et al.*, 1991). They may also be of importance at larger scales (Pitchford, 1981), although this aspect has received little attention.

At temperatures above 30°C, snail mortality increases (Pflüger, 1980; Joubert *et al.*, 1984) and fecundity decreases through the retardation of gametogenesis and gonad development (Appleton and Eriksson, 1984). Below 16°C snails tend to die before the cercariae mature from sporocysts (Pflüger, 1980; Joubert *et al.*, 1986). Temperature also has an effect on the rate of development, known as the 'intrinsic rate of natural increase' (r_m) (Shiff, 1964; Jordan and Webbe, 1993), which is suggested to be maximized at temperatures broadly similar (25 ± 5°C) for both *Bulinus* spp. and *Biomphalaria* spp. (Sturrock, 1993; Appleton, 1978).

High temperatures have been used to explain the absence of *Biomphalaria* spp. from coastal East Africa (Sturrock, 1966), and is confirmed by the virtual absence of *S. mansoni* infections in these coastal areas (Diesfeld and Hecklau, 1978; Doumenge *et al.*, 1987). A similar exclusion has been observed in southern Africa, where a high mortality of *Bi. pfeifferi* in South Africa is associated with periods of continuous high temperatures (25–27°C) (Pitchford and Visser, 1965; Appleton, 1977). Although this may be an example of an important limiting factor, several other factors may also convincingly explain the absence of this species in given areas (McCullough, 1972). Nonetheless, a valuable example of temperature-based prediction is provided by Pitchford (1981), who related composite measures of temperature to outdoor experimental data concerning the life-cycle of the schistosomes and transmission. High temperatures caused reductions in the survival of the snail host, development of cercariae and the numbers of shedding snails. It was suggested that monthly means of daily minimum (MDn) of 17°C for 7 months were sufficient to have adverse effects on the development of the parasite, while MDns of 1°C caused snails exposed to *S. haematobium* miracidia to die, thus setting the lower temperature cut-off. For *S. mansoni* this level was raised to 3°C. Based on these limits, the predicted distribution was shown to match the observed distribution well.

4.1.2. *Water Body Type*

It has been suggested that *Bulinus* spp. are adapted more to temporary water bodies because of their relatively high r_m and ability to colonize habitats rapidly (Appleton, 1978). Studies of snail distribution confirm this suggestion and reveal that *Bulinus* spp. usually occurs in seasonal, temporary bodies of water (McCullough, 1972; Brown et al., 1981), whereas Bi. pfeifferi populates predominantly perennial streams, irrigation channels and lakes (Webbe, 1962; McCullough, 1972; Brown et al., 1981; Marti et al., 1985).

The observation that different snail species occupy different types of water bodies raises the possibility of defining patterns of schistosome infection based on the permanence of water bodies (Appleton, 1978) and their distance from communities. It has been well recognized for 30 years that in the lake areas of East Africa there is an inverse relationship between prevalence of S. mansoni infection and the distance along a transect from the lake's edge (Nelson, 1958; McCullough, 1972; Kabaterine et al., 1996; Lwambo et al., 1999), whereas S. haematobium may be absent on the lakeshore (McCullough, 1972). Permanent water bodies can be assessed using both RS and radar technologies (Hay, this volume) and GIS can be used to measure the distance between each community and water bodies (Lwambo et al., 1999).

4.1.3. Rainfall

That different snail species inhabit different types of water bodies suggests that the influence of rainfall will vary between species, with the greater effect on species exploiting temporary water bodies (Marti et al., 1985). It has been suggested that since Bi. pfeifferi is not usually present in seasonal water bodies it is unlikely to occur in areas of low rainfall (Brown et al., 1981); heavy rainfall may, however, wash away snail populations (Davies et al., 2000). Thus, it would appear that the effect of rainfall on the spatial variation of snail distributions might be modified by temporal variation in rainfall. The relationship between rainfall and the temporal dynamics of both Bulinus spp. and Biomphalaria spp. populations is a well-studied topic and for detailed treatment of this area the reader is referred to Southgate and Rollinson (1987) and Sturrock (1993).

4.1.4. Water Velocity

An important way in which rainfall may influence geographical patterns of snail species is through its effect on water velocity. Most snail species are known to tolerate only a narrow range of water velocity, up to a maximum flow of 0.3 m/s (Appleton, 1978; Brown, 1994). Thus, in areas of heavy rainfall the rapidly flowing river systems may prevent the establishment of snail populations in many parts of Africa.

Another useful proxy for water velocity may be slope, since this determines, in part, the rate of water flow. Slope can be derived from altitude surfaces using GIS, based on the maximum rate of change in values of a

Table 1 Altitudinal limits (m) of snail intermediate hosts and schistosomiasis.

Species	Country	Lower limit	Upper limit	Source
Intermediate hosts				
Bi. pfeifferi	Tanzania	200		Beirrie (1970)
Bi. pfeifferi	Kenya	300	2000	Brown *et al.* (1981)
Bi. pfeifferi	Kenya	300	1800	Highton (1974)
Bi. pfeifferi	South Africa		1200	Van Eeden and Combrinck (1966)
Bu. africanus	Kenya	0	1800	Highton (1974)
Bulinus spp.	South Africa	NR	1000	Van Eeden and Combrinck (1966)
Schistosomiasis				
S. haematobium	Kenya	NR	2000	Diesfeld (1969)
S. mansoni	Kenya	200	1800	Diesfeld (1969)
S. mansoni	Uganda	NR	1700	Nelson (1958)
S. mansoni	Ethiopia	500–1000	2000–2200	Kloos *et al.* (1978)

NR, not recorded.

digital elevation model (Burrough and McDonnell, 1998). A recent development in earth observations has been the successful use of Synthetic Aperture Radar (SAR) in describing hydrologic landscapes and for detecting water bodies (Hay, this volume), which offers a potential application for the prediction of water velocities that would restrict snail populations.

4.1.5. Altitude

A final factor that may be useful in predicting schistosome distributions is altitude. This affects both temperature and rainfall, crucial factors in the life-cycles of both the schistosome parasite and the intermediate host. Several studies indicate that altitude serves to restrict the distribution of snail species by providing an upper and lower limit for population and transmission dynamics (Table 1). This factor is particularly attractive for predicting schistosomiasis distribution since it is a measure that is readily available from digital elevation models. The distribution limits, however, will differ according to region due to differences in the relationship between altitude and temperature. This emphasizes a need to consider altitude used in combination with measures of rainfall and/or temperature in a multivariate approach.

4.2. Direct Life-cycle Geohelminths

The main geohelminths that infect man are hookworms (*Ancylostoma duodenale* and *Necator americanus*), *A. lumbricoides* and *T. trichiura* (Pawlowski, 1984a,b; Schad and Banwell, 1984). The life-cycles of geohelminths are direct, involving sexually maturing parasites in the human host and free-living stages. Uninfective stages of geohelminths are passed in human faeces, which are often deposited in the soil environment where development to the infective stage occurs. Since geohelminth eggs are non-motile and hookworm larvae stages have limited motility, their rate of development and survival are dependent on the surrounding environmental humidity and temperature (Crompton, 1994). It follows therefore that these factors, and indirectly related factors such as rainfall, soil type and altitude, may influence transmission success (Anderson, 1993) and hence spatial patterns of infection. Parasitiologists have typically used animal models to explore experimentally the effect of environmental variables on the survival and development of infective stages. These models have included the cat and dog hookworm (*A. tubaeforme* and *A. caninum*), the pig roundworm (*A. suum*), and mouse trichuriasis (*T. muris*) and pig trichuriasis (*T. suis*) and are detailed further below.

4.2.1. Temperature

Several laboratory studies have investigated the effect of temperature on the embryonation and hatching rates of the geohelminth infective stages. In these studies the infective stages were maintained under controlled conditions so that embryonation and hatching could be recorded for a range of temperatures, and the minimum, optimum and maximum temperatures for hatching and embryonation documented (Table 2). The two species of human hookworm are said to differ in their susceptibility to temperature, such that the thermal tolerance is greater for *A. duodenale* than for *N. americanus* (Matsusaki, 1963; Hoagland and Schad, 1978). These differing climatic responses may cause partial separation of the species, making the prediction of spatial patterns problematic. Several studies have also pointed to an apparent difference in temperature thresholds for *Ascaris* spp. and *Trichuris* spp., with the latter having a higher optimal threshold (Brown, 1927; Rukmono, 1980; Sargent, 1971, cited in Larsen and Roepstorff, 1999), which may be important in the differential distribution of these species.

The problem with transferring these findings to larger spatial scales is that it is often difficult to apply the results of laboratory experiments to field conditions (Larsen and Roepstorff, 1999). Existing knowledge of thermal limits of geohelminth distribution is, at present, inadequate and more

Table 2 The optimum, minimum and maximum temperature (°C) limits for hatching and embryonation of geohelminths.

Species	Optimum temperature	Minimum and maximum temperature	Source
A. tubaeforme	20–30	15–37	Nwosu (1978)
A. caninum		15–37	McCoy (1930)
A. duodenale	15–35	7–45	Hoagland and Schad (1978)
N. americanus	20–35	15–40	Hoagland and Schad (1978); Udonsi and Atata (1987)
A. suum	31	16–34	Seamster (1950); WHO (1967); Arene (1986)
A. lumbricoides	28–32	NR	Crompton and Pawlowski (1985)
Trichuris spp.	32–35	5–45	Beer (1971, 1973, 1976)

NR, not recorded.

detailed geographical data are needed to assess the potential of using temperature to predict spatial patterns of geohelminths.

4.2.2. *Soil Moisture and Humidity*

Soil moisture and humidity are thought to be important for the successful development of infective ova and larvae. At low humidity (atmospheres less than 80% saturation) the ova of human and pig ascaris do not embryonate (Otto, 1929). By contrast, there does not appear to be an upper lethal limit since *A. lumbricoides* ova will develop at 95% relative humidity (Seamster, 1950), although the precise effect of humidity will be modified by temperature (Otto, 1929). It has been stated that the moisture requirement of *T. trichiura* is lower than that for *A. lumbricoides* (Spindler, 1929), and thus the former species is more affected by differences in soil moisture. There is also evidence that for individuals to be infected with hookworm, the soil needs to be sufficiently moist so that the larvae remain on the soil surface to penetrate the human skin (Augustine, 1922). Studies have shown that as the soil dries, hookworm migrate downwards (Augustine, 1922), to depths of 20 cm (Udonsi *et al.*, 1980), where larvae would be unavailable for human infection. Development of hookworm eggs is also stopped in waterlogged soils (Payne, 1922).

Vegetation may be a useful proxy for soil moisture, since a large amount of vegetation tends to prevent evaporation and conserve soil moisture. An RS indicator of vegetation is provided by the Normalized Difference Vegetation Index (NDVI) (Hay, this volume), which has been shown to be closely associated with saturation deficit (Rogers and Randolph, 1991).

Alternatively, since rainfall increases humidity and soil moisture, which enhances larval and ova development, it is probable that the total rainfall in an area and its seasonal distribution might help to explain observed patterns of infection. Wetter areas are usually associated with increased transmission of *T. trichiura* (Sweet, 1924; Chandler, 1928; Cort *et al.*, 1929), *A. lumbricoides* (Prost, 1987) and hookworm (Stott, 1960; Sturrock, 1967; Hall *et al.*, 1982; De Clercq *et al.*, 1995). Other data from South Africa (Appleton and Gouws, 1996; Appleton *et al.*, 1999) suggest a negative association that presumably reflects regional differences in the dynamic interplay of several environmental factors, such as altitude and temperature.

4.2.3. Soil Type

In considering the effect of soil type for geohelminths the focus has been on their texture and porosity. Under experimental conditions, sandy soil harbours a higher percentage of infective hookworm larvae than does clay soil (Stoll, 1923; Augustine and Smillie, 1926; Vinayak *et al.*, 1979), because of its large particle size and good aeration. So although infective larvae quickly die on the surface of sandy soil in direct sunlight, they are able to migrate rapidly into the soil and during the rains are able to migrate vertically as moisture permits (Beaver, 1953). In contrast, clay soils are less well aerated and less porous, which restricts vertical migration, causing the larvae to die from desiccation (Beaver, 1953). Other evidence, however, suggests that *A. suum* eggs survive better in less permeable clay soils, because eggs are dispersed in the soil by water and clay soils have greater water dispersion properties (Mizgajska, 1993).

Early American studies suggested that hookworm infection was more prevalent in areas with sandy soils than with clay soil (Stiles, 1903; Smith, 1903; Augustine and Smillie, 1926; Keller *et al.*, 1940; Scott, 1945). This is often quoted as evidence that hookworms are more abundant in areas with sandy soils than with clay soils. None of these studies, however, has been truly quantitative since each involved observational associations. Increased abundance of hookworms in sandy soils has been shown in Liberia (Hsieh, 1990) but not Papua New Guinea. Recent studies indicate that infective stages disappear from the soil over a period of 2 to 6 months (Mizgajska, 1993; Larsen and Roepstorff, 1999), suggesting that temperature and soil moisture may be of greater importance.

4.2.4. Altitude

It is often observed in Africa that the prevalence of hookworm is negatively associated with altitude. This effect is widespread and has been reported in Ethiopia (McConnell and Armstrong, 1976; Jemaneh and Tedla, 1984;

Jemaneh, 1998), Kenya (Diesfeld and Hecklau, 1978; Hall *et al.*, 1982) and Madagascar (Hanitrasoamampionona *et al.*, 1998). In particular, low temperatures and humidity may exclude the parasite at high altitudes (Tedla, 1986), while the hot and humid conditions of coastal areas provide suitable environments for transmission (Hall *et al.*, 1982). Survey results from South Africa and Madagascar showed that the prevalence of *A. lumbricoides* and *T. trichiura* tends to be higher at greater altitude (Appleton and Gouws, 1996; Hanitrasoamampionona *et al.*, 1998), whilst Ethiopian studies have produced conflicting results (McConnell and Armstrong, 1976; Jemaneh, 1998).

4.3. Filariasis

Lymphatic filariasis is caused by infection with any three closely related parasitic nematodes—*Wuchereria bancrofti*, *Brugia malayi* or *B. timori*. Like schistosomes, filarial parasites also have an indirect life-cycle, involving mosquito vectors. Infective third-stage larvae are transmitted to humans during blood feeding by mosquitoes. These organisms are deposited from the mouth parts of the mosquito in the vicinity of the skin puncture wound, from where they penetrate through the dermis to enter the local lymphatics. There, these larvae then moult twice and over 9 months develop into sexually mature male and female worms. Once mated, fecund female parasites release large numbers of first-stage larvae called microfilariae. These frequently have a nocturnal periodicity whereby most are present in the peripheral human host blood circulation between late night and 06:00, with few or none present in the circulation during the day. When absent from the peripheral circulation, microfilariae are sequestered in deep vascular beds of the lung and other organs. This behaviour appears to be an example of adaptation to local ecological conditions in that the time at which peak parasitaemia occurs coincides with the time when the local female mosquitoes take their blood meal, which in most areas of the world occurs during the night. The cycle is completed when microfilariae ingested via blood meals develop over a period of 10 days to 2 weeks within the vectors to become third-stage larvae capable of infecting another human.

As discussed above, environmental factors that may govern filariasis distribution may be expected to affect the life history parameters of both the mosquito vector and the parasite. Traditionally, attention in this area has focused primarily on temperature, humidity and altitude (Raghavan, 1957, 1969; Sasa, 1976; Lindsay *et al.* 1984; Attenborough *et al.* 1997).

4.3.1. Temperature

Temperature is known to have several effects on both filarial parasites and the mosquito vectors. As with schistosomes, filarial parasites have well defined minimum, optimum, and maximum temperatures for their development or replication in the vector. The time from imbibing an infected meal until the vector is able to transmit to a vertebrate host is called the extrinsic incubation period, which is shorter at higher temperatures. The optimal temperature range for the development of the filarial parasites may be as wide as 15.5°C to 32.2°C (Basu and Rao, 1939; see also Omori, 1958), although more recent experiments carried out on the development of larval *W. bancrofti* in *Aedes polynesiensis* suggest that the relationship between parasite development and temperature is likely to be non-linear, with temperatures above 31.5°C limiting for most larval stages (Lardeux and Cheffort, 1997). Temperature also has an effect on the rapidity of development, the feeding interval and the survival rate of the vectors (Lindsay *et al.* 1984). At high temperatures adult mosquito mortality is usually increased, which may counteract any increased vectorial capacity due to more rapid development (Shope, 1999).

4.3.2. *Humidity*

Lymphatic filariasis is limited to the humid areas of the tropics. High humidity (i.e. a relative humidity above 60%) is thought to be important both for the successful development of infective larvae (Acton and Rao, 1931; Basu and Rao, 1939) and for larval penetration into the skin (McGreevy *et al.* 1974). The latter conclusion stems from observations that successful penetration of infective larvae into bite wounds depends on the enveloping haemolymph secreted by the mosquito as it feeds. At low humidity, the protective drop of haemolymph may evaporate faster thereby reducing the success of larval penetration and hence parasite transmission (Lindsay *et al.* 1984). The problem with using humidity as the sole influence on transmission is highlighted by the fact that, paradoxically, filariasis is also endemic in areas where the rate of evaporation is greatest. This may be explained by the fact that this parasite is transmitted predominantly by mosquitoes which feed at night, when evaporation is reduced. This emphasizes the need to take a multi-scale approach when quantifying and modelling the impacts of such effects on parasite distribution.

4.3.3. *Altitude*

Altitude is another factor that has long been known to influence filariasis distribution (Raghavan, 1969). In particular, filariasis is expected to be more prevalent in valleys and other low altitude areas where the warmer temperatures (and higher humidity) favour both mosquito survival and faster parasite development through the extrinsic incubation period

(Attenborough *et al.* 1997). Although data are limited, early work suggested that infections due to *W. bancrofti* and *B. malayi* may be limited to an altitude of about 600 m in India and *W. bancrofti* in Nepal and Africa up to 1300 m (Raghavan, 1969; Jordan, 1956). Similarly, the proportion of anopheline and culicine mosquitoes with infective larvae of *W. bancrofti* also decreased from 170 to 650 m in Papua New Guinea (Attenborough *et al.* 1997).

4.4. Prospects

The above evidence suggests that the distribution of helminth infections may be predicted using environmental information (see Table 3). This approach requires careful consideration of several issues.

A major problem is that developing predictive models and maps on large spatial scales cannot take into account the micro-scale spatial patterns. For example, microhabitats, influenced by local housing and sanitation, may provide suitable transmission foci within unsuitable areas for geohelminth transmission and vice versa (Spindler, 1929; Sturrock, 1967; Bundy and Cooper, 1989; Crompton, 1994). Similarly, the distribution of schistosomiasis is influenced by several small-scale factors that are not available from RS data (Huang and Manderson, 1992; Kloos *et al.*, 1997). Moreover, many of the relations between the environment and spatial patterns of infection will vary in a scale-dependent fashion (Walsh *et al.*, 1999).

A further complication is that the importance of each environmental factor in determining the distribution of snail hosts and schistosome infections may vary from area to area so that it may be necessary to develop separate prediction models. Much is known about the geographical distribution of snail species in Africa (Rollinson and Simpson, 1987; Sturrock, 1993; Brown, 1994), although this information is not yet available digitally.

Finally, many observations linking helminth prevalence and climatic variables have not been quantified in relation to a single environmental factor. Seasonal changes in environmental factors, and their interactions, have the potential to induce temporal fluctuations in prevalence and intensity of infection by modifying parasite transmission success (Anderson, 1993). Hookworm infection levels, for example, shows marked seasonal variation (e.g. Sturrock, 1967; Nawalinski *et al.*, 1978; Nwosu and Anya, 1980; Udonsi, 1983) especially in relation to rainfall. Such changes may be of little consequence to the long-term success of observed infection patterns for helminths, however, since the longest lived developmental stages in the life-cycle of these parasites are typically much longer (*N. americanus*, for example, has an estimated life span of 3.5 years, while the mean life span of *Wuchereria bancrofti* is estimated between 5 and 8 years) than the periods in the year during which the parasite reproductive number (R_0) is less than unity (Anderson, 1993). This clearly highlights the need to understand biology, as opposed to seeking statistical associations when attempting to model parasite distributions using environmental variables. It also emphasizes a need to take a multi-factorial approach. Also of relevance is the importance of spatial autocorrelation and spatial structure in infection and environmental data (Rossi *et al.*, 1992; Bailey and Gatrell, 1995), since this may affect the confidence intervals placed on the significance of results (Thomson *et al.*, 1999).

The potential of RS in filling the gaps in current helminth atlases will, however, remain undefined until large scale studies are conducted that consider some of these issues. Some studies of this type have already been undertaken and are discussed in the next section.

5. APPLICATIONS OF REMOTE SENSING FOR PREDICTING HELMINTH DISTRIBUTIONS

Since 1980s RS technologies have increasingly been used to study the spatial and temporal patterns of infectious diseases (Rogers, this volume; Randolph, this volume; Hay *et al.*, this volume). For helminth infections, these applications have only been attempted for schistosomiasis and filariasis.

An early elegant study used a combination of weather and Landsat-3 Multispectral Spectral Scanner (MSS) data (bands 4–7) to define the risk of schistosomiasis in the Caribbean and in the Philippines (Cross and Bailey, 1984; Cross *et al.*, 1984). By using discriminant analysis, it was possible to predict the absence or presence of schistosomiasis using weather data with an accuracy of 87.1% in the Caribbean and 93.2% in the Philippines (Cross and Bailey, 1984). Further study using Landsat MSS data produced a disease distribution map based on the probability of infection using initial analysis of weather data and mean values of Landsat bands in each the 52 survey sites, although the accuracy of this analysis was not reported.

Malone *et al.* (1994, 1997) used RS data on diurnal temperature differences (dT) in conjunction with spatial data on *S. mansoni* infection prevalence. These dTs indicate surface and subsurface moisture contained in soil and plant canopy and hence may act as a surrogate for the abundance of the snail vector, *Bi. alexandrina,* whereby wetter and more suitable habitats for *Bi. alexandrina* corresponded to lower dT values. This is because water acts as a temperature buffer for diurnal fluctuations of surface temperature, and *Bi. alexandrina* is known to be more sensitive to temperature variation

Table 3 Potential environmental factors for the prediction of spatial distribution of helminths using remote sensing and environmental data.

Factor	Possible indicator	Species	Comment
Temperature	Land surface temperature from NOAA-AVHRR. Interpolated meteorological surfaces	Schistosomes/ filariasis	Acts to define the upper and lower limit for transmission. Can be used to predict geo-graphical limits of snail populations
		Geohelminths	Wide tolerance range of survival and development. Undefined potential
Distance to water bodies	Calculated GIS distances	*S. mansoni*	Predictive in areas where the species is known to be endemic
Soil moisture/ humidity	NDVI from NOAA-AVHRR, rainfall	Schistosomes	No apparent effect
		Geohelminths/ filariasis	Critical factor in the survival and development of infective stages
Rainfall	Cold Cloud Duration from Meteosat. Interpolated meteorological surfaces	Geohelminths/ schistosomes	Acts to modify seasonal distribution. Possible effect of excess rainfall on water velocity
Water velocity	SAR radar, rainfall	Schistosomes	Alters soil moisture and humidity predictively
Altitude	Digital elevation surfaces	Schistosomes	Defines the upper and lower limit for transmission
		Hookworm/ filariasis	Limits transmission
		Ascaris and *Trichuris*	Conflicting evidence

(Continued)

Table 3 Continued.

Factor	Possible indicator	Species	Comment
Soil type	Digital soil maps	Geohelminths	Experimental evidence suggests important role, but its usefulness in predicting large scale distribution is not yet established

NOAA-AVHRR, National Oceanic and Atmospheric Administration–Advanced Very High Resolution Radiometer.

and does not survive drying out of water bodies. Studies in Egypt have shown that low values of dT are associated with increased snail abundance in wet areas with a slow current flow (Abdel-Rahman *et al.*, 1997), and this is closely mirrored by the patterns of *S. mansoni* prevalence (Malone *et al.*, 1994, 1997).

Thompson *et al.* (1996) also used dT maps in conjunction with spatial data on case prevalence from 297 villages within the southern Nile delta, and showed that this environmental variable may underlie the observed spatial distribution of lymphatic filariasis at least within their Egyptian study region. It is suggested that dTs act as a surrogate for the abundance of the mosquito vector, *Culex quinquefasciatus*. Satellite image data from NOAA-AVHRR were analysed to determine dTs for the southern Nile delta, while the case prevalence and locational data for each of the 297 villages were entered into a GIS. The digitized filariasis prevalence data were superimposed on the dT map and assigned to each of four prevalence categories of 0.5%, 5%, 15% and 25% respectively. The association between village dT value and prevalence category was investigated using stepwise polychotomous logistic regression, which indicated a significant relationship between the two variables.

Hassan *et al.* (1998a,b) examined the relationship between NDVI and moisture derived from the Landsat Thematic Mapper (TM) indices and the distribution of filariasis in Egypt. Discriminant analysis showed that marginal vegetation, water, wetness and moisture indices were the most important predictors of disease risk, such that it was possible to identify accurately 77% of high and low prevalence villages.

While the potential of RS techniques has been demonstrated for helminth infections in Egypt, the country provides a relatively unique ecosystem for snail populations and it remains unclear how this relationship could be extended to other areas of Africa. Recently, Lindsay and Thomas (2000) examined the relationship between environmental factors (as assessed by interpolated meteorological maps) and the distribution of filariasis at the continental level in Africa. Logistic regression analysis of the climate variables predicted with 76% accuracy whether sites had microfilaraemic patients or not. Based on these analyes they were able to develop a risk map for Africa and a separate one for Egypt, where both *Cx. pipiens* and *Cx, quinquefasciatus* occur. Although this work belies the fine-scale spatial heterogeneity within countries, it represents a first large area model and is useful in identifying broad transmission patterns where field data are unavailable.

The applications of RS have not been restricted to studying infection prevalence and vector populations. There also exists potential for understanding more fully the relationship between geographical distribution and genetic diversity. For example, Krüger *et al.* (2000) investigated the relationship between intra- and inter-genetic population structure of *B. globosus* and *Bi. pfeifferi* using random amplified polymorphic DNA (RAPD) markers and RS data in the Zimbabwe highveld. Analyses indicated that geographic and genetic distances were significantly correlated for both species. NOAA-AVHRR derived NDVI was positively correlated with intra-population genetic diversity, since high NDVI indicates rainfall and is suggestive of population mixing after rainfall and population extinction during the dry season. While molecular malacology remains a developing area of investigation, further studies may benefit from the integration of genetic and spatial data. Furthermore little information is available in the literature concerning the spatio-temporal dynamics of population genetics and the environment, although the importance of such data has been recently recognized (Davies *et al.*, 2000).

5.1. Two Case Studies

Section 4 outlined those environmental factors that have been shown to be associated with helminth distribution in Africa. To emphasize the message of the potential for RS in predicting helminth distribution, we examine in detail two examples of the relationships between RS derived environmental variables and the prevalence of helminth infection.

5.1.1. A Case Study in Cameroon: A. lumbricoides *and* T. trichiura *infection*

A nationwide survey of geohelminth infection (*A. lumbricoides*, *T. trichiura*, and hookworms), the largest in Africa, took place betwen 1985 and 1987 in Cameroon (Ratard *et al.*, 1991). The distribution of infection has been discussed but not yet analysed in relation to environmental variation (Ratard

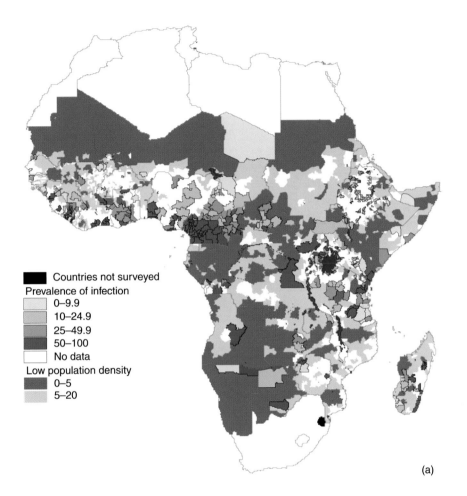

Legend:
Countries not surveyed
Prevalence of infection
0–9.9
10–24.9
25–49.9
50–100
No data
Low population density
0–5
5–20

(a)

Plate 11 (a) Distribution of geohelminths in sub-Saharan Africa, showing the highest prevalence of either *Ascaris*, *Trichuris* or hookworm in each district. Of the 2304 administrative units (typically districts) in 39 countries, data are available for geohelminths in 343 (14.9%). The population in Africa is not homogeneously distributed, and many of the districts for which no data are currently available also have low-density populations. To highlight the high population densities areas of Africa we have created a template (dotted shading) that masks out all districts with population density less than 5 persons/km^2 and between 5 and 20 persons/km^2. The areas of white on the map indicate high population density areas for which we have no data. See Brooker and Michael (this volume).

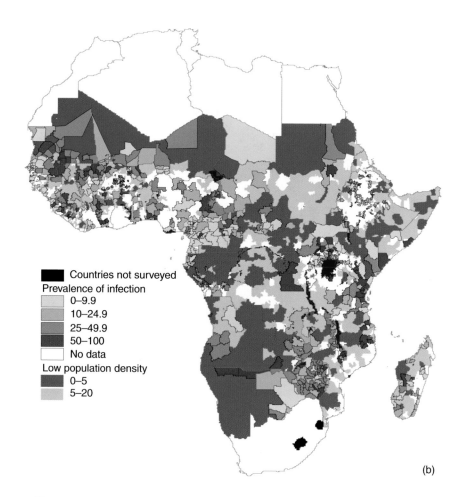

Legend:

Countries not surveyed

Prevalence of infection
- 0–9.9
- 10–24.9
- 25–49.9
- 50–100
- No data

Low population density
- 0–5
- 5–20

(b)

Plate 11 (contd) (b) Distribution of schistosomiasis in sub-Saharan Africa, showing the highest prevalence of *Schistosoma mansoni* or *S. haematobium* in each district. Of the 2304 administrative units (typically districts) in 39 countries, data are available for schistosomiasis in 684 (29.7%). The population in Africa is not homogeneously distributed, and many of the districts for which no data are currently available also have low-density populations. To highlight the high population densities areas of Africa we have created a template (dotted shading) that masks out all districts with population density less than 5 persons/km^2 and between 5 and 20 persons/km^2. The areas of white on the map indicate high population density areas for which we have no data. See Brooker and Michael (this volume).

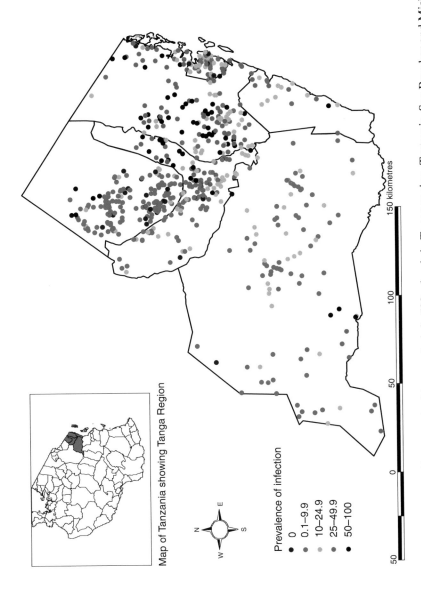

Plate 12 The distribution of self-reported schistosomiasis in 598 schools in Tanga region, Tanzania. See Brooker and Michael (this volume).

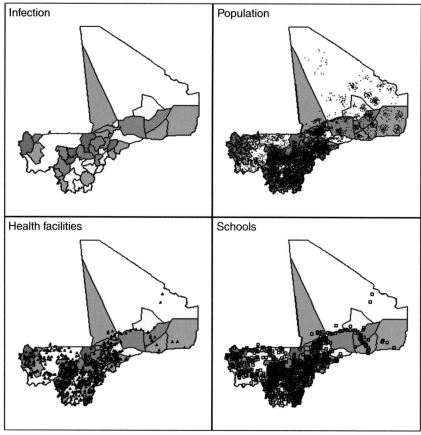

Prevalence of infection

	0–9.9
	10–24.9
	25–49.9
	50–100
	No data

Plate 13 The integration of epidemiological and infrastructure information that could be used to plan a school-based helminth control programme in Mali. It can be seen that few schools (represented by squares) serve the north-eastern part of the country and health centres (open boxes)). This type of analysis allows policymakers to focus school health resources where both the problem of schistosomiasis is prevalent and where schools and populations are concentrated. The example of Mali is kindly provided by HealthMap, which has been working with the government and the local UNICEF office in the development of a dynamic public health atlas for the country. See Brooker and Michael (this volume).

et al., 1991; Crompton, 1994). In collaboration with Raoult Ratard we have explored these data with respect to making predictions of prevalence based on RS and meteorological variables.

Figure 4 shows that there exists marked geographical variation in the prevalence of *A. lumbricoides* and *T. trichiura*, both being highest in the equatorial south of the country. The relationships between disease prevalence and the NDVI, land surface temperature (LST) obtained from Pathfinder AVHRR data and rainfall are shown in Figure 5. For both *A. lumbricoides and T. trichiura*, these are markedly non-linear. The relationships suggest that a threshold of 1400 m rainfall is required for prevalence >10% (Figure 5a, b) and confirms earlier work in West Africa (Prost, 1987). In those areas with rainfall greater than 1500 mm rainfall, prevalence typically exceeds 50%. A similar threshold is apparent for LST, where an annual mean temperature of less than 37°C is required for prevalences above 10% (Figure 5c, d). Above this threshold there is much scatter in the relationship, but it does appear that prevalence increases with increasing LST.

Many of the environmental factors are highly intercorrelated, and therefore it is important to undertake multivariate analysis to control for this. Subsequent analysis using generalized linear models showed that there is a strong relationship between environmental variables and the prevalence of infection. This preliminary work requires extensive validation of the predictions to check the model's spatial robustness. Furthermore, whilst multivariate statistics are a convenient method of analysis, they generally assume spatial independence between the prevalence of infection and environmental data (Rossi *et al.*, 1992; Robinson, this volume). These preliminary results suggest, however, that spatial patterns of *A. lumbricoides* and *T. trichiura* can be predicted, for the first time, on the basis of RS data. This is a focus of current research.

5.1.2. A Case Study in Tanzania: Urinary Schistosomiasis

Recently, together with the Partnership for Child Development, we have examined the relationships between RS data and the distribution of the prevalence of self-reported schistosomiasis in Tanga Region, Tanzania (see Plate 12). Self-reported schistosomiasis is a useful tool to identify high risk schools/communities (Red Urine Study Group, 1995); it typically underestimates the actual prevalence of infection, but by a consistent amount (Ansell *et al.*, 1997; Partnership for Child Development, 1999). Preliminary analyses indicated no consistent relationships between prevalence of reported haematuria, termed locally *kichocho,* and any environmental variable (Figure 6). The data do suggest, however, that there are specific thresholds necessary

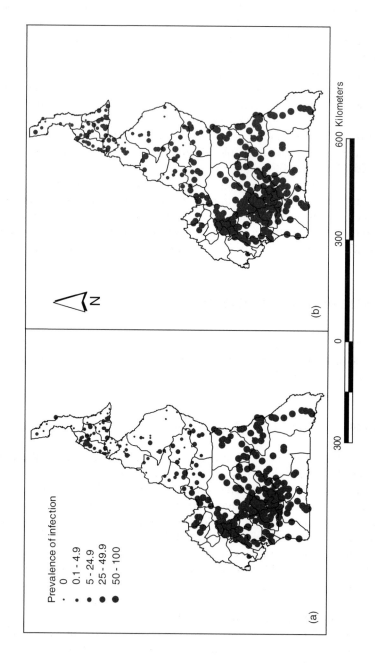

Figure 4 The geographical distribution of (a) *Ascaris lumbricoides* and (b) *Trichuris trichiura* by school in Cameroon in 1987 national survey. Data from Raoult Ratard.

for any significant transmission of urinary schistosomiasis. Above 1500 m no school has a prevalence of *kichocho* of >10%, and only three schools above 1000 m have a prevalence of *kichocho* >10% (Figure 6a). Similarly, a minimum mean LST threshold of 26°C is necessary to yield a *kichocho* prevalence >20% (Figure 6b). In contrast, there is no relationship between mean annual rainfall or NDVI and the prevalence of reported schistosomiasis (Figures 6c, d). Thus it seemed possible to identify schools that had a prevalence of self-reported schistosomiasis >30%, which is equivalent to an infection prevalence >50% (Ansell *et al.*, 1997) and warrants mass treatment with praziquantel. A number of logistic regression models were subsequently fitted to a 50% random subsample of schools classified according to the 30% threshold. The best-fit models were produced from a subset of three environmental variables, chosen from the original ten so as to minimize inter-correlation. In order to validate the predictions of the model they were tested in Tanga region using the remaining 50% of schools not selected originally to develop the model. Validation of the model showed that it correctly classified 76% of schools with a reported *kichocho* prevalence >30% in Tanga region.

6. PUBLIC HEALTH APPLICATIONS

This section builds on the previous sections and attempts to provide examples of how the perspectives offered by RS/GIS technologies can potentially be used to develop and guide helminth control programmes.

6.1. Maps and the Planning of Control Efforts

The development of any spatial database is meaningless unless there is a clear identification of the goals and definitions for using the information for informed decision-making. The recent mapping initiatives outlined in Section 3 aim to provide information on the spatial distribution of helminth infections. Such information can identify high-risk areas and can provide a very useful tool for planning, targeting and monitoring of control measures (Mott *et al.*, 1995; Nuttall *et al.*, 1999). For instance, the WHO/UNICEF Joint Programme on Health Mapping and GIS, HealthMap, has been working with national governments and key agencies in promoting and implementing the use of mapping and GIS for planning, monitoring and managing public health programmes. GIS provides a means for integrating multi-sectoral data and allows the easy visualization of the extent of the problem in relation to surrounding environment and existing health and social infrastructures such as health facilities, schools and roads. Such

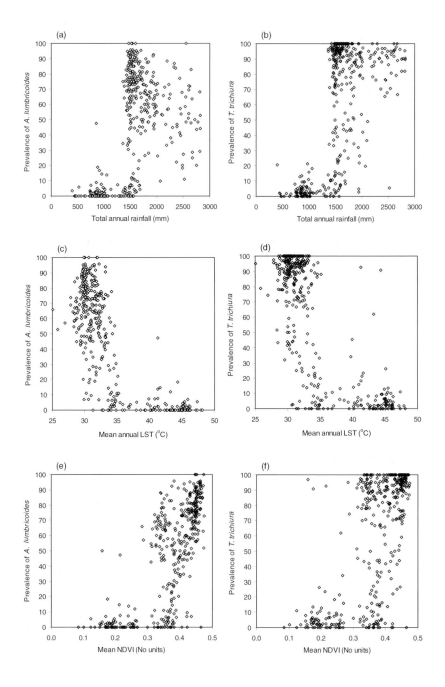

information, when mapped, creates a powerful tool for monitoring and managing control programmes and allows the more efficient targeting of resources to those communities that might be most in need. Plate 13 provides an illustration of such a potential approach. By combining epidemiological and demographic data, it can also allow an estimation of the total population at risk (Brooker *et al.*, 2000).

6.2. Targeting Schistosomiasis Control

School-based questionnaires have been identified as an effective approach to help locate high-risk communities/schools requiring mass treatment (Red Urine Study Group, 1995). If the questionnaire were to be done elsewhere it would probably be best to identify the areas where urinary schistosomiasis is not prevalent (WHO, 1995). Routine health statistics provide a potential source of such information, but in many countries these are not collated nationally. An alternative approach, basing the selection of areas on ecological criteria, has been suggested; however, there are currently no guidelines for the health planner to identify these zones (WHO, 1995). In Section 5.1 we highlighted the relationship between prevalence of schistosome infection, as indicated by self-reported blood in urine, and environmental factors. These relationships could then be used to develop a risk map for a country showing the predicted areas of where *S. haematobium* might be a public health problem, and areas where transmission is limited and would not be priority areas for a questionnaire approach.

7. CONCLUSIONS

The use of GIS and RS is in its infancy for helminth epidemiology and control. This chapter has highlighted the increasing use of these approaches and their potential as tools for visualization and analysis, leading to a better

---

Figure 5 Relationship between prevalence of infection with *Ascaris lumbricoides* and *Trichuris trichiura* and key RS environmental variables in Cameroon. The RS data were contemporary Pathfinder AVHRR data at an 8 × 8 km spatial resolution (provided by TALA, University of Oxford). These data were further processed to produce environmental proxies (Hay *et al.*, 1996; Hay and Lennon, 1999): NDVI which is an index of photosynthetically active vegetation amount; and LST. Other environmental data were obtained from the Spatial Characterization Tool (SCT) (Corbett and O'Brien, 1997) that included interpolated climate surfaces for Africa.

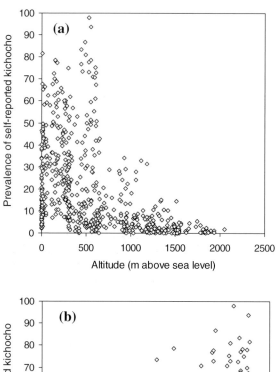

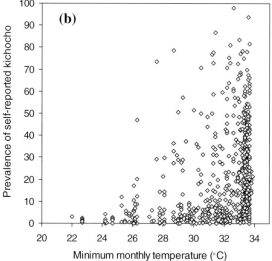

Figure 6 Relationship between prevalence of self-reported schistosomiasis and key RS environmental variables in Tanga region, Tanzania. An interpolated digital elevation model (DEM) of Africa was obtained from the Global Land Information System (GLIS) of the United States Geological Survey (EROS Data Center, 1996).

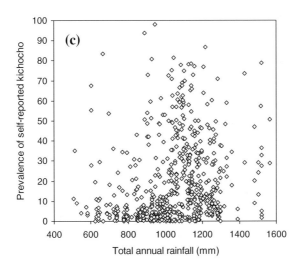

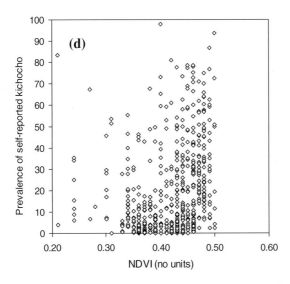

Figure 6 Continued.

understanding of the spatial extent of infection and disease. The present challenge is to build upon these developments and to exploit to the full the power of these approaches in helminth disease research and control. In particular, we identify the following issues that need to be addressed.

First, there is a pressing need for future work to recognize and address the central importance of scale in spatial analysis, both in understanding aetiological factors underlying patterns at different spatial scales and also the availability and resolution of GIS data, which ideally should match the scale at which disease variations occur. Although current experimental evidence provides some indication of the importance of environmental variables in determining spatial patterns of infection, little is currently known regarding the relative importance of different factors at different spatial scales. Future epidemiological studies will also need to introduce statistical techniques that take adequate account of spatial autocorrelation, multicolinearity and non-linear processes. In addition, issues related to the accuracy of GIS-derived exposure information in such studies need to be more fully addressed. These include: (i) the aggregate/ecological nature of the data, which may be relevant at the individual level; (ii) the quality of the data that are input into the GIS (e.g. accuracy and completeness); (iii) the appropriateness of combining multiple databases; and (iv) the relevance of the map layers to the exposure/outcome relationship of interest (e.g. timing of the data collection). The latter is a particular problem in helminthological research given that infection survey data typically span several decades, while map layers for candidate exposure variables are more recent.

In conclusion, we suggest that the GIS/RS approach is likely to become an essential tool in gaining a better understanding of the epidemiology and control of helminths. It already provides an integrated technology for providing a composite data view, and as improved exploratory and statistical data analysis and spatial modelling tools are developed, we predict that its value is likely to become more manifest.

ACKNOWLEDGEMENTS

We wish to thank many people for the numerous helpful conversations and for the valuable comments they provided. Particular thanks are due to Bernhard Bakker, María-Gloria Basáñez, David Brown, Don Bundy, Christl Donnelly, Helen Guyatt, Andrew Hall, Simon Hay, Brad Lobitz, Nicholas Lwambo, Sarah Randolph, Andrew Roddam, David Rogers, Vaughan Southgate, Joanne Webster, Willy Wint, and Byron Woods. We also wish to thank our many collaborators. First and foremost to Raoult Ratard for allowing us to present the Cameroon data, which were collected as part of a

USAID funded project. We also wish to thank Charles Kihamia, Wahab Issae and members of the Tanzania Partnership for Child Development for allowing us to present the Tanzania school data. Thanks also go to collaborators of the WHO mapping initiatives, and include Kathy O'Neill, Isabelle Nuttall and Jean-Pierre Meert of the WHO/UNICEF HealthMap Programme, Communicable Disease Surveillance and Response, WHO and Dirk Engels, Lester Chitsulo, Antonio Montresor and Lorenzo Savioli of WHO. This work was supported by the Wellcome Trust (SB) and the Medical Research Council (EM).

REFERENCES

Abdel Rahman, M.S., El Bahy, M.M., El Bahy, N.M. and Malone, J.B. (1997). Development and validation of a satellite based geographic information system (GIS) model for epidemiology of Schistosoma risk assessment on snail level in Kafr El-Sheikh governorate. *Journal of the Egyptian Society of Parasitology* **27**, 299–316.

Acton, H.W. and Rao, S.S. (1931). Diagnosis of lymphatic obstruction of filarial origin. *Indian Medical Gazette* **66**, 11–17.

Allen, T.F.H. and Hoekstra, T.W. (1992). *Toward a Unified Ecology*. New York: Columbia University Press.

Anderson, R.M. (1993). Epidemiology. In: *Modern Parasitology* (F.E.G. Cox, ed.), pp. 75–116. Oxford: Blackwell Science.

Anderson, R.M. and May, R.M. (1979). Prevalence of schistosome infections within molluscan populations: observed patterns and theoretical predictions. *Parasitology* **79**, 63–94.

Anderson, R.M. and May, R.M. (1991). *Infectious Diseases of Humans: Dynamics and Control*. Oxford: Oxford University Press.

Anderson, R.M., Mercer, J.G., Wilson, R.A. and Carter, N.P. (1982). Transmission of *Schistosoma mansoni* from man to snail: experimental studies of miracidial survival and infectivity in relation to larval age, water temperature, host size and host age. *Parasitology* **85**, 339–360.

Ansell, J., Guyatt, H., Hall, A. *et al.* (1997). The reliability of self-reported blood in urine and schistosomiasis as indicators of *Schistosoma haematoibum* infection in school children: a study in Muheza District, Tanzania. *Tropical Medicine & International Health* **2**, 1180–1189.

Appleton, C.C. and Eriksson, I.M. (1984). The influence of fluctuating above-optimal temperature regimes on the fecundity of *Biomphalaria pfeifferi* (Mollusca: Planorbidae). *Transactions of the Royal Society of Tropical Medicine and Hygiene* **78**, 49–54.

Appleton, C.C. (1977). The influence of temperature on the life-cycle and distribution of *Biomphalaria pfeifferi* (Krauss 1948) in south-eastern Africa. *International Journal for Parasitology* **7**, 335–345.

Appleton, C.C. (1978). Review of literature on abiotic factors influencing the distribution and life-cycles of Bilharziasis intermediate host snails. *Malacological Review* **11**, 1–25.

Appleton, C.C. and Gouws, E. (1996). The distribution of common intestinal nematodes along an altitudinal transect in Kwa-Zulu Natal, South Africa. *Annals of Tropical Medicine and Parasitology* **90**, 181–188.

Appleton, C.C., Maurikungirire, M. and Gouws, E. (1999). The distribution of helminth infections along the coastal plains of Kwa Zulu–Natal province, South Africa. *Annals of Tropical Medicine and Parasitology* **93**, 859–868.

Arene, F.O. I. (1986). *Ascaris suum*: influence of embryonation temperature on the viability of the infective larva. *Journal of Thermal Biology* **11**, 9–15.

Attenborough, R.D., Burkot, T.R. and Gardner, D.S. (1997). Altitude and the risk of bites from mosquitoes infected with malaria and filariasis among the Mianmin people of Papua New Guinea. *Transactions of the Royal Society of Tropical Medicine and Hygiene* **91**, 8–10.

Augustine, D.L. (1922). Investigations on the control of hookworm disease. VIII. Experiments on the migration of hookworm larvae in soils. *American Journal of Hygiene* **2**, 162–187.

Augustine, D.L. and Smillie, W.G. (1926). The relation of the type of soils of Alabama to the distribution of Hookworm Disease. *American Journal of Hygiene* **6**, 36–62.

Bailey, T.C. and Gatrell, A.C. (1995). *Interactive Spatial Data Analysis*. Harlow: Longman.

Barnish, G. and Ashford, R.W. (1990). *Strongyloides* cf. *Fuelleborni* and other intestinal helminths in Papua New Guinea: distribution according to environmental factors. *Parassitologia* **32**, 245–263.

Basu, P.C. and Rao, S.S. (1939). Studies on filariasis transmission. *Indian Journal of Medical Research* **27**, 233–249.

Beaver, P.C. (1953). Persistence of hookworm larvae in soil. *American Journal of Tropical Medicine and Hygiene* **2**, 102–108.

Beer, R.J. (1971). Whipworms of domestic animals. *The Veterinary Bulletin* **41**, 343–349.

Beer, R.J. (1976). The relationship between *Trichuris trichiura* (Linnaeus 1758) of man and *Trichuris suis* (Schrank 1788) of the pig. *Research in Veterinary Science* **20**, 47–54.

Beer, R.J. S. (1973). Studies on the biology of the life-cycle of *Trichuris suis* Schrank, 1788. *Parasitology* **67**, 253–262.

Berrie, A.D. (1970). Snail problems in African schistosomiasis. *Advances in Parasitology* **8**, 43–96.

Brooker, S., Rowlands, M., Haller, L., Savioli, L. and Bundy, D.A.P. (2000a). Towards an atlas of helminth infection in sub-Saharan Africa: the use of geographical information systems (GIS). *Parasitology Today* **16**, 303–307.

Brooker, S., Conelly, C.A. and Guyatt, H. (2000b). A practical application of an index to estimate the number of cases of intestinal nematode infections and schistosomiasis within a country: Cameroon as a case study. *Bulletin of the World Health Organization*, in press.

Brown, D.S. (1994). *Freshwater Snails of Africa and their Importance*. London: Taylor and Francis.

Brown, D.S., Jelnes, J.E., Kinoti, G.K. and Ouma, J. (1981). Distribution in Kenya of intermediate hosts for *Schistosoma*. *Tropical and Geographical Medicine* **33**, 95–103.

Brown, H.W. (1927). Studies on the rate of development and viability of eggs of *Ascaris lumbricoides* and *Trichuris trichiura* under field conditions. *Journal of Parasitology* **14**, 1–15.

Bundy, D.A.P., Chan M.S., Medley, G.F., Jamison, D. and Savioli, L. (2000). Intestinal nematode infections. In: *The Global Epidemiology of Infectious Diseases* (C. J.L. Murray and A.D. Lopez, eds). Cambridge: Harvard University Press, in press.

Bundy, D.A.P., Chandiwana, S.K., Homeida, M.M.A., Yoon, S. and Mott, K.E. (1991). The epidemiological implications of a multiple-infection approach to the control of human helminth infections. *Transactions of the Royal Society of Tropical Medicine and Hygiene* **85**, 274–276.

Bundy, D.A.P. and Cooper, E.S. (1989). *Trichuris* and trichuriasis in humans. *Advances in Parasitology* **28**, 107–173.

Burrough, P.A. and McDonnell, R.A. (1998). *Principles of Geographical Information Systems*. Oxford: Oxford University Press.

Chandler, A.C. (1928). The prevalence and epidemiology of hookworm and other helminthic infections in India. 12. General summary and conclusions. *Indian Journal of Medical Research* **15**, 696–743.

Cliff, A.D. and Haggett, P. (1988). *Atlas of Disease Distributions. Analytical Approaches to Epidemiological Data*. Oxford: Blackwell.

Corbett, J.D. and O'Brien, R.F. (1997). *The Spatial Characterization Tool—Africa v 1.0*. Texas Agricultural Experiment Station, Texas A and M University, Blackland Research Center Report No. 97–03, Documentation and CD-ROM.

Cort, W.W., Schapiro, L., Riley, W.A. and Stoll, N.R. (1929). A study of the influence of the rainy season on the level of helminth infestations in a Panama Village. *American Journal of Hygiene* **10**, 626–634.

Crompton, D.W.T. and Pawlowski, Z.S. (1985). Life history and development of *Ascaris lumbricoides* and the persistence of human ascariasis. In: *Ascariasis and its Public Health Significance* (D.W.T. Crompton, M.C. Nesheim and Z.S. Pawlowski, eds), pp. 9–23. London: Taylor and Francis.

Crompton, D.W.T. (1994). *Ascaris lumbricoides*. In: *Parasitic and Infectious Diseases: Epidemiology and Ecology* (M.E. Scott and G. Smith, eds), pp. 175–196. New York: Academic Press.

Crompton, D.W.T. and Tulley, J.J. (1987). How much ascariasis is there in Africa. *Parasitology Today* **3**, 123–127.

Cross, E.R. and Bailey, R.C. (1984). Prediction of areas endemic for schistosomiasis through use of discriminant analysis of environmental data. *Military Medicine* **149**, 28–30.

Cross, E.R., Perrine, R., Sheffield, C. and Pazzaglia, G. (1984). Predicting areas endemic for schistosomiasis using weather variables and a Landsat data base. *Military Medicine* **149**, 542–543.

Davies, C.M., Webster, J.P., Kruger, O., Munatsi, A., Ndamba, J. and Woolhouse, M.E.J. (1999). Host-parasite population genetics: a cross-sectional comparison of *Bulinus glosus* and *Schistosoma haematobium*. *Parasitology* **119**, 295–302.

Davies, C.M., Webster, J.P., Munatsi, A., Krüger, O., Ndamba, J., Noble, L.R. and Woolhouse, M.E.J. (2000). Schistosome host–parasite population genetics in the Zimbabwean Hiighveld. In: *A Status of Research on Medical and Veterinary Malacology in Africa* (H. Madsen, T.K. Kristensen and M. Chimbari, eds). Charlottenlund: DBL.

De Clercq, D., Sacko, M., Behnke, J.M., Traore, M. and Vercruysse, J. (1995). *Schistosoma* and geohelminth infections in Mali, west Africa. *Annales de la Societé Belge de Médicine Tropicale* **75**, 191–199.

Deichmann U. (1996). *African Population Database. Digital Database and Documentation*. United Nations Environmental Programme. Santa Barbara, CA:

National Center for Geographic Information and Analysis (http://
    grid2.cr.usgs.gov/globalpop/africa).
Diesfeld, H.J. (1969). Die Befundhaufigkeit von Helminthen beim Menschen in
    Kenia in Beziehung zu Umweltfaktoren [Incidence of helminth infestations in
    humans in Kenya and their relation to environmental factors]. *Zeitschrift für
    Tropenmedizin und Parasitologie* **20**, 310–333.
Diesfeld, H.J. and Hecklau, H.K. (1978). *Kenya: Regional Studies in Geographical
    Medicine*. Berlin: Springer-Verlag.
Doumenge, J.P., Mott, K.E., Cheung, C. *et al.* (1987). *Atlas of the Global Distribution
    of Schistosomiasis*. Bordeaux: Presses Universitares de Bordeaux.
EROS Data Center. (1996). GTOPO30 Documentation. Universal Resourse
    Locator: (http://edcwww.cr.usgs.gov/landdaac/gtopo30/). Sioux Falls, SD: Global
    Land Information System, EROS Data Center.
Guyatt, H., Brooker, S. and Donnelly, C.A. (1999). Can prevalence of infection in
    school-aged children be used as an index for assessing community prevalence.
    Parasitology **118**, 257–268.
Hall, A., Latham, M.C., Crompton, D.W.T., Stephenson, L.S. and Wolgemuth, J.C.
    (1982). Intestinal parasitic infections of men in four regions of rural Kenya.
    Transactions of the Royal Society of Tropical Medicine and Hygiene **76**, 728–733.
Hanitrasoamampionona, V., Brutus, L., Hebrard, G. *et al.* (1998). Epidemiological
    study of main human intestinal helminthiasis in the middle west of Madagascar.
    Bulletin de la Societé de Pathologie Exotique **91**, 77–80.
Hassan, A.N., Beck, L.R. and Dister, S. (1998a). Prediction of villages at risk for
    filariasis transmission in the Nile delta using remote sensing and geographic
    information system technologies. *Journal of the Egyptian Society of Parasitology* **28**,
    75–87.
Hassan, A.N., Dister, S. and Beck, L. (1998b). Spatial analysis of lymphatic filariasis
    distribution in the Nile delta in relation to some environmental variables using
    geographical information system technology. *Journal of the Egyptian Society of
    Parasitology* **28**, 119–131.
Hay, S.I. and Lennon, J.J. (1999). Deriving meteorological variables across Africa for
    the study and control of vector-borne disease: a comparison of remote sensing and
    spatial interpolation of climate. *Tropical Medicine and International Health* **4**,
    58–71.
Hay, S.I., Tucker, C.J., Rogers, D.J. and Packer, M.J. (1996). RS surrogates of
    meteorological data for the study of the distribution and abundance of arthropod
    vectors of disease. *Annals of Tropical Medicine and Parasitology* **90**, 1–19.
Hay, S.I., Packer, M.J. and Rogers, D.J. (1997). The impact of remote sensing on the
    study and control of invertebrate intermediate hosts and vectors for disease.
    International Journal of Remote Sensing **18**, 2899–2930.
Hay, S.I., Snow, R.W. and Rogers, D.J. (1998). From predicting mosquito habitat to
    malaria seasons using RS data: Practice, problems and perspectives. *Parasitology
    Today* **14**, 306–313.
Highton, R.B. (1974). Schistosomiasis. In: *Health and Disease in Kenya* (L.C. Vogel,
    A.S. Muller, R.S. Odingo, Z. Onyango and A.De Geus, eds), pp. 347–355. Nairobi:
    East African Literature Bureau.
Hoagland, K.E. and Schad, G.A. (1978). *Necator americanus* and *Ancylostoma
    duodenale*: life history parameters and epidemiological implications of two
    sympatric hookworms of humans. *Experimental Parasitology* **44**, 36–49.
Hsieh, H.C., Stoll, N.R., Reber, E.W., Chen, E.R., Kang, B.T. and Kuo, M. (1972).
    Distribution of *Necator americanus* and *Ancylostoma duodenale* in Liberia. *Bulletin
    of the World Health Organization* **47**, 317–324.

Huang, Y. and Manderson, L. (1992). Schistosomiais and the social patterning of infection. *Acta Tropica* **51**, 175–194.

Jemaneh, L. and Tedla, S. (1984). The distribution of *Necator americanus* and *Ancylostoma duodenale* in school populations, Gojam and Gondar administrative regions. *Ethiopian Medical Journal* **22**, 87–90.

Jemaneh, L. (1998). Comparative prevalences of some common intestinal helminth infections in different altitudinal regions in Ethiopia. *Ethiopian Medical Journal* **36**, 1–8.

Jordan, P. (1956). Filariasis in the Eastern, Tanga and Northern provinces of Tanganyika. *East African Medical Journal* **33**, 225–233.

Jordan, P. and Webbe, G. (1993). Epidemiology. In: *Human Schistosomiasis* (P. Jordan, G. Webbe and R.F. Sturrock, eds), pp. 87–158. Wallingford: CAB International.

Joubert, P.H., Pretorius, S.J., de Knock, K.N. and van Eeden, J.A. (1986). Survival of *Bulinus africanus* (Krauss), *Bulinus globosus* (Morelet) and *Biomphalaria pfeifferi* (Krauss) at constant high temperatures. *South African Journal of Zoology* **21**, 85–88.

Joubert, P.H., Pretorius, S.J., de Kock, K.N. and van Eeden, J.A. (1984). The effect of low temperatures on the survival of *Bulinus africanus* (Krauss), *Bulinus globosus* (Morelet) and *Biomphalaria pfeifferi*(Krauss). *South African Journal of Zoology* **19**, 314–316.

Kabaterine, N.B., Kazibwe, F. and Kemijumbi, J. (1996). Epidemiology of schistosomiasis in Kampala, Uganda. *East African Medical Journal* **73**, 795–800.

Kareiva, P. (1990). Population dynamics in spatially complex environments: theory and data. *Philosophical Transactions of the Royal Society of London. Series B* **330**, 175–190.

Keller, A.E., Leathers, W.S. and Densen, P.M. (1940). The results of recent studies of hookworm in eight southern states. *American Journal of Tropical Medicine and Hygiene* **20**, 493–509.

Kloos, H., Fulford, A.J., Butterworth, A.E. *et al.* (1997). Spatial patterns of human water contact and *Schistosoma mansoni* transmission and infection in four rural areas in Machakos District, Kenya. *Social Science and Medicine* **44**, 949–968.

Kloos, H., Lemma, A. and Desole, G. (1978). *Schistosoma mansoni* distribution in Ethiopia: a study in medical geography. *Annals of Tropical Medicine and Parasitology* **72**, 461–470.

Krüger, O., Ndamba, J. and Webster, J.P. (2000). Genetic analysis of the population structure of the two sympatric freshwater snails *Bulinus globosus* and *Biomphalaria pfeifferi* from the Zimbabwean highveld incorporating satellite remotely-sensed data. In: *Medical and Veterinary Malacology in Africa* (H. Madsen, T.K. Kristerson and P. Ndlovu, eds). Charlottelund: DBL.

Lardeux, F. and Cheffort, J. (1997). Temperature threshold and statistical modelling of larval *Wuchereria bancrofti* (Filariidea: Onchocercidae) developmental rates. *Parasitology* **114**, 123–134.

Larsen, M.N. and Roepstorff, A. (1999). Seasonal variation in development and survival of *Ascaris suum* and *Trichuris suis* eggs on pastures. *Parasitology* **119**, 209–220.

Levin, S.A. (1992). The problem of pattern and scale in ecology. *Ecology* **73**, 1943–1967.

Levin, S.A. and Pacala, S.W. (1997). Theories of simplification and scaling of spatially distributed processes. In: *Spatial Ecology. The Role of Space in Population Dynamics and Interspecific Interactions* (D. Tilman and P. Kareiva, eds), pp. 271–295. Princeton, NJ: Princeton University Press.

Lindsay, S.W., Denham, D.A. and McGreevy, P.B. (1984). The effect of humidity on the transmission of *Brugia pahangi* infective larvae to mammalian hosts by *Aedes aegypti*. *Transactions of the Royal Society of Tropical Medicine and Hygiene* **78**, 19–22.

Lindsay, S.W. and Thomas, C.J. (2000). Mapping and estimating the population at risk from lymphatic filariasis in Africa. *Transactions of the Royal Society of Tropical Medicine and Hygiene* **94**, 37–45.

Lwambo, N.J. S., Siza, J.E., Brooker, S., Bundy, D.A.P. and Guyatt, H. (1999). Patterns of concurrent infection with hookworm and schistosomiasis in school children in Tanzania. *Transactions of the Royal Society of Tropical Medicine and Hygiene* **93**, 497–502.

Malone, J.B., Abdel Rahman, M.S., El Bahy, M.M., Huh, O.K., Shafik, M. and Bavia, M. (1997). Geographic information systems and the distribution of *Schistosoma mansoni* in the Nile delta. *Parasitology Today* **13**, 112–119.

Malone, J.B., Huh, O.K., Fehler, D.P. *et al.* (1994). Temperature data from satellite imagery and the distribution of schistosomiasis in Egypt. *American Journal of Tropical Medicine and Hygiene* **50**, 714–722.

Marti, H.P., Tanner, M., Degrémont, A.A. and Freyvogel, T.A. (1985). Studies on the ecology of *Bulinus globosus*, the intermediate host of *Schistosoma haematobium* in the Ifakara area, Tanzania. *Acta Tropica* **42**, 171–187.

Matsusaki, G. (1963). Influence of low temperature, sunshine, desiccation and human excreta upon the eggs of *Ancylostoma duodenale* and *Necator americanus* Part I. Low temperature. *Yokohama Medical Bulletin* **14**, 73–79.

McConnell, E. and Armstrong, J. C. (1976). Intestinal parasitism in fifty communities on the central plateau of Ethiopia. *Ethiopian Medical Journal* **14**, 159–165.

McCoy, O.R. (1930). The influence of temperature, hydrogen-ion concetration, and oxygen tension on the development of the eggs and larvae of the dog hookworm, *Ancylostoma caninum*. *American Journal of Hygiene* **11**, 413–448.

McCullough, F.S. (1972). The distribution of *Schistosoma mansoni* and *S. haematobium* in East Africa. *Tropical and Geographical Medicine* **24**, 199–207.

McGreevy, P.B., Theis, J.H., Lavoipierre, M.M.J. and Clark, J. (1974). Studies on filariasis. III. *Dirofilaria immitis*: emergence of infective larvae from the mouth parts of *Aedes aegypti*. *Journal of Helminthology,* **48**, 221–228.

Michael, E. and Bundy, D.A.P. (1997). Global mapping of lymphatic filariasis. *Parasitology Today* **13**, 472–476.

Michael, E., Bundy, D.A.P. and Grenfell, B.T. (1996). Re-assessing the global prevalence and distribution of lymphatic filariasis. *Parasitology* **112**, 409–428.

Michael, E. and Bundy, D.A.P. (1998). Herd immunity to filarial infection is a function of vector biting rate. *Proceedings of the Royal Society of London, Series B* **265**, 855–860.

Mizgajska, H. (1993). The distribution and survival of eggs of *Ascariasis suum* in six different natural soil profiles. *Acta Parasitologica* **38**, 170–174.

Mott, K.E., Nuttall, I., Desjeux, P. and Cattand, P. (1995). New geographical approaches to control of some parasitic zoonoses. *Bulletin of the World Health Organization* **73**, 247–257.

Nawalinski, T., Schad, G.A. and Chowdhury, A.B. (1978). Population biology of hookworms in children in rural West Bengal II. Acquisition and loss of hookworms. *American Journal of Tropical Medicine and Hygiene* **27**, 1162–1173.

Nelson, G.S.E. (1958). *Schistosoma mansoni* infection in the West-Nile district of Uganda. Part II. The distribution of *S. mansoni* with a note on the probable vectors. *East African Medical Journal* **35**, 335–344.

Nuttall, I., O'Neill, K. and Meert, J.P. (1998). Systemes d'information géographique et lutte contre les maladies tropicales. *Médecine Tropicale* **58**, 221–227.

Nwosu, A.B.C. (1978). Investigations into the free-living phase of the cat hookworm life-cycle. *Zeitschrift für Parasitenkunde* **56**, 243–249.

Nwosu, A.B.C. and Anya, A.O. (1980). Seasonality in human hookworm infection in an endemic area of Nigeria, and its relationship to rainfall. *Tropenmedizin und Parasitologie* **31**, 201–208.

Omori, N. (1958). Experimental studies on the role of the house mosquito, *Culex pipiens pallens* in the transmission of Bancroftian filariasis 5. On the distribution of infective larvae in the mosquito and the effects of parasitism of filariae upon the host insect. *Nagasaki Medical Journal* **33**, 143–155.

Openshaw, S. (1996). Geographical information systems and tropical diseases. *Transactions of the Royal Society of Tropical Medicine and Hygiene* **90**, 337–339.

Ottesen, E.A., Duke, B.O.L., Karam, M. and Behbehani, K. (1997). Strategies and tools for the control/elimination of lymphatic filariasis. *Bulletin of the World Health Organization* **75**, 491–503.

Otto, G.F. (1929). A study of the moisture requirements of the eggs of the horse, the dog, human and pig ascarids. *American Journal of Hygiene* **10**, 497–520.

Partnership for Child Development (1999). Self-diagnosis as a possible basis for treating urinary schistosomiasis: a study of schoolchildren in a rural area of the United Republic of Tanzania. *Bulletin of the World Health Organization* **77**, 477–483.

Pavlovsky, E.P. (1966). *Natural Nidality of Transmissible Diseases—with Special Reference to the Landscape Epidemiology of Zooanthroponoses*. Urbana: University of Illinois Press.

Pawlowski, Z.A. (1984a). Ascariasis. In: *Tropical and Geographical Medicine* (K.S. Warren and A.A.F. Mahmouud, eds), pp. 369–378. New York: McGraw-Hill.

Pawlowski, Z.A. (1984b). Trichuriasis. In: *Tropical and Geographical Medicine* (K.S. Warren and A.A.F. Mahmouud, eds), pp. 380–384. New York: McGraw-Hill.

Payne, F.K. (1922). Investigations on the control of hookworm disease. XI. Vertical migration of infective hookworm larvae in the soil. *American Journal of Hygiene* **11**, 254–263.

Pflüger, W. (1980). Experimental epidemiology of schistosomiasis I. The prepatent period and cercarial production of *Schistosoma mansoni* in *Biomphalaria* snails at various constant temperatures. *Zeitschrift für Parasitenkunde* **63**, 159–169.

Pitchford, R.J. (1981). Temperature and schistosome distribution in South Africa. *South Africa Journal of Science* **77**, 252–261.

Pitchford, R.J. and Visser, P.S. (1965). Some further observations on schistosome transmission in the Eastern Transvaal. *Bulletin of the World Health Organization* **32**, 83–104.

Prost, A. (1987). L'ascaridiose en Afrique de l'ouest. Revue épidémiologique. *Annales de Parasitologie Humaine et Comparée* **62**, 434–455.

Raghavan, N.G.S. (1957). Epidemiology of filariasis in India. *Bulletin of the World Health Organization* **16**, 553–579.

Raghavan, N.G.S. (1969). Clinical manifestations and associated epidemiological factors of filariasis. *Journal of Communicable Disease* **1**, 75–102.

Ratard, R.C., Kouemeni, L.E., Bessala, M.M.E., Ndamkou, C.N., Sama, M.T. and Cline, B.L. (1991). Ascariasis and trichuriasis in Cameroon. *Transactions of the Royal Society of Tropical Medicine and Hygiene* **85**, 84–88.

Red Urine Study Group (1995). *Identification of High Risk Communities for Schistosomiasis in Africa: A Multi-country Study*. Social and Economic Research Project Reports, No. 15. Geneva: World Health Organization.

Rogers, D.J. and Randolph, S.E. (1991). Mortality rates and population density of tsetse flies correlated with satellite imagery. *Nature* **351**, 739–741.

Rogers, D.J. and Randolph, S.E. (1993). Distribution of tsetse and ticks in Africa: past, present and future. *Parasitology Today* **9**, 266–271.

Rogers, D.J. and Williams, B.G. (1993). Monitoring trypanosomiasis in space and time. *Parasitology* **106**, S77-S92.

Rollinson, D. and Simpson, A.J.G. (1987). *The Biology of Schistosomes. From Genes to Latrines.* London: Academic Press.

Rossi, R.E., Mulla, D.J., Journel, A.G. and Franz, E.H. (1992). Geostatistical tools for modeling and interpreting ecological spatial dependence. *Ecological Monographs* **62**, 277–314.

Rukmono, B. (1980). Infection route of roundworm and hookworm (with reference to the development and viability in soil of infective stages). In: *Collected Papers on the Control of Soil-transmitted Helminthiases*, Vol. 1, pp. 125–128. Tokyo: Asian Parasite Control Organisation.

Sasa, M. (1976). *Human Filariasis—A Global Survey of Epidemiology and Control.* Tokyo: University of Tokyo.

Savioli, L., Renganathan, E., Montresor, A., Davis, A. and Behbehani, K. (1997). Control of schistosomiasis—a global picture. *Parasitology Today* **13**, 444–448.

Schad, G.A. and Banwell, J.G. (1984). Hookworms. In: *Tropical and Geographical Medicine* (K.S. Warren and A.A.F. Mahmouud, eds), pp. 369–378. New York: McGraw-Hill.

Scott, J.A. (1945). Hookworm disease in Texas. *Texas Reports of Biology and Medicine* **3**, 558–568.

Seamster, A.P. (1950). Developmental studies concerning the eggs of *Ascaris lumbricoides* var. *suum*. *The American Midland Naturalist* **43**, 450–468.

Shiff, C.J. (1964). Studies on *Bulinus* (*Phyopsis*) *globosus* in Rhodesia. I. The influence of temperature on the intrinsic rate of natural increase. *Annals of Tropical Medicine and Parasitology* **58**, 94–105.

Shiff, C.J., Evans, A., Yiannakis, C. and Eardley, M. (1975). Seasonal influence on the production of *Schistosoma haematobium* and *S. mansoni* cercariae in Rhodesia. *International Journal of Parasitology* **5**, 119–123.

Shope, R.E. (1999). Factors influencing geographic distribution and incidence of tropical infectious diseases. In: *Tropical Infectious Diseases. Principles, Pathogens and Practice* (R.L. Guerrant, D.H. Walker and P.F. Weller, eds), pp. 16–21. London: Churchill Livingstone.

Smith, A.J. (1903). Uncinariasis in Texas. *American Journal of Medicine* **126**, 768–798.

Southgate, V.R. and Rollinson, D. (1987). Natural history of transmission and schistosome interactions. In: *The Biology of Schistosomiasis. From Genes to Latrines* (D. Rollinson and A.J.G. Simpson, eds), pp. 347–378. London: Academic Press.

Spindler, L.A. (1929). The relation of moisture to the distribution of human trichuris and ascaris. *American Journal of Hygiene* **10**, 476–496.

Stephenson, L.S. (1987). *Impact of Helminth Infection in Human Nutrition.* New York: Taylor and Francis.

Stiles, C.W. (1903). Report upon the prevalence and geographic distribution of hookworm disease (uncinariasis or anchylostomiasis) in the United States. *Hygienic Laboratory Bulletin* **10**, 1–10.

Stoll, N.R. (1923). Investigations on the control of hookworm disease. XXIV. Hookworm cultures with humus, sand, loam and clay. *American Journal of Hygiene* **6**, 1–36.

Stoll, N.R. (1947). This wormy world. *Journal of Parasitology* **33**, 1–18.

Stott, G. (1960). Hookworm infection and anaemia in Mauritius. *Transactions of the Royal Society of Tropical Medicine and Hygiene* **55**, 20–25.

Sturrock, R.F. (1966). The influence of temperature on the biology of *Biomphalaria pfeifferi* (Krauss), an intermediate host of *Schistosoma mansoni. Annals of Tropical Medicine and Parasitology* **60**, 100–105.

Sturrock, R.F. (1967). Hookworm studies in Tanganyika (Tanzania): the results of a series of surveys on a group of primary schoolchildren and observations on the survival of hookworm infective larvae exposed to simulative field conditions. *East African Medical Journal* **44**, 143–149.

Sturrock, R.F. (1993). The intermediate hosts and host–parasite relationships. In: *Human Schistosomiasis* (P. Jordan, G. Webbe and R.F. Sturrock, eds), pp. 33–85. Wallingford: CAB International.

Sweet, W.C. (1924). The intestinal parasites of man in Australia and its dependencies as found by the Australian hookworm campaign. *The Medical Journal of Australia* **1**, 405–407.

Tedla, S. (1986). Hookworm infections in several communities in Ethiopia. *East African Medical Journal* **63**, 134–139.

Thompson, D.F., Malone, J.B., Harb, M. *et al.* (1996). Bancroftian filariasis distribution and diurnal temperature differences in the southern Nile delta. *Emerging Infectious Diseases* **2**, 234–235.

Thomson, M.C., Connor, S.J., Dalessandro, U. *et al.* (1999). Predicting malaria infection in Gambian children from satellite data and bed net use surveys: the importance of spatial correlation in the interpretation of results. *American Journal of Tropical Medicine and Hygiene* **61**, 2–8.

Udonsi, J.K., Nwosu, A.B.C. and Anya, A.O. (1980). *Necator americanus*: population structure, distribution, and fluctuations in population densities of infective larvae in contaminated farmlands. *Zeitschrift für Parasitenkunde* **63**, 251–259.

Udonsi, J.K. (1983). *Necator americanus* infection: a longitudinal study of an urban area in Nigeria. *Annals of Tropical Medicine and Parasitology* **77**, 305–310.

Udonsi, J.K. and Atata, G. (1987). *Necator americanus*: temperature, pH, light, and larval development, longevity, and desiccation tolerance. *Experimental Parasitology* **63**, 136–142.

Utroska, J.A., Chen, M.G., Dixon, H. *et al.* (1989). *An Estimate of the Global Needs for Praziquantel within Schistosomiasis Control Programmes.* Geneva: World Health Organization.

van Eeden, J.A. and Combrinck, C. (1966). Distributional trends of four species of freshwater snails in South Africa with special reference to the intermediate hosts of bilharzia. *Zoologica Africana* **2**, 95–109.

Vinayak, V.K., Chitkara, N.L. and Chhuttani, P.N. (1979). Soil dynamics of hookworm larvae. *Indian Journal Medical Research* **70**, 609–614.

Walsh, S.J., Evans, T.P., Welsh, W.F., Entwisle, B. and Rindfuss, R.R. (1999). Scale-dependent relationships between population and environment in northeastern Thailand. *Photogrammetric Engineering & Remote Sensing* **65**, 97–105.

Warren, K.S., Bundy, D.A.P., Anderson, R.M. *et al.* (1993). Helminth infections. In: *Disease Control Priorities in Developing Countries* (D.T. Jamison, W.H. Mosley, A.R. Measham and J.L. Bobadilla, eds), pp. 131–160. Oxford: Oxford University Press.

Watkins, W.E. and Pollitt, E. (1997). 'Stupidity or Worms': do intestinal worms impair mental performance? *Psychological Bulletin* **121**, 171–191.

Webbe, G. (1962). The transmission of *Schistosoma haematobium* in an area of Lake Province, Tanganyika. *Bulletin of the World Health Organization* **27**, 59–85.

WHO (1967). *Control of Ascariasis*. Report of a WHO Expert Committee. Geneva: World Health Organization.

WHO (1984). *Lymphatic Filariasis*. Fourth Report of the WHO Expert Committee on Filariasis. WHO Technical Report Series, 702. Geneva: World Health Organization.

WHO (1995) *The Schistosomiasis Manual*. Social and Economic Research Project Reports, No. 3. Geneva: World Health Organization.

WHO (1999). *Report on the WHO Informal Consultation on Schistosomiasis Control*. Geneva, 2–4 December 1999. WHO/CDS/CPC/SIP/99.2. Geneva: World Health Organization.

Wiens, J.A. (1989). Spatial scaling in ecology. *Functional Ecology* **3**, 385–397.

Wiens, J.A., Stenseth, N.C., Van Horne, B. and Ims, R.A. (1993). Ecological mechanisms and landscape ecology. *Oikos* **66**, 369–380.

Woolhouse, M.E.J. (1994). Epidemiology of human schistosomes. In: *Parasitic and Infectious Diseases: Epidemiology and Ecology*, (M.E. Scott and G. Smith, eds), pp. 197–217. New York: Academic Press.

Woolhouse, M.E.J., Watts, C.H. and Chandiwana, S.K. (1991). Heterogeneities in transmission rates and the epidemiology of schistosome infection. *Proceedings of the Royal Society London, Series B* **245**, 109–114.

Advances in Satellite Remote Sensing of Environmental Variables for Epidemiological Applications

S.J. Goetz, S.D. Prince and J. Small

Department of Geography, University of Maryland, College Park, Maryland, MD 2074–8225,USA

ABSTRACT

Earth-observing satellites have provided an unprecedented view of the land surface but have been exploited relatively little for the measurement of environmental variables of particular relevance to epidemiology. Recent advances in techniques to recover continuous fields of air temperature, humidity, and vapour pressure deficit from remotely sensed observations have significant potential for disease vector monitoring and related epidemiological applications. We report on the development of techniques

ADVANCES IN PARASITOLOGY VOL 47
0065-308-X $30.00

to map environmental variables with relevance to the prediction of the relative abundance of disease vectors and intermediate hosts. Improvements to current methods of obtaining information on vegetation properties, canopy and surface temperature and soil moisture over large areas are also discussed. Algorithms used to measure these variables incorporate visible, near-infrared and thermal infrared radiation observations derived from time series of satellite-based sensors, focused here primarily but not exclusively on the Advanced Very High Resolution Radiometer (AVHRR) instruments. The variables compare favourably with surface measurements over a broad array of conditions at several study sites, and maps of retrieved variables captured patterns of spatial variability comparable to, and locally more accurate than, spatially interpolated meteorological observations. Application of multi-temporal maps of these variables are discussed in relation to current epidemiological research on the distribution and abundance of some common disease vectors.

1. INTRODUCTION

There have been rapid and widespread advances in the application of Earth-observing satellite sensor data since the launch of the first 'land satellite' (LandSat) in 1972. Satellite remote sensing observations are now routinely used to monitor land use change, estimate rangeland and crop production, improve weather predictions, and drive ecosystem, energy balance and ecophysiology models. Interdisciplinary Earth system science, which has emerged in research programmes worldwide, has advanced rapidly with the advent of Earth-observing satellites. The use of satellite observations has met with reticence in some disciplines, which is not unwarranted, but this has diminished as the utility of the observational data sets has become more evident in a diverse array of applications. Recently the National Aeronautics and Space Administration (NASA) has established an initiative focused on epidemiological applications of satellite sensor data and made it a component of their Earth Science Enterprise Program (Myers *et al.*, this volume). Epidemiology is of interest owing to the impacts of vector-borne disease on millions of people worldwide. In particular, remote sensing provides observations useful for monitoring environmental conditions favourable to the reproductive success, development, dispersal, and survival of disease vectors, primarily arthropods and particularly insects, of which mosquitoes are the most prevalent and deleterious (Snow *et al.*, 1999). Information on environmental conditions has proved useful for predicting the relative abundance of vectors, thus the potential for spread of pathogens to, and episodic outbreaks of disease among, human populations (Onori and Grab, 1980; Hay *et al.* 2000).

Satellite observations have been applied to epidemiological applications in a number of different ways (Hay *et al.*, 1997; Rogers, this volume; Hay *et al.*, this volume; Randolph and Brooker, this volume; Michael, this volume). For the most part, those who have used Advanced Very High Resolution Radiometer (AVHRR) have relied on information related to the spatial and temporal variability of spectral vegetation indices (SVIs, such as the Normalized Difference Vegetation Index, NDVI) and apparent land surface temperature (LST) (e.g. Rogers *et al.*, 1996; Hay *et al.*, 1998a). There are, however, a number of environmental variables in addition to those inferred from NDVI (e.g. proportional vegetation cover) and LST that are important to epidemiological research, particularly for improving predictive models of disease vector spatial distributions and their sensitivity to a range of interacting variables. These include atmospheric humidity, vapour pressure deficit, soil moisture and air temperature, among others.

In reality, few of the environmental variables that affect the abundance of disease vectors are independent. Rather than limiting the applicability of remote sensing to epidemiology, however, this allows the use of SVIs as surrogates for several related variables. Vegetation cover, for example, is affected by the many variables listed above and, in fact, frequently varies in relation to them in a predictable manner. Thus remote sensing can be particularly useful for monitoring regions or localized areas where temporal variability in weather conditions, which affect vegetation cover, result in epidemics. This most frequently occurs in areas where historical transmission levels, thus collective immunity, are relatively low (e.g. semi-arid and mountainous zones of Africa) (Hay *et al.*, 2000). The relative importance of any single environmental variable, nevertheless, cannot be dismissed and may change rapidly if the breeding habitat is altered or a threshold of relevance to larval development, for example, is reached. This may occur at short time scales even if correlated variables, such as vegetation cover, remain invariant. Thus it is important to consider both integrative variables, such as vegetation indices, as well as surface environmental variables that vary more rapidly.

We describe methods to obtain environmental variables of relevance to mapping and predicting the presence and relative abundance of disease vectors from satellite observations, and discuss areas of epidemiological research that may benefit from recent advances in this area.

2. SATELLITE RECOVERY OF LAND SURFACE VARIABLES

A wide range of land surface variables, both environmental and biophysical, can be inferred from satellite observations. We briefly review the methodologies used to recover a suite of such variables and, where possible, provide accuracy metrics based on comparison with field measurements. It is important to keep in mind that field measurements are rarely available at a spatial scale that incorporates the range of spatial variability in surface reflectance or thermal emission captured by satellite sensors. Thus sample variability and error present in field measurements limit their utility as 'ground truth' (an unfortunate misnomer), in some cases perhaps as much as the 'errors' in satellite-based recoveries of land surface variables introduced by variability in atmospheric constituents, sensor calibration, and geometric rectification (each of which are research issues beyond the scope of this chapter).

2.1. Vegetation Cover and Type

A substantial proportion of the literature on remote sensing of the land surface has been focused on vegetation type classification, partly because of the long history of classification work in plant ecology and partly because many modern ecosystem models are based on plot-scale studies that are specific to the type of vegetation in which they were conducted. In epidemiology and more generally, vegetation type may be most relevant in that it reflects and modifies land surface processes such as energy or materials exchange. In terrestrial carbon exchange modelling, for example, there is a trend towards de-emphasis of species composition and a focus on rate-limiting factors associated with nutrient availability, resource scaling, and carbon allocation (Goetz and Prince, 1999). Similarly, there are frequently robust associations of disease and vector abundance with the amount and density, rather than the species compostion of vegetation cover (e.g. Rejmankova *et al.*, 1991; Hay *et al.*, 1998b). Typically this results from the co-variation of vegetation cover with a suite of biophysical variables (temperature, energy environment, habitat diversity). Satellite sensor observations of the land surface are uniquely suited to mapping the density of photosynthetically active material (i.e. foliage) and its spatial and temporal variations. The wide array of SVIs that have proliferated since the advent of land remote sensing all exploit the divergence of light absorption and reflection between the visible and near-infrared wavelengths, a property indicative of vegetation amount sometimes expressed as the fraction of photosynthetically active radiation absorbed (Fpar), leaf area index or green biomass.

There have been numerous studies linking the spatio-temporal dynamics of SVIs with the presence of disease vectors. The literature on these studies has been summarized elsewhere (Rogers, this volume; Hay *et al.*, this volume; Randolph, this volume; Brooker and Michael, this volume). Thus, rather than emphasizing vegetation dynamics *per se*, we focus on those environmental variables that have been identified as important for epidemiological research but which have only recently been measured over large areas from remotely sensed observations. We note here that there are limits to the utility of SVIs alone for the study of landscape factors related to epidemiology, largely because of rapidly varying land surface variables critical to vector breeding success and developmental productivity (Rogers *et al.*, 1996). One of the most commonly utilized and epidemiologically relevant of these is surface temperature.

2.2. Land Surface Temperature

In epidemiological applications there is interest in both absolute and relative estimates of LST. We note that LST is a term that has been used rather ambiguously to include a wide range of 'surface temperature' variables including kinetic, thermodynamic, radiometric, canopy, and even air temperature (Norman and Becker, 1995). Satellite sensors measure a signal that is a combination of the radiant temperature of the land surface and the intervening atmosphere. Moreover, the land surface emits radiance differently across the thermal spectrum, and the emitted radiance is affected by the composition of the surface constituents, particularly spectral emissivity (ε, the ratio of surface emission to that of a perfect emitter at the same temperature). Partially vegetated areas with soils containing a substantial quartz content, for example, will have significantly lower ε than surrounding vegetated areas (Salisbury and D'Aria, 1992). The effect of ε on the determination of the actual radiometric temperature can be substantial. A difference in ε from 1.0 to 0.99 typically results in an increase of 1–2°C (Coll *et al.*, 1994; Goetz, 1997; Schmugge *et al.*, 1998). Disease vector or parasite survival, development or oviposition cycles can also be sensitive to variations of this magnitude.

From this point on we use LST to refer to a general index of the apparent environmental temperature (whether soil or vegetation), and radiometric surface temperature (T_s) as a well-defined variable described by Planck's law, including the effects of ε and the atmosphere (Li and Becker, 1993; Goetz *et al.*, 1995):

$$T_s = \frac{K_2}{\ln(1 + K_1 T\varepsilon/L - L_a - L_r)} \tag{1}$$

where

T is atmospheric transmission,
L is surface radiance,

L_a is atmospheric radiance,
L_r is the reflected portion of downward irradiance, and
K_1 and K_2 are parameters derived from Planck's equation and the spectral calibration of the sensors (after Markham and Barker, 1986).

There have been numerous methods devised to estimate and account for the effects of surface emissivity and the contribution of the atmosphere to at-sensor radiance. Generalized global emissivity maps have been available since the 1970s (Prabhakara and Dalu, 1976), and a more recent map has been produced by the Clouds and Earth Radiation Energy Systems (CERES) project using measurements by Salisbury and D'Aria (1992). Several methods have also recently been developed to generate maps of ε from multispectral imagery using variations of emitted radiance in different thermal channels (Schmugge *et al.*, 1991; Coll *et al.*, 1994; Casalles *et al*, 1997). A similar methodology is used to account for atmospheric attenuation of thermal radiance detected by satellite sensors. In this case an atmospheric

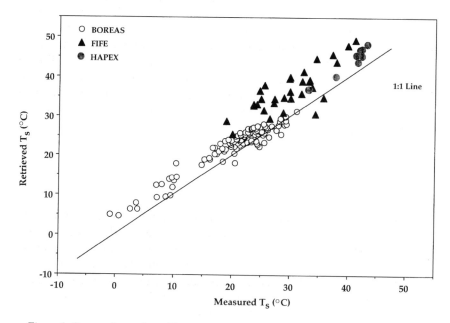

Figure 1 Comparison of satellite retrieved and near-surface measured radiometric surface temperatures (°C) for three study areas including the Boreal Ecosystem Atmosphere Study (BOREAS) in central Canada, the First International Satellite Land Surface Climatology Project Field Experiment (FIFE) in central Kansas, and the Hydrology–Atmosphere Pilot Experiment in the Sahel (HAPEX-Sahel). The satellite-based estimates accounted for 89% of the variability in 150 near-coincident surface measurements.

'split-window' approach uses the fact that the different signal in two thermal bands that are observing the same area of the surface is caused by the differential absorption of radiation in the atmosphere (Price, 1984; Dalu, 1986; Harris and Mason, 1992). Nearly 20 split-window algorithms have been developed using semi-empirical approaches that relate radiative transfer model simulations of T_s to remotely sensed observations. These algorithms perform with varying degrees of success depending on the range of conditions for which they were developed (e.g. Asrar et al., 1988; Becker and Li, 1990; Kalluri and Dubayah, 1995). More recent split-window algorithms have attempted to develop generalized coefficients that account for a wider range of atmospheric and surface conditions, as well as bandwidth variations between different sensors (Czajkowski et al., 1998; Ouaidrari et al., 2000). An assessment of the accuracy of a recent split-window T_s algorithm (Ouaidrari et al, 2000) for 150 measurements over a broad range of surface conditions in disparate study areas (Prince et al., 1998) resulted in a root mean square error (RMSE) of 3.48°C for a range of 48.0°C (Figure 1).

The intricacies of T_s determination may be less relevant to epidemiologists when used as a relative index of conditions favourable to vector abundance. Determination of spatial and temporal variability in LST, for example, may be used as a correlative index of vector abundance (e.g. Malone et al., 1994; Rogers et al., 1996). The utility of LST to disease monitoring may be significantly enhanced in studies of vector survival and reproductive success if, rather than using LST as a statistical predictor of spatio-temporal variability in relative abundance, absolute values of temperature were used.

2.3. Air Temperature

Limitations in the applicability of LST to epidemiological applications are not restricted to issues of apparent versus actual radiometric surface temperature. Some of the more relevant and useful variables of interest to disease vector monitoring include near-surface air temperature (T_a) and relative humidity (h). Methods have been devised to obtain these variables using remote sensing. These innovative techniques use a contextual combination of vegetation indices and split-window estimates of T_s. Inference of T_a is based on an assumption that the radiometric temperature of a fully vegetated canopy is in equilibrium with ambient air temperature, because of the similar heat capacity of dense vegetation and the surrounding air (Goward et al., 1994; Prihodko and Goward, 1997). A high SVI value (e.g. 0.65 NDVI for calibrated AVHRR radiances) is used to set the equilibrium threshold (Figure 2). The contextual relationship between vegetation index and split-window T_s values is accomplished with ordinary least squares regression, where the sample size is set by the dimensions of a moving window passed over the images. The regression coefficients are used to solve for $T_s = T_a$ at the threshold SVI value:

$$T_a = a \times SVI_{max} + b \qquad (2)$$

where

SVI$_{max}$ is set between 0.6 and 0.9 depending on the specific SVI and sensor used, and

a and b are ordinary least squares regression coefficients that change with the moving window contextual approach.

Application of this method, known as TvX, at the global scale with AVHRR data sets of 8 km (64 km^2) spatial resolution has produced maps of T_a that are comparable to climatological maps in terms of their magnitude, spatial patterns, and temporal variability (see Plate 14). Comparisons of TvX-recovered T_a values with 712 field measurements at a broad range of study sites indicate that T_a can be estimated using this methodology to within 3.93°C RMSE over a range of 36.0°C. These comparisons, however, included both 1 km^2 and 64 km^2 AVHRR observations. The 1 km^2 results alone were more accurate (RMSE 2.98°C over a comparable temperature range)

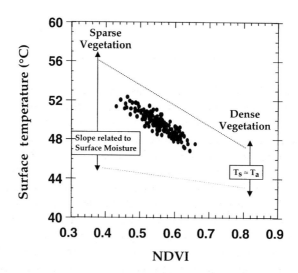

Figure 2 The contextual TvX approach used to recover air temperature and surface wetness from radiometric surface temperature and a spectral vegetation index, in this case the NDVI. The approach uses a moving window, where the resolution of each window element is dependent on the spatial characteristics of the sensor.

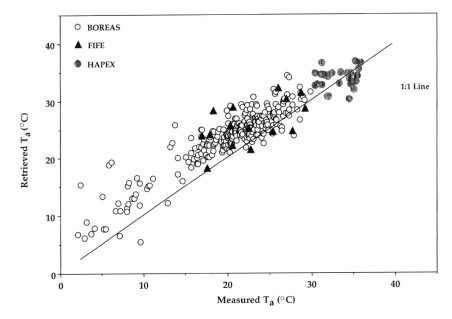

Figure 3 Same as Figure 1 but for air temperature (T_a). The satellite-based T_a estimates accounted for 83% of the variability in 314 near-coincident surface measurements.

because the areas captured in higher resolution satellite sensor data were more comparable to that characterized by the meteorological stations (Figure 3).

The timing of the satellite observations can have a substantial effect on the recovery of T_a using the TvX approach (Goetz, 1997). In most situations T_s increases shortly after sunrise, peaks after solar noon owing to thermal inertia, and decreases through the remainder of the diurnal cycle until the following sunrise. T_a follows a similar cycle but lags T_s as a result of evaporative cooling and partitioning of sensible and latent heat fluxes. In the case of the AVHRR, a mid-afternoon overpass time typically results in estimates of T_a that approximate the daily maximum rather than a daily T_a average. This was evident in comparisons of T_a recovered using the TvX approach with gridded T_a fields spatially interpolated from meteorological stations for continental Africa (Figure 4). The image depicts the mean residual of satellite-derived T_a maps and monthly averaged maximum daily temperature fields, prepared from climate station data by the Climate Research Unit (CRU, East Anglia, UK), for 96 months between 1982 and 1989. The image shows close agreement between the two measures across the continent (mean difference = 2.6°C), and agreement within 1°C over

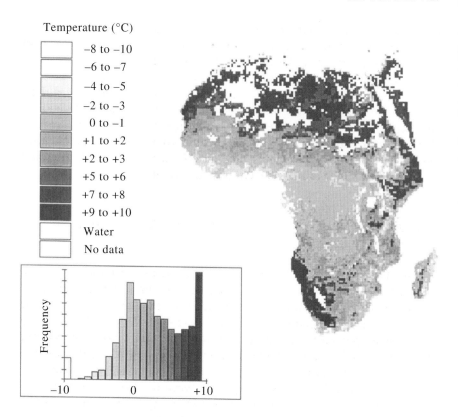

Figure 4 Difference between air temperature fields derived from AVHRR satellite sensor observations (TvX) and interpolated meteorological station observations (CRU), based on monthly residuals averaged over a 96-month period. The 64 km² satellite data were degraded to match the 0.5° latitude/longitude grid of interpolated station measurements.

most vegetated regions, but systematically higher T_a from the TvX method over very sparsely vegetated areas. The reasons for these differences are explored elsewhere (Hay and Lennon 1999; Goetz *et al.*, 2000), but include dependence of the TvX method on the presence of a sufficient range of SVI and T_s conditions to infer T_a in equilibrium with T_s. In sparsely vegetated regions this equilibrium occurs infrequently. Interpolation of meteorological stations across regions with sparse measurements add further uncertainty to comparisons between these different approaches to mapping T_a over large areas.

The advantage of the maps derived from the satellite observations is that the image data were acquired continuously across the landscape, thus providing higher resolution than can be accomplished through the interpolation of meteorological stations, particularly in remote areas with few stations or where maintenance is infrequent. We note, for example, that the comparisons shown in Figure 4 required degradation of the satellite sensor observations, both spatially (64 km^2 to 2500 km^2) and temporally (daily to monthly) to match that available from the spatially interpolated climate fields. The temperature maps derived from the satellite observations clearly capture spatial variability more effectively than is possible with interpolated field measurements (see also Prince et al., 1998; Hay and Lennon, 1999; Goetz et al., 2000; Green and Hay, 2000).

2.4. Atmospheric and Near-surface Humidity, Vapour Pressure Deficit

Near-surface atmospheric humidity may be expressed in terms of mass of water vapour to mass of air (specific humidity, q), mass of water vapour to volume of air (absolute humidity, χ), the ratio of actual to saturation vapour pressures (relative humidity, h), or the difference between saturation and actual vapour pressures–the vapour pressure deficit (D, kPa). D, an index of 'drying power' of air, is a humidity measure particularly relevant to disease vector populations. Tsetse fly mortality, for example, has been found to be inversely related to D (Rogers, 1991). Until recently it has not been possible to estimate D over large areas other than by interpolating meteorological station observations of wet and dry-bulb temperatures, or using correlative indices as surrogates for D.

Using fields of T_a calculated with the TvX technique, and a split-window estimate of precipitable water vapour content of the total atmospheric column (U, g cm^3), it is possible to calculate D over large areas with satellite observations. The technique was developed to provide input to ecosystem models (e.g. Goward et al., 1994; Prince and Goward, 1995; Goetz et al., 1999), building on a series of developments in the measurement of surface physical variables with remote sensing. The value of U is estimated at the surface using a logarithmic approximation of the vertical distribution of water vapour, assuming it increases towards the surface as a result of boundary layer dynamics (after Smith, 1966). The near-surface water vapour amount is then used to calculate an estimate of saturated vapour pressure based on dewpoint temperature (T_d), and is differenced with vapour pressure estimated at T_a (Monteith and Unsworth, 1990; Prince and Goward, 1995):

$$D = 0.611 \times \left[\exp\left(17.27 \times \frac{T_a}{T_a + 237} \right) - \exp\left(17.27 \times \frac{T_d}{T_d + 237} \right) \right] \tag{3a}$$

where

$$T_d = \frac{\ln(\lambda+1)+\ln(U)-0.1133}{0.0393} \tag{3b}$$

λ = coefficient to adjust for latitude and season (Smith, 1966) and

$$U = \left[17.32 \times \left(\frac{\Delta T - 0.0683}{T_s - 291.97}\right)\right] + 0.546 \tag{4}$$

ΔT is the radiometric temperature difference between the two thermal channels of the AVHRR (the split-window).

Maps of D calculated using this approach at the global scale (see Plate 15) show patterns of spatial and temporal variability that are in general agreement with known climate patterns and with spatial interpolation of meteorological observations (Prince *et al.*, 1998; Green and Hay, 2000). Additional assessment of recovered values of D, based on comparison with 699 field measurements at a broad range of study sites, suggest an RMSE of 10.9 mbar over a range of 58 mbar (Prince *et al.*, 1998). These comparisons, again, included both 1 km^2 and 64 km^2 AVHRR observations. The 1 km^2 results alone had an RMSE of 6.1 mbar with a range of 36 mbar (Figure 5), which was comparable to that found by Hay and Lennon (1999) based on 200 stations in Africa.

The errors in D recovered using this approach tend to be larger than errors in T_a or U alone, owing to the compounding of errors in each of these terms, both of which contribute to the calculation of D. Nevertheless, the estimated values of D were accurate within the above-specified limits. The fields of D mapped using this technique provide what are probably the best high temporal and spatial resolution estimates currently available over large areas. We note again that map comparisons of climate fields, such as those shown in Figure 4, have required the spatial and temporal degradation of the satellite data to match the low spatial resolution provided by interpolated climate fields.

2.5. Surface Wetness and Soil Moisture

The final variable of importance to epidemiological applications that we consider is surface wetness, an index of soil moisture. The contextual approach used to measure T_a can also be used to estimate this variable, one that is particularly relevant to mosquito borne diseases (Beck *et al.*, 1994; Linthicum *et al.*, 1999), but also to the development and breeding success of other vectors including, for example, a range of tsetse fly (Rogers, 1991) and tick species (Randolph 1993, 1994).

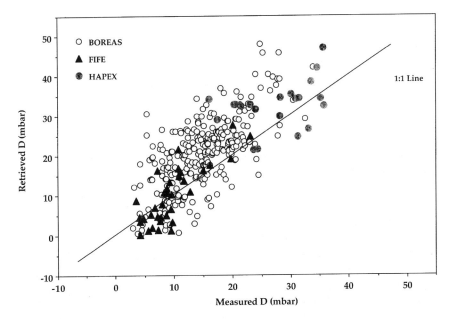

Figure 5 Same as Figure 1 but for vapour pressure deficit (*D*). The satellite-based estimates of *D* accounted for 45% of the variability in 307 near-coincident surface measurements.

Owing to its importance in crop growth, a great deal of research has been undertaken to develop maps of surface soil moisture from remotely sensed observations. Microwave techniques to estimate soil moisture have been favoured over optical and thermal methods, primarily because the longer wavelengths of microwaves penetrate materials, unlike thermal and optical wavelengths. Moreover, microwaves are less attenuated by clouds and vegetation cover. Much of this work has been focused on passive microwave measurements, which are sensitive to the relative moisture content of materials as expressed through their dielectric properties (e.g. Mattikalli *et al.*, 1998; Schmugge, 1998). Active microwave remote sensing has also been used, specifically radio detection and ranging (radar) (e.g. Wang *et al.*, 1997; Owe *et al.*, 1999). No widely available method exists to estimate surface moisture routinely with microwave remote sensing, largely because many other properties of the surface affect the signal and so additional information is needed to enable the moisture effect to be isolated (Engman and Chauran, 1995; Hall *et al.*, 1995).

Surface radiometric temperature is also sensitive to surface moisture but is also affected by the degree of vegetation cover. Thus, when coupled with SVI estimates of vegetation cover, T_s can be used to provide an estimate of

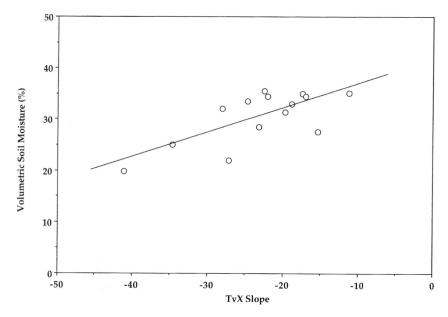

Figure 6 The slope of the TvX relationship accounted for 66% of the variability in surface (25 mm depth) volumetric soil moisture, averaged over a 225 km² area, at the FIFE site. There were 13 days with near-coincident measurements available for comparison, none of which fell below the 20% 'wilting point' threshold which induces a rapid rise in T_s, hence the TvX slope (see Goetz, 1997).

surface moisture conditions. The TvX technique provides a contextual approach to combine optical and thermal observations for soil moisture estimation. The TvX slope tends to be strongly negative in dry conditions owing to increased T_s and decreased SVI over sparsely vegetated areas, and closer to zero in densely vegetated areas (see Figure 2). This phenomenon has been linked to the effect of surface soil moisture on evapotranspiration rates and energy balance terms (e.g. Goward *et al.*, 1985; Hope *et al.*, 1987; Nemani and Running, 1989; Nemani *et al.*, 1993; Friedl and Davis, 1994; Carlson *et al.*, 1995). In a tallgrass prairie region of the central United States, for example, Goetz (1997) found the TvX slope explained 66% of the variability in surface soil moisture (<20 mm) (Figure 6) compared with 41% of the evaporative fraction (i.e. the amount of net radiation expressed as latent heat of evaporation).

There are a number of factors that can modify the sensitivity of the TvX slope to surface moisture conditions, including wind speed, soil properties, surface roughness and spectral emissivity, among others (Goward *et al.*, 2000). Thus further work is required before TvX can provide a routine measure of surface moisture, as is also the case for those techniques based on microwave observations. Nevertheless, several approaches, including TvX, can provide a relative index of moisture conditions over a given site, thus enabling epidemiologists to monitor environmental conditions that may affect the breeding success and survival rates of numerous disease vectors.

3. CONCLUSIONS AND FUTURE DIRECTIONS

We have provided a brief overview of some of the recent developments in satellite-based mapping of environmental variables that are relevant to the establishment, relative reproductive success, and survival of disease vectors. Advances are being made on a number of fronts that will increase the accuracy and precision of environmental variable measurement from remote sensing. These include the reliability of the satellite data sets with which they are driven, calibration, stability and other features of the sensors that provide the data, and the physical bases of the algorithms used to estimate the variables of use to epidemiological applications. The development of new sensors is proceeding at a rapid pace. Hyperspectral and very high spatial resolution satellite sensor imagery have been little explored and may provide useful methods for the prediction of conditions favourable to various disease vectors. The thermal split-window techniques used in the inference of surface temperature and atmospheric water vapour, for example, can now be applied at high spatial resolution using a series of new Earth-observing sensors. Satellite missions currently under development include microwave sensors sensitive to soil moisture, radically improved atmospheric sounders, and laser instruments that have the capability to measure canopy vertical structure (Prince, 1999). Observational data sets from these sensors will make it possible, for example, to map fine-scale surface topography, perhaps including the monitoring of ephemeral pools of standing water that are critical to the prediction of many mosquito species, the most ubiquitous and deadly of disease vectors.

At the very least, these new or proposed sensors will provide a wealth of data for existing techniques, as well as those yet to be developed, to improve the accuracy at which environmental variables are mapped. Resource limitations of many sorts, including data handling and storage requirements, will continue to limit the applicability of fine scale observations over large areas, thus scaling methodologies and sampling schemes will continue to play an important role in remote-sensing applications to epidemiology. At the same time, the inevitability of improved surface condition maps will continue to provide invaluable resources for the monitoring of disease vectors and, perhaps more importantly, the prediction of disease incidence under dynamic land surface conditions associated with changes in climate.

Predicting the spread of disease and the potential for epidemics undoubtedly depends on numerous factors other than the environmental conditions discussed here (see related chapters in this issue), but advances in satellite remote sensing and associated spatial information technologies will clearly play a critical role in advancing disease prediction necessitated by a changing Earth environment.

ACKNOWLEDGEMENTS

We acknowledge the editorial comments of Simon Hay, David Rogers and Sarah Randolph, and extend our thanks to Nathan Morrow for conducting the comparisons summarized in Figure 4. This work was partially supported by NASA grants from the Terrestrial Ecology (NAG56421) and Land Cover/ Land Use Change programs (NAG54811).

REFERENCES

Asrar, G., Cooper, D.I. and Harris, T.R. (1988). Surface radiative temperature of the burned area in a tall grass prairie. *Remote Sensing of Environment* **24**, 447–457.

Beck, L.R., Rodriguez, M.H., Dister, S.W. *et al.* (1994). Remote sensing as a landscape epidemiologic tool to identify villages at high risk for malaria transmission. *American Journal of Tropical Medicine and Hygiene* **51**, 270–281.

Becker, F. and Li., Z.L. (1990). Towards a local split window method over land surfaces. *International Journal of Remote Sensing* **11**, 1509–1522.

Carlson, T.N., Gillies R.R. and Schmugge, T.J. (1995). An interpretation of methodologies for indirect measurement of soil water content. *Agricultural and Forest Meteorology* **77**, 191.

Casalles, V., Coll C. and Rubio, E. (1997). Thermal band selection for the PRISM instrument: analysis and comparison of the existing atmospheric and emissivity correction methods for land surface temperature recovery. *Journal of Geophysical Research, Atmospheres* **102**, 19611–19629.

Coll, C., Caselles, V. and Schmugge, T.J. (1994). Estimation of land surface emissivity differences in the split-window channels of AVHRR. *Remote Sensing of Environment* **48**, 127–134.

Czajkowski, K.P., Goward, S.N. and Ouaidrari, H. (1998). Impact of AVHRR filter function on surface temperature estimation from the split window approach. *International Journal of Remote Sensing* **19**, 2007–2012.

Dalu, G. (1986). Satellite remote sensing of atmospheric water vapour. *International Journal of Remote Sensing* **7**, 1089–1097.

Engman, E.T. and Chauhan, N. (1995). Status of microwave soil moisture measurements with remote sensing. *Remote Sensing of Environment* **51**, 189–198.

Friedl, M.A. and Davis, F.W. (1994). Sources of variation in radiometric surface temperature over a tallgrass prairie. *Remote Sensing of Environment* **48**, 1–17.

Goetz, S.J. (1997). Multi-sensor analysis of NDVI, surface temperature, and biophysical variables at a mixed grassland site. *International Journal of Remote Sensing* **18**, 71–94.

Goetz, S.J. and Prince, S.D. (1999). Modeling terrestrial carbon exchange and storage: evidence and implications of functional convergence in light use efficiency. In: *Advances in Ecological Research* (A.H. Fitter and D. Raffaelli, eds), pp. 57–92. San Diego: Academic Press.

Goetz, S.J., Halthore, R., Hall, F.G. and Markham, B.L. (1995). Surface temperature retrieval in a temperate grassland with multi-resolution sensors. *Journal of Geophysical Research* **100**, 25397–25410.

Goetz, S.J., Prince, S.D., Goward, S.N., Thawley M. and Small, J. (1999). Satellite remote sensing of primary production: an improved production efficiency modeling approach. *Ecological Modeling* **122**, 239–255.

Goetz, S.J., Small, J., Morrow, N. and Prince, S.D. (2000). Mapping global fields of surface air temperature and humidity with AVHRR satellite observations. *Remote Sensing of Environment*, in press.

Goward, S.N., Cruickshanks, G.D. and Hope, A.S. (1985). Observed relation between thermal emission and reflected spectral radiance of a complex vegetated landscape. *Remote Sensing of Environment* **18**, 137–146.

Goward, S.N., Waring, R.H. and Dye, D.G. (1994). Ecological remote sensing at OTTER: macroscale satellite observations. *Ecological Applications* **4**, 322–343.

Goward, S.N., Xue, Y. and Czajkowski, K. (2000). Evaluating land surface moisture conditions from remotely sensed temperature and vegetation index measurements: an exploration employing the simplified simple biosphere model. *Remote Sensing of Environment*, in press.

Green, R.H. and Hay, S.I. (2000). Mapping of climate variables across tropical Africa and temperate Europe using meteorological satellite data. *Remote Sensing of Environment*, in press.

Hall, F.G., Townshend, J.T. and Engman, E.T. (1995). Status of remote sensing algorithms for estimation of land surface state parameters. *Remote Sensing of Environment* **51**, 138–156.

Harris, A.R. and Mason, I.M. (1992). An extension to the split-window technique giving improved atmospheric correction and total water vapour. *International Journal of Remote Sensing* **13**, 881–892.

Hay, S.I. and Lennon, J.J. (1999). Deriving meteorological variables across Africa for the study and control of vector-borne disease: a comparison of remote sensing and spatial interpolation of climate. *Tropical Medicine & International Health* **4**, 58–71.

Hay, S.I., Packer, M.J. and Rogers, D.J. (1997). The impact of remote sensing on the study and control of invertebrate intermediate hosts and vectors for disease. *International Journal of Remote Sensing* **18**, 2899–2930.

Hay, S.I., Snow, R.W. and Rogers, D.J. (1998a). From predicting mosquito habitat to malaria seasons using remotely sensed data: practice, problems and perspectives. *Parasitology Today* **14**, 306–313.

Hay, S.I., Snow, R.W. and Rogers, D.J. (1998b). Prediction of malaria seasons in Kenya using multi-temporal meteorological satellite sensor data. *Transactions of the Royal Society of Tropical Medicine and Hygiene* **92**, 12–20.

Hay, S.I., Rogers, D.J., Myers, M.F. and Snow, R.W. (2000). Malaria early warning in Kenya. *Parasitology Today*, in press.

Hope, A.S., Petzold, D.E., Goward, S.N. and Ragan, R.M. (1987). Simulating canopy reflectance and thermal infrared emissions for estimating evapotranspiration. *Water Resources Bulletin* **22**, 1011–1019.

Kalluri, S.N.V. and Dubayah, R.O. (1995). A comparison of atmospheric correction models for thermal bands of AVHRR over FIFE. *Journal of Geophysical Research* **100**, 25411–25418.

Li, Z.-L. and Becker, F. (1993). Feasibility of land surface temperature and emissivity determination from AVHRR data. *Remote Sensing of Environment* **43**, 67–85.

Linthicum, K.J., Anyamba, A., Tucker, C.J., Kelly, P.W., Myers, M.F and Peters, C.J. (1999). Climate and satellite indicators to forecast Rift Valley Fever epidemics in Kenya. *Science* **285**, 397–400.

Malone, J.B., Huh, O.K., Fehler, D.P. *et al.* (1994). Temperature data from satellite imagery and the distribution of schistosomiasis in Egypt. *American Journal of Tropical Medicine and Hygiene* **50**, 714–722.

Markham, B.L. and Barker, J.L. (1986). Landsat MSS and TM post calibration dynamic ranges, exoatmospheric reflectances, and at-satellite temperatures. *Landsat User Notes* **1**, 3–8.

Mattikalli, N.M., Engman, E.T. and Ahuja, L.R. (1998). Microwave remote sensing of temporal variations of brightness temperature and near-surface soil water content during a watershed-scale field experiment, and its application to the estimation of soil physical properties. *Water Resources Research* **34**, 2289–2299.

Monteith, J.L. and Unsworth, M.H. (1990). *Principles of Environmental Physics*. Edward Arnold: London.

Nemani, R.R. and Running, S.W. (1989). Estimation of regional surface resistance to evapotranspiration from NDVI and thermal-IR AVHRR data. *Journal of Applied Meteorology* **28**, 276–284.

Nemani, R., Pierce, L., Running, S. and Goward, S. (1993). Developing satellite-derived estimates of surface moisture status. *Journal of Applied Meteorology* **32**, 548–557.

Norman, J.M. and Becker, F. (1995). Terminology in thermal infrared remote sensing of natural surfaces. *Remote Sensing Reviews* **12**, 159–173.

Onori, E. and Grab, B. (1980). Indicators for the forecasting of malaria epidemics. *Bulletin of the World Health Organization* **58**, 91–98.

Ouaidrari, H., Czajkowski, K. and Goward, S. (2000). Land surface temperature estimation from AVHRR thermal infrared measurements: an assessment for the AVHRR Land Pathfinder II data set. *Journal of Geophysical Research, Atmospheres*, in press.

Owe, M., Van de Griend, A.A. and Engman, E.T. (1999). Estimating soil moisture from satellite microwave observations: past and ongoing projects and relevance to GCIP. *Journal of Geophysical Research, Atmospheres* **104**, 19735–19748.

Prabhakara, C. and Dalu, G. (1976). Remote sensing of surface emissivity at 9 μm over the globe. *Journal of Geophysical Research* **81**, 3719–3724.

Price, J.C. (1984). Land surface temperature measurements from the split window channels of the NOAA advanced very high resolution radiometer. *Journal of Geophysical Research* **89**, 7231–7237.

Prihodko, L. and Goward, S.N. (1997). Estimation of air temperature from remotely sensed observations. *Remote Sensing of Environment* **60**, 335–346.

Prince, S.D. (1999). What practical information about land-surface function can be determined by remote sensing? Where do we stand? In: *Integrating Hydrology, Ecosystem Dynamics, and Biogeochemistry in Complex Landscapes* (J.D. Tenhunen and P. Kabat, eds), pp. 39–60. Chichester: Wiley & Sons.

Prince, S.D. and Goward, S.J. (1995). Global primary production: a remote sensing approach. *Journal of Biogeography* **22**, 815–835.

Prince, S.D., Goetz, S.J., Czajkowski, K., Dubayah, R. and Goward, S.N. (1998). Inference of surface and air temperature, atmospheric precipitable water and vapour pressure deficit using AVHRR satellite observations: validation of algorithms. *Journal of Hydrology* **212/213**, 231–250.

Randolph, S.E. (1993). Climate, satellite imagery and the seasonal abundance of the tick *Rhipicephalus appendiculatus* in southern Africa: a new perspective. *Medical and Veterinary Entomology* **7**, 243–258.

Randolph, S.E. (1994). Population dynamics and density-dependent seasonal mortality indexes of the tick *Rhipicephalus appendiculatus* in eastern and southern Africa. *Medical and Veterinary Entomology* **8**, 351–368.

Rejmankova, E., Saveage, H.M., Rejmanek, M., Roberts, D.R. and Arrendondo-Jimenez, J.I. (1991). Multivariate analysis of relationships between habitats, environmental factors and occurrence of *anophelene* mosquito larvae (*Anopheles albimanus* and *A. pseudopuntipennis*) in southern Chiapas, Mexico. *Journal of Applied Ecology* **28**, 827–841.

Rogers, D.J. (1991). Satellite imagery, tsetse and trypanosomiasis in Africa, *Preventive Veterinary Medicine* **11**, 201–220.

Rogers, D.J., Hay, S.I. and Packer, M.J. (1996). Predicting the distribution of tsetse flies in West Africa using temporal Fourier processed meteorological satellite data. *Annals of Tropical Medicine and Parasitology* **90**, 225–241.

Salisbury, J.W. and D'Aria, D.M. (1992). Emissivity of terrestrial materials in the 8–14 μm atmospheric window. *Remote Sensing of Environment* **42**, 83–106.

Schmugge, T. (1998). Applications of passive microwave observations of surface soil moisture. *Journal of Hydrology* **212–213**, 187–197.

Schmugge, T.J., Becker, F. and Li, Z.-L. (1991). Spectral emissivity variations observed in airborne surface temperature measurements. *Remote Sensing of Environment* **34**, 95–104.

Schmugge, T., Hook, S.J. and Coll, C. (1998). Recovering surface temperature and emissivity from thermal infrared multispectral scanner data. *Remote Sensing of Environment* **65**, 121–131.

Smith, W.L. (1966). Note on the relationship between precipitable water and surface dew point. *Journal of Applied Meteorology* **5**, 726–727.

Snow, R.W., Craig, M., Deichmann, U. and Marsh, K. (1999). Estimating mortality, morbidity and disability due to malaria among Africa's non-pregnant population. *Bulletin of the World Health Organization* **77**, 624–640.

Wang, J.R., Hsu, A. and Engman, E.T. (1997). A comparison of soil moisture retrieval models using SIR-C measurements over the Little Washita River Watershed. *Remote Sensing of Environment* **59**, 308–322.

Forecasting Disease Risk for Increased Epidemic Preparedness in Public Health

M.F. Myers,[1] D.J. Rogers,[2] J. Cox,[3] A. Flahault[4] and S.I. Hay[2]

[1] *Human Health Initiative, NASA—Goddard Space Flight Center, Code 902, Bldg 32/S130E, Greenbelt, Maryland, MD 20771, USA;*
[2] *Trypanosomiasis and Land use in Africa (TALA) Research Group, Department of Zoology, University of Oxford, South Parks Road, Oxford OX1 3PS, UK;*
[3] *Department of Infectious and Tropical Diseases, London School of Hygiene and Tropical Medicine, Keppel Street, London WC1E 7HT, UK;*
[4] *Institut National de la Santé et de la Recherche Médicale (INSERM) Unité 444, WHO Collaborating Centre for Electronic Disease Surveillance, Faculté de Médecine Saint-Antoine, 27 Rue Chaligny, F-75571 Paris Cedex 12, France*

ADVANCES IN PARASITOLOGY VOL 47
0065-308-X $30.00

ABSTRACT

Emerging infectious diseases pose a growing threat to human populations. Many of the world's epidemic diseases (particularly those transmitted by intermediate hosts) are known to be highly sensitive to long-term changes in climate and short-term fluctuations in the weather. The application of environmental data to the study of disease offers the capability to demonstrate vector–environment relationships and potentially forecast the risk of disease outbreaks or epidemics. Accurate disease forecasting models would markedly improve epidemic prevention and control capabilities. This chapter examines the potential for epidemic forecasting and discusses the issues associated with the development of global networks for surveillance and prediction. Existing global systems for epidemic preparedness focus on disease surveillance using either expert knowledge or statistical modelling of disease activity and thresholds to identify times and areas of risk. Predictive health information systems would use monitored environmental variables, linked to a disease system, to be observed and provide prior information of outbreaks. The components and varieties of forecasting systems are discussed with selected examples, along with issues relating to further development.

1. INTRODUCTION

Environmental change, human demography, international travel, microbial evolution and the breakdown of public health facilities have all contributed to the changing spectrum of infectious diseases with which the global community is challenged (Bryan *et al.*, 1994). Existing mechanisms for infectious disease surveillance and response are inadequate to meet the increasing needs for prevention, detection, reporting and response (CDC, 1994; CISET, 1995). The ability to predict epidemics will provide a mechanism for governments and health-care services to respond to outbreaks in a timely fashion, enabling the impact to be minimized and limited resources to be saved (LaPorte, 1993; Wilson, 1994). For many infectious diseases, particularly those transmitted by arthropod vectors, advanced surveillance and modelling technologies incorporating environmental data create the potential to predict the temporal and spatial risk of epidemics. When combined with communication technologies, these techniques can provide important tools that are both cost-effective and timely (Susser and Susser, 1996). As disease boundaries shift and expand to threaten new populations, there is increasing need to develop operational models with predictive capacity: 'As more experience is gained in linking changes detected by global imaging with changes in disease patterns, geographical information systems are likely to play an increasingly important role in forecasting outbreaks, especially those of vector-borne diseases such as malaria' (Greenwood, 1998).

Advances in disease surveillance systems, epidemiological modelling and information technology have generated the expectation that early warning systems are not only feasible but necessary tools to combat the re-emergence and spread of infectious diseases. While many of the environmental data used in these systems are available free or at low cost, the quality and availability of epidemiological data vary enormously. The length and spatial extent of the epidemiological data series are particularly important for investigating annual and inter-annual patterns of disease. Elsewhere in this volume the evolution of remote sensing instrumentation (Hay, this volume) and the application of satellite data to problems of disease risk prediction are reviewed and discussed (Rogers, this volume; Hay *et al.*, this volume; Randolph, this volume; Brooker and Michael, this volume) and will not be reiterated here. This chapter focuses particularly on techniques being developed with the view to predicting diseases in both time and space. Figure 1 illustrates the terminology that is used in this chapter. Briefly, surveillance and early detection refer to the monitoring of reported case data; disease forecasting is a medium term warning of suitable conditions for a disease (e.g. increased rainfall for malaria); epidemic warning and prediction are more short-term indications of risk with more specificity in time and space.

2. DISEASE FORECASTING

2.1. Historical Early Warning Systems

Efforts to use environmental data for epidemic prediction and response began in the early 1920s in India, when nearly half a century of meteorological data and 30 years of records of tropical diseases had been amassed by province and district (Rogers, 1925a). Risk maps were developed by combining meteorological and health records for diseases such as leprosy (Rogers, 1923), pneumonia (Rogers, 1925b) and smallpox (Rogers, 1926). These maps offered predictions with a 2–3 month lead-time to allow government response. Gill (1921) was also able to link rainfall and river flooding to subsequent outbreaks of malaria. Further work by Gill (1923) used rainfall and a series of health and demographic inputs to establish 4-month and 2-month malaria predictive warnings for two decades in the

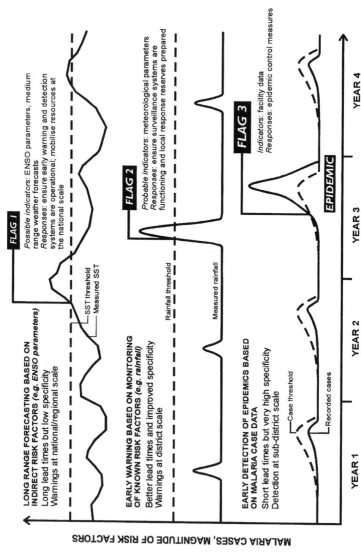

LONG RANGE FORECASTING BASED ON INDIRECT RISK FACTORS (*e.g. ENSO parameters*)
Long lead times but low specificity
Warnings at national/regional scale

FLAG 1

Possible indicators: ENSO parameters, medium range weather forecasts
Responses: ensure early warning and detection systems are operational; mobilise resources at the national scale

SST threshold
Measured SST

EARLY WARNING BASED ON MONITORING OF KNOWN RISK FACTORS (*e.g. rainfall*)
Better lead times and improved specificity
Warnings at district scale

FLAG 2

Probable indicators: meteorological parameters
Responses: ensure surveillance systems are functioning and local response reserves prepared

Rainfall threshold

Measured rainfall

EARLY DETECTION OF EPIDEMICS BASED ON MALARIA CASE DATA
Short lead times but very high specificity
Detection at sub-district scale

FLAG 3

Indicators: facility data
Responses: epidemic control measures

EPIDEMIC

Case threshold
Recorded cases

MALARIA CASES, MAGNITUDE OF RISK FACTORS

YEAR 1   YEAR 2   YEAR 3   YEAR 4

Figure 1 A three-tiered approach for epidemic forecasting, early warning and detection. Each tier is associated with specific indicators and responses. In this simplified example for malaria epidemics, a first warning flag is raised at the regional level after Sea Surface Temperature (SST) anomalies suggest an impending El Niño event. Subsequent excess rainfall is monitored directly as part of an early warning system and Flag 2 is raised once a critical threshold is reached. Malaria cases are monitored at the individual (sentinel) facility level and an epidemic declared once a defined threshold has been reached. (Cox *et al.*, 1999).

various districts of the Punjab, until the province ceased to exist with the partitioning of India in 1947. Swaroop (1949) provides an excellent review of these investigations.

2.2. The Components of an Early Warning System

A framework for early warning systems (EWSs) for epidemic preparedness was developed in the mid-1980s, during the collaborative efforts to design and build EWSs for famine prediction in Africa (Walsh, 1988). Davies *et al.* (1991) defined an EWS for famine as:

> a system of data collection to monitor people's access to food, in order to provide timely notice when a food crisis threatens and thus to elicit appropriate responses.

Reviews of these based on remote sensing operational famine EWSs can be found in Hutchinson (1991) and Hielkema and Snijders (1994). Early warning for epidemics refers to risk formulation or modelled projections of potential outbreaks based on systematically collected information from the monitored site(s) to allow appropriate and timely actions for mitigation and response.

There are three components of an EWS. These are (1) routine surveillance of the targeted disease; (2) modelling the disease risk based on historical surveillance and contemporary environmental data; and (3) forecasting future risk through the use of predictive models and continued epidemiological and environmental surveillance. During the 1990s, technological advances have made disease EWSs more feasible. These include the development and global penetration of the Internet (Valleron *et al.*, 1986), online electronic communication between health care providers and epidemiologists (CISET, 1995), and improvements in satellite imaging that allow improved environmental characterization of sentinel sites (Hay *et al.*, 1996; Hay and Lennon, 1999; Goetz *et al.*, this volume). We acknowledge the obvious challenge to make such technological developments available to all.

2.2.1. Component 1: Disease Surveillance

A sentinel network is an interactive disease surveillance system that involves the collection of health data on a routine basis, usually by health care professionals over a wide (usually country level) area (Valleron *et al.*, 1986; Girard, 1997; Fourquet and Drucker, 1997). In most industrialized nations, notification of many infectious diseases is a statutory requirement. Rapid collection of data and assessment of regional and national statistics leads to early detection of changes in the incidence of infections (CDC, 1994; Greenwood, 1998; Heymann and Rodier, 1998). The database also provides information for the planning and implementation of intervention(Choi, 1998; Kafadar and Stroup, 1992). The growth of such sentinel systems, from independent national networks to co-ordinated international information systems, has generated a demand for health information systems capable of forecasting disease (Flahault *et al.*, 1998; Nabarro and Tayler, 1998).

When teleinformatics was introduced into public health fields the potential for its use as a rapid and early warning system soon became evident. France developed its sentinel system (Valleron *et al.*, 1986; Fourquet and Drucker, 1997), New York State implemented Healthcom, and the Center for Disease Control (CDC) set up an electronic network between the various US states' Departments of Health. As data were collected it became possible to identify evolving temporal and spatial patterns, such as growing or lessening risks of reported diseases, seasonality, clustering, and so on. This spawned a huge literature on detection of anomalies in disease surveillance data (for example, Zeng *et al.*, 1988; Stroup *et al.*, 1989; Watier *et al.*, 1991; Frisén, 1992; Nobre and Stroup, 1994; Stern and Lightfood, 1999; Vanbrackle and Williamson, 1999).

Increasingly powerful platforms for data collection and manipulation developed alongside these information networks. Relational databases were undergoing rapid evolution, permitting manipulation and analysis of huge fluxes of information. Client–server architecture developments also provided for more rapid remote access, allowing multiple simultaneous sessions to be open on a unique application for analysts and data providers. For the first time it became possible not only to provide baseline health statistics in near real time, but also to archive and mine the data electronically and to add additional information from other sources. Profiles of the outbreak and spread of a disease could also be created quickly enough so that the information could be returned to public health organizations charged with containing the disease, increasing their ability to respond. This led to the present understanding that a facility-based sentinel surveillance system can play an important role in providing information for monitoring communicable diseases, guiding further investigation, evaluating control measures, and predicting epidemics (Berkelman, 1994, 1998; Shalala, 1998). Some of the epidemiological insights that can be drawn from the archival and systematic long-term collection of disease data are identified by Cliff *et al.* (1998).

(a) *The French Sentinel System.* France took a technological lead in electronic disease surveillance, with a national telecommunications programme instituted in 1983 that provided videotext home terminals free of charge to French citizens. In 1984, the Institut National de la Santé et de la Recherche Médicale (INSERM), in collaboration with the Ministry of Health, initiated a program to provide for electronic monitoring of communicable diseases. Today, a volunteer sample of about 1% of French general practitioners (GPs) remotely enters reports on several diseases on a weekly basis (Flauhault *et al.*, 1998). This system is the basis for one of the largest databases of individual cases (including time-to-onset and geographical location) of diseases such as influenza-like illness, acute diarrhoea and chickenpox. Bonabeau *et al.* (1998) demonstrate some of the detailed insights into the geographical spread of disease that can be derived from such large databases. The contemporaneous systems of the United States' Centers for Disease Control (CDC) in Atlanta, as well as the Royal College of GPs in the UK, still used a system based on index cards.

(b) *The US Surveillance System.* In the United States, individual states determine which diseases are reported internally, and together determine which diseases will be reported to the federal government on a voluntary countrywide basis (Berkelman, 1998). Currently, all states use a standardized weekly form submitted by e-mail to the CDC on the National Electronic Telecommunications System for Surveillance (NETSS). In return, the CDC established the Public Health Network (PHNET), a tool to return information alerts to state departments of health (SHD) (Halperin *et al.*, 1992). For a long time, this system was based on an electronic text-message system of the Morbidity and Mortality Weekly Report (MMWR) which SHDs received earlier than other subscribers. Currently, using the graphic capabilities of the Internet, up-to date maps and graphs are now increasingly available.

(c) *Development of a Global Network for Disease Surveillance.* Other sentinel systems have been set up in Europe (Snacken *et al.*, 1992, 1998; Szecsenyi *et al.*, 1995; Fleming and Cohen, 1996; Gylys *et al.*, 1998). Following the establishment and operation of national surveillance systems, international cooperation is resulting in the development of global surveillance systems for targeted diseases. Working through the trans-Atlantic Agenda, the US and the European Union (EU) are negotiating to share surveillance data on a variety of diseases (Berkelman, 1998; Heyman and Rodier, 1998). Collaborative surveillance efforts also exist between US agencies and the World Health Organization (WHO) for the establishment of regional centres for monitoring disease as well as for improving communications infrastructure for future efforts (Shalala, 1998).

2.2.2. *Component 2: Developing a Model*

Disease forecasting involves modelling, which may be based either on statistical relationships established between past case numbers and environmental predictors (the 'statistical approach'), or on sets of equations that attempt to capture the biology of the transmission processes (the 'biological approach'), both reviewed by Rogers (this volume). Briefly, the statistical approach requires samples from as wide a range of environmental conditions as possible: predictions arising from this approach assume that the future will be the same as the past, i.e. that the relationships already established between case numbers and environmental variables will persist into the future. The biological approach requires details on all the parameters and variables considered to be important in transmission (these may sometimes be estimated by *post hoc* analysis of disease data sets): predictions arising from this approach are in theory able to incorporate the effects of environmental changes, or interventions, as long as the impacts of each of these changes on the key transmission parameters are established.

It should follow from the above that in the absence of full knowledge of all the transmission pathways for any particular diseases, only the statistical approach is possible. This explains why much of the early epidemiology of poorly-understood diseases such as cancer adopted the statistical route. Statistical models can be extremely powerful, but should be only a temporary substitute for the biological process-based models, whose development exposes our full ignorance of the systems we study. It is only by addressing this ignorance that real progress will be made.

There is, however, an important dilemma in the statistical/biological model debate. It seems likely that case numbers in many diseases arise from a combination of factors, some of which are intrinsic to the disease and its various hosts and others which are extrinsic, or environmental. The intrinsic factors include herd susceptibility, infection and immunity etc., which change over time through the normal processes of disease transmission. The extrinsic factors are often due to climate, which affects the average amount of transmission in any area, and weather which influences its seasonality. Biological models should describe the intrinsic factors well, but will be rendered more or less inaccurate by the extrinsic factors, unless they are explicitly included. Statistical models exploit the relationships of case numbers to these extrinsic factors, but are unable to cope with the intrinsic factors easily. Thus, what is the 'signal' for one approach is the statistical 'noise' of the other. Biological models tend to cope with their statistical noise by drawing demographic parameters from predefined frequency distributions that may or may not be linked to seasonality: statistical models cope with their noise by introducing spatial or temporal autoregressive terms that patently acknowledge, in a rather ill-defined way, the biological fact that present case numbers depend on nearby, or past, case numbers. We believe that progress will be made in this field by combining the best of the statistical and biological approaches, and warn against the exclusive use of one or the other. Many examples of this epidemiological 'exploration' process are reviewed by Rogers (this volume), Hay *et al.* (this volume), Randolph (this volume) and Brooker and Michael (this volume).

Developing an appropriate model is one of the most crucial steps in determining the robustness of any early warning system and is discussed with respect to malaria early warning systems (MEWS) in Africa (Hay *et al.* 2000a). Africa displays considerable spatial heterogeneity in its climate and ecology (see Plate 16). It follows that malaria distribution will reflect this environmental heterogeneity in space and in time. Without elaborating the quantitative epidemiology (Hay *et al.*, 2000a), it is plain that in extremely arid areas malaria will be limited by rainfall, which provides habitats in which mosquitoes can oviposit. For example, the relationship between malaria cases and rainfall in Wajir, a town in the arid north of Kenya, is shown in Plate 17. Positive rainfall anomalies in Wajir are a good indicator of malaria cases 3 months into the future; the lag presumably reflects the time for the mosquito population to establish. Following on from this, a system to monitor rainfall anomalies for the arid and semi-arid areas of Africa is shown in Plate 18. It is particularly important to monitor these arid areas, because the usual lack of malaria means that the population has little immunity to the disease and is therefore very susceptible to infection. Epidemics in these areas can have a devastating impact across all age groups (Brown *et al.*, 1998). Satellite sensors can provide timely remotely sensed data on which to base such monitoring systems (Hay *et al.*, 1996, 1997; Hay, 1997; Hay, this volume).

Similarly, positive temperature anomalies in cold areas (usually at high elevation in Africa) can be reliable predictors of malaria, since low temperatures normally limit parasite development within the vectors. Thus simple systems to monitor temperature anomalies could be important for 'epidemic' warning in these locations.

In more endemic malaria areas, such as Kericho in western Kenya, human and parasite population dynamic effects complicate relationships between malaria cases and climate. Malaria is always present, so children who survive to adulthood develop a functional immunity and epidemics can only occur when the non-immune population has grown, through recovery, births and immigration (Shanks *et al.*, 2000). Climate therefore acts in concert with the population dynamics of malaria in endemic areas (Hay *et al.*, 2000b) and will have to be considered when developing MEWS for such zones.

The above contrasts show some of the complexity inherent in developing a predictive malaria model. A single biological model should be capable of describing all malarious situations, but is not yet available because the interaction between extrinsic and intrinsic factors in the expression of malaria is not completely understood. For the time being, therefore, we are left with several (location-specific) statistical models.

2.2.3. *Component 3: Disease Forecasting and Prediction*

At the heart of early warning is a basic trade-off between the specificity of predictions (in space and time) and the lead times which those predictions can provide. In general, long-range forecasts give the least specific warnings, but have the advantage of providing planners with relatively long lead times. At the other extreme, systems based on early detection of cases provide highly specific information on the timing and location of outbreaks, but allow little time for implementing remedial measures. Any prediction of risk should include an estimate of its reliability (Frisén, 1992). This is particularly important from a health-planning standpoint, as resources will only be mobilized once a 'critical level' of confidence has been exceeded. While some elements of intervention (such as allocating extra resources within health budgets) may require relatively low confidence predictions, other activities may only proceed once more specific predictions are available (and the danger of a false alarm is less likely).

Epidemic prevention and control activities usually involve a chain of events and it is important to recognize the potential usefulness of a wide range of indicators, which may be combined to create an integrated prediction strategy. Such a hierarchical system has recently been proposed for tracking malaria epidemics in highland areas of Africa (Cox *et al*., 1999). It combines elements of long and medium range forecasting as well as the early detection of malaria outbreaks through direct epidemiological surveillance (see Figure 1). These elements provide warning signals that can be thought of as a series of 'flags', which correspond to increasing degrees of alarm, and trigger activities of increasing degrees of urgency. Each flag relates to a specific set of indicators, and leads to a specific set of responses following predefined procedures. These responses also anticipate the next level of indicators with an increased sensitivity. As shown in Figure 1, successive flags carry increasing weight. From a planning perspective, it is important that the higher weight flags are implemented first.

3. TYPES OF EARLY WARNING SYSTEM

The criteria for selection of any EWS are: the information requirements of the health community; the scale of analysis (i.e. local, regional, national); and the technological requirements for modelling or prediction. EWSs comprise one or more of the following types of activity: (i) reportorial, involving collected reports of outbreaks from health care professionals; (ii) risk maps or indicators based on seasonality or changes in environment of the vector; (iii) threshold alerts indicating changes in acceptable ranges derived from ongoing surveillance systems; and (iv) EWSs modelled from ongoing disease surveillance and operational environmental monitoring of sentinel sites.

3.1. Reportorial Systems

Formal and informal systems provide case data for reporting and investigation. Formal networks include the US CDC, the UK Public Health Services, the French Instituts Pasteur and the Training in Epidemiology and Public Health Network, among others (Parsons *et al.*, 1996). These networks provide laboratory-confirmed reports of outbreaks of new diseases and shifts in patterns of endemic ones (Bryan *et al.*, 1994; Berkelman, 1998). Most are, or will become, part of the WHO Collaborating Centre network (Heymann and Rodier, 1998). While falling primarily under the category of early detection, there is also potential within this system for providing long-range risk forecasting of disease events by expert surveillance and laboratory confirmation.

Informal networks may be national, such as SentiWeb, or regional, such as the Pacific Network (PacNet). Important informal networks include the Program for Monitoring Emerging Diseases (ProMed) that provides open postings of outbreaks of both familiar and new diseases to 18 000 subscribers in 150 countries via email (Chase, 1996). The Global Public Health Information Network (GPHIN), meanwhile actively trawls the Internet looking for reports of communicable diseases in news groups, wire service postings and other listings, and reports its findings to WHO for response and verification (Cribb, 1998).

While remote sensing is used little, if at all, in reportorial systems, high-resolution satellite imaging can be an important tool for increasing the efficiency of notification of population networks, particularly for emerging problems in unmapped areas. The use of remote sensing for population surveillance may see greater use in reportorial systems as WHO implements a strategy of geographically identified surveillance centres and ProMed develops its proposal of selected surveillance centres for monitoring responsibilities (FAS, 1999).

3.2. Risk Mapping Systems

Disease data, however collected, may be turned into static maps of risk. They may also be used to develop new statistical approaches to risk mapping, to test the association of weather anomalies with disease outbreaks and to test biological models that give rise to risk predictions. This sequence is one of increasing complexity, and therefore increasing uncertainty. Each of the many links in the chain of causation must be accurately described before a biologically based model can accurately describe disease risk through both space and time.

3.2.1. *Static Risk Maps*

Mapping hot spots for disease was one of the earliest methods to identify risk areas for epidemiology. As discussed earlier, Rogers (1923, 1925a,b, 1926) and Gill (1921, 1923) were able to use historic environmental and epidemiological data to develop risk maps for a wide variety of diseases in the first quarter of the twentieth century; these maps were subsequently used for decades. Risk maps of malaria in Africa were developed by experts beginning in the 1950s (Hay *et al.*, 1998) and their production is among the primary aim of the Mapping Malaria Risk in Africa/Atlas du Risque de la Malaria en Afrique (MARA/ARMA) collaboration (Hay *et al.*, this volume). Data collation within a geographical information system (GIS) will identify disease hot spots and these may be targeted for long-term control. No further attempt need be made to understand the reasons for the hot spots in the first place: like a fire risk map, a static disease risk map tells us where, and perhaps when, to expect an outbreak, but not why.

3.2.2. *Statistical Risk Maps*

The GIS disease data may be related to ancillary data, such as satellite sensor information, soil and water types, human agricultural activities etc., using a variety of regression or maximum likelihood methods (Curran *et al.*, this volume; Robinson, this volume; Rogers, this volume). The relationships established between the predictor (e.g. satellite) and the predicted (e.g. disease) data may then be used to predict risk in previously unsurveyed areas. Seasonally varying satellite data may also be used to describe seasonally varying risk. Anomalies from the usual patterns of satellite data in both space and time can be associated with varying risks, to improve the accuracy of short-term risk map forecasts.

Just as statistical analysis may be used to predict spatial variation in risk, so different sorts of statistical analyses can predict variation through time. A whole variety of time series analytical methods is available, from spectral analysis to autoregression methods (Chatfield, 1975; Diggle, 1990). It is possible to show the fundamental similarities between many of these techniques. As outlined in the introduction to this section, it is assumed that patterns in past data can be projected into the future to make predictions of future case numbers. Uptake of such systems will depend upon both the reliability of such forecasts and the lead time they give for sensible mitigating responses. One example of this approach is the development of a dengue early warning system by NASA's Inter-agency Research Partnership for Infectious Diseases (IntRePID).

IntRePID began life as a US federal agency working group in 1996, investigating whether technologies and data from NASA's suite of earth observing satellites could be applied to the development of early warning systems for infectious diseases. The dengue early warning system (DEWS) is a prototype system which is undergoing testing to validate the accuracy of the predictions in real-time. The prototype is designed to receive data from Bangkok and the four main regions of Thailand and is based on previous Thai systems for malaria 'epidemic' surveillance (Cullen *et al.*, 1984) and responds to calls for such a facility (Gunakasem *et al.*, 1981). When fully validated, the system will form an international EWS for dengue. The system comprises several models.

The surveillance model allows new case data to be compared against the long-term average case data. As additional disease data are received they are plotted against the long-term average for that month. The user is then able to determine the severity of the current outbreak against historical conditions. The line of two standard deviations above the long-term mean is also drawn to help with this comparison: if cases exceed two standard deviations from the normal, there is significant cause for concern. This is flag 3 of Figure 1.

The DEWS risk map module, again using Thailand as a prototype, displays historical records from separate administrative units (i.e. changwats) to show the spatial distribution of dengue cases on a national basis from 1982 to 1997. These data refer to relatively severe cases requiring a visit to a local clinic or hospital, although it is not possible to distinguish in them severe dengue from dengue haemorrhagic fever. Estimated populations over the same period of time, are used to turn disease cases into disease incidence per 100 000 population. Incidence was related to National Oceanic and Atmospheric Administration (NOAA) Advanced Very High Resolution Radiometer (AVHRR) satellite data using maximum likelihood methods, and the full resolution satellite data were used to produce country-wide risk maps. The analysis helps to identify crucial environmental variables determining local variation in risk, and the risk maps may be updated in near real time as recent satellite data are incorporated into the risk map model.

The DEWS forecasting module is based on a time-series analysis of past case numbers of this disease. It was initially developed for Bangkok, followed by the four main regions of Thailand (the north, northeast, central and southern regions). Inspection of monthly case data over many years shows that there are not only within-year cycles of variation in dengue in Bangkok, but considerable between-year variation as well. Temporal Fourier analysis (Rogers, this volume) of the de-trended time series splits the case data into a series of regular cycles (the 'harmonics') with frequencies ranging from one complete cycle every 2 months to one complete cycle only once in the entire duration of the records. The between-year cycles are included without modification in the model predictions. To these are added both a fitted trend line and a description of within-year variation. The latter are predicted from relationships established between monthly temperature and the within-year residuals 4 months later; biologically this implies that annual temperature changes trigger a series of processes that result in changing case numbers in the future, with a peak at the 4 month mark.

The descriptive skill of this model is excellent, with past dengue records being described with acceptable accuracy. The three components of the model can also be projected into the future to make disease predictions. The first two component projections are based on extending the trend or between-year harmonics and the third uses the observed mean monthly temperatures for previous years. These predictions are reasonably good for non epidemic years, but are not yet able to capture the full extent of epidemic cycles which occur irregularly.

3.2.3. *Anomaly Risk Maps*

There is increasing evidence that longer term climatic events, such as the aperiodic El Niño southern oscillation (ENSO) that affect local patterns of rainfall, have some impact on vector-borne diseases (among others, Nicholls 1993; Bouma *et al.*, 1997; Baylis *et al.*, 1999; Maelzer *et al.*, 1999). The effect of El Niño on local rainfall varies spatially; in some places rainfall increases, in others it decreases. The effect also varies temporally; some El Niños cause an increase in local rainfall, others a decrease in the same areas, and disease outbreaks may only be associated with one of these changes (Baylis *et al.*, 1999). Our ability to detect the early signs of developing El Niño conditions has increased dramatically in recent years. If we could be confident of predicting the climatic consequences of these events, El Niño-associated disease outbreaks could be anticipated and mapped.

Factors intrinsic to the disease system can also generate periodic outbreaks, however, and these may be difficult to distinguish from extrinsically driven cycles. There is strong evidence that these intrinsic cycles have periods approaching, but not quite matching, those of El Niño (Hay *et al.*, 2000b). Teasing apart the intrinsic and extrinsic influences in such cases is technically difficult, and remains controversial.

3.2.4. *Biological Risk Maps*

These maps exploit what is known about the biological relationships between organisms in the period leading up to disease outbreaks. In many cases a particular event, or chain of events, triggers an eventual outbreak of disease.

(a) *Hantavirus Pulmonary Syndrome.* Glass *et al.* (2000) uses Landsat thematic mapper (TM) data to establish annual risk predictions for hantavirus pulmonary syndrome (HPS). This is a disease of humans caused by infection with members of the viral genus *Hantavirus,* which are carried in the US by certain native rodent species (Engelthaler *et al.,* 1999). The disease was first recognized in the US in 1993, following the 1991–1992 El Niño event. The presumptive chain of events leading to the original outbreak involves increased precipitation during the winter and spring, leading to increased vegetation and insect population growth, which in turn provide food and shelter for the rodent reservoirs. When climatic conditions return to normal, vegetation growth declines, forcing the rodents into human habitation in search of food and shelter. This leads to increased contact with humans and transmission of the hantavirus. The entire sequence has been termed the 'trophic cascade hypothesis' (TCH), and is currently being tested (Glass *et al.,* 2000).

Surveillance of HPS is based on the assumption that environmental conditions favouring the rodents precede, by a substantial time, the increase in rodent populations and their subsequent movement into houses. To generate an efficient surveillance algorithm, locations where cases of disease occurred in 1993 were compared with locations where people had not contracted HPS. Positive sites were characterized using Landsat-TM data and these characteristics were used to evaluate risk for the subsequent year for each pixel in a 105 000 km^2 region. Areas of high, medium and low risk were defined. The algorithm predicted the extent and timing of HPS risk in 1994, 1996, 1998, and 1999 using satellite imagery from each of the preceding years. More than 90% of cases in these years occurred in the predicted medium to high risk areas. Risk maps are now produced annually in conjunction with the American Indian Health Service.

(b) *Nasal Bot Fly.* The nasal bot fly, *Oestrus ovis,* is an insect pest of livestock in Namibia. It develops at shallow depths in the soil and the timing of its emergence is directly dependent on the number of degree days, i.e. the summed product of soil temperatures above a developmental threshold and the time over which they apply (Flasse *et al.,* 1998). Meteosat satellite sensor imagery provides accumulated soil temperature information to identify trigger conditions conducive to outbreaks and provides a broader warning capability than ground surveillance alone.

(c) *Rift Valley Fever.* Recent work has identified a complex relationship between Rift Valley fever (RVF) in Kenya with sea surface temperature change in the Indian Ocean (Linthicum *et al.,* 1999). RVF affects domestic animals and humans throughout Africa and results in widespread livestock losses and frequent human mortality. Virus outbreaks in East Africa, from 1950 to May 1998, followed periods of abnormally high rainfall, and previous work used Normalized Difference Vegetation Index (NDVI) data derived from the NOAA-AVHRR to detect conditions associated with the earliest stages of an RVF epizootic (Linthicum *et al.,* 1987, 1990). Identification of potential outbreak areas was refined using higher resolution Landsat-TM,

Satellite pour l'Observation de la Terre (SPOT), and air-borne Synthetic Aperture Radar data to identify mosquito habitats. By incorporating both Pacific and Indian Ocean sea surface temperature anomaly data together, recent studies have successfully predicted each of the three RVF outbreaks that occurred between 1982 and 1998, without predicting any false RVF events (Linthicum *et al.*, 1999).

(d) *Dengue and Dengue Haemorrhagic Fever.* Focks *et al.* (1995) have developed a two-part predictive model for dengue that incorporates entomological and human population data with weather data. The two parts are named CIMSiM (container-inhabiting mosquito simulation model) and DENSiM (dengue simulation model). The entomological model (CIMSiM) is a dynamic life-table simulation model that produces mean-value daily estimates of various parameters for all cohorts of a single species of *Aedes* mosquito within a representative 1-hectare area. The model takes account of breeding container type and its relative abundance in the environment, and predicts adult production from these variables. Because microclimate is an essential determinant of survival and development for all stages, CIMSiM also contains an extensive database of daily weather information.

DENSiM is essentially the corresponding account of the dynamics of a human population driven by country- and age-specific birth and death rates. The entomological variables output from CIMSiM are input into DENSiM, which follows the individual infection history of the modelled human population.

Parameters estimated by DENSiM include demographic, entomological, serological, and infection information on a human age group and/or time basis. As in the case of CIMSiM, DENSiM is a stochastic model. The DENSiM/CIMSiM model combination has been validated in many locations and is currently being used to model dengue risk in Brownsville, New Orleans and Puerto Rico.

3.2.5. *Developing a Global Mapping Capability*

The joint WHO/UNICEF programme HealthMap is a data management, mapping and GIS system for public health. Initially created in 1993, HealthMap was initiated to support management and monitoring of the Guinea Worm Eradication Programme (GWEP). Since 1995, it has grown in response to the increasing demand to include mapping and GIS activities for other disease control programmes, including malaria, onchocerciasis, African trypanosomiasis and lymphatic filariasis (HealthMap, 2000).

Maps for the GWEP combine village-level epidemiological maps with entomological data maps to track and visualize local prevalence trends, dependencies such as access to health resources, and social infrastructures.

The system today incorporates more socioeconomic and environmental variables to provide a broader-access mapping display designed to enable policy makers to target scarce resources better at communities at greatest risk. HealthMap archives historical data and can be considered to provide predictive data through its trending information displays. As part of the WHO international programme Roll Back Malaria, HealthMap is developing relational geo-referenced databases to determine the various types of malaria transmission. These use global positioning systems, ground and satellite sensor information, including rainfall, elevation and temperature, and epidemiological data. The maps will serve as an operational tool for planning and target control interventions including bed nets and spray operations (HealthMap, 2000).

3.3. Threshold Alert Systems

These systems are based on time series of surveillance data and were discussed in Section 2.2.1.

3.4. Environmental Early Warning Systems

The objective of a forecasting system is to predict the future course of disease case numbers, giving health care workers sufficient warning for them to deal with unexpectedly high (or low) case numbers, or else to implement control measures to prevent disease outbreaks from happening in the first place. In general, epidemic forecasting is most useful to health services when it predicts case numbers 2 to 6 months into the future, allowing tactical responses to be made when disease risk is predicted to increase. Longer-term forecasting is required when strategic control of diseases is the objective (e.g. as in WHO's Onchocerciasis Control Programme to reduce river blindness in parts of West Africa), something which is possible only with a very clear understanding of the transmission dynamics of the disease being controlled. Both spatial and temporal changes in environmental conditions may be important determinants of vector-borne disease transmission.

4. CONCLUSIONS: WHAT MAKES A GOOD PREDICTION?

Several factors can be identified as important components in establishing a good prediction for risk of epidemic or disease. Primary among these are the accuracy of prediction, as well its geographical scale and temporal duration.

Processes that should be incorporated into EWS implementation include broad validation of the model, application of models at scales appropriate to public health managers, and regular reassessment of data reliability, all coupled with expert review. Predictions should be linked with response initiatives so that they can be updated based on these actions. In this way, officials responsible for containing an outbreak can determine the reliability of predictions, the effectiveness of their responses and the level of effort required for an ongoing outbreak. Further issues include international cooperation in sharing sometimes sensitive surveillance data, as well as the burden of prediction validation.

The previous experiences of the famine EWS suggests that the impact of the system is often less related to the accuracy of the prediction, than to the fact that EWS information is not routinely used by the relevant decision makers. There were many, principally related to political and institutional factors and to logistical obstacles to launching adequate, timely response (Buchanan-Smith and Downing, 1995; Buchanan-Smith, 1996). These authors found that the international relief system 'responds to famine once it is underway but is ill-equipped to provide genuinely *early* warning'. This situation will not simply be changed by proving that an EWS is reliable. If policy-makers cannot easily determine the human or economic value of an early warning, the likelihood of implementation is small. Information therefore needs to be provided in a way that can be easily interpreted and in such a way that it influences the decision making process. These are areas that require investigation and forethought.

ACKNOWLEDGEMENTS

The authors would like to acknowledge the generous support of the National Aeronautics and Space Administration's Earth Science Enterprise Environment & Health Initiative and The Innovation Fund of the National Performance Review, whose support for the Interagency Partnership for Infectious Diseases (INTREPID) made this review possible. We are also grateful to Bob Snow and Sarah Randolph for comments on this manuscript. SIH is an Advanced Training Fellow funded by the Wellcome Trust (No. 056642).

REFERENCES

Baylis, M., Mellor, P.S. and Meiswinkel, R. (1999). Horse sickness and ENSO in South Africa. *Nature* **397**, 574.

Berkelman, R.L. (1994). Emerging infectious diseases in the United States. *Annals of the New York Academy of Science* **740**, 346–361.

Berkelman, R.L. (1998). The public health response to emerging infectious diseases: are current approaches adequate? In: *New And Resurgent Infections: Prediction, Detection and Management of Tomorrow's Epidemics* (B. Greenwood and K. De Cock, eds). London: John Wiley.

Bonabeau, E., Toubiana, L. and Flahault, A. (1998). The geographical spread of influenza. *Proceeedings of the Royal Society of London B* **265**, 2421–2425.

Bouma, M.J., Poveda, G., Rojas, W. *et al.* (1997). Predicting high risk years for malaria in Colombia using parameters of El Niño Southern Oscillation. *Tropical Medicine and International Health* **2**, 1122–1127.

Brown, V., Issak, M.A., Rossi, M., Barboza, P. and Paugam, A. (1998). Epidemic of malaria in North Eastern Kenya. *Lancet* **352**, 1356–1357.

Bryan, R.T., Pinner, R.W., and Berkelman, R.L. (1994). Emerging infectious diseases in the United States. *Annals of the New York Academy of Science* **740**, 346–361.

Buchanan-Smith, M. and Downing, T.E. (1992). Drought and famine in Africa: time for effective action. *Food Policy* **17**, 465–467.

Buchanan-Smith, M. (1996). Early warning systems: contrasting micro and macro approaches. *Appropriate Technology* **23**, 21–22.

Centers for Disease Control and Prevention (CDC) (1994). *Addressing Emerging Infectious Disease Threats: A Prevention Strategy for the United States*. Atlanta: US Department of Health and Human Services, Public Health Service.

Chase, V. (1996). ProMED: a global early warning system for disease. *Environmental Health Perspectives* **104**, 699.

Chatfield, C. (1975). *The Analysis of Time-series: Theory and Practice*. London: Chapman and Hall.

Choi, B.C. (1998). Perspectives on epidemiological surveillance in the 21st century. *Chronic Diseases in Canada* **19**, 145–151.

Cliff, A.P., Haggett, P and Smallman-Raynor, M. (1998). *Deciphering Global Epidemics—Analytical Approaches to the Disease Records of World Cities, 1888–1912*. Cambridge: Cambridge University Press.

Committee on International Science, Engineering, and Technology (CISET) (1995). *Infectious Diseases: A Global Health Threat*. CISET Working Group on Emerging and Re-emerging Infectious Diseases, National Science and Technology Council. Washington, DC: Government Printing Office.

Cox, J.S., Craig, M.H., Le Sueur, D. and Sharp, B. (1999). *Mapping Malaria Risk in the Highlands of Africa*. Durban: MARA/HIMAL Technical Report.

Cribb, R. (1998). Docs use net as disease detector. *Wired News*.

Cullen, J.R., Chitprarop, U., Doberstyn, E.B. and Sombatwattanangkul, K. (1984). An epidemiological early warning system for malaria control in northern Thailand. *Bulletin of the World Health Organization* **62**, 107–114.

Davies, S., Buchanan-Smith, M. and Lambert, R. (1991). *Early Warning in the Sahel and the Horn of Africa—The State of the Art: A Review of the Literature*. Brighton: Institute of Development Studies, University of Sussex.

Diggle, P.J. (1990). *Time Series: A Biostatistical Introduction*. Oxford: Oxford University Press.

Engelthaler, D.M., Mosley, D.G., Cheek, J.E. *et al.* (1999). Climatic and environmental patterns associated with hantavirus pulmonary syndrome, Four Corners region, United States. *Emerging Infectious Diseases* **5**, 87–94.

Federation of American Scientists (FAS) (1999). *Global Monitoring of Emerging Diseases—Design for Demonstration Program*. Federation of American Scientists.

Flahault, A., Dias-Ferrao, V., Chaberty, P., Esteves, K., Valleron, A.-J. and Lavanchy, D. (1998). Flunet as a tool for global monitoring of influenza on the web. *Journal of the American Medical Association* **280**, 1330–1332.

Flasse, S., Walker, C., Biggs, H., Stephenson, P. and Hutchinson, P. (1998). Using remote sensing to predict outbreaks of *Oestrus ovis* in Namibia. *Preventive Veterinary Medicine* **33**, 31–38.

Fleming, D.M. and Cohen, J.M. (1996). Experience of European collaboration in influenza surveillance in the winter of 1993–1994. *Journal of Public Health and Medicine* **18**, 133–142.

Focks, D.A., Daniels, E., Haile, D.G. and Keesling, J.E. (1995). A simulation model of the epidemiology of urban dengue fever - literature analysis, model development, preliminary validation, and samples of simulation results. *American Journal of Tropical Medicine and Hygiene* **53**, 489–506.

Fourquet, F. and Drucker, J. (1997). Communicable disease surveillance: the sentinel network. *Lancet* **349**, 794–795.

Frisén, M. (1992). Evaluation of methods for statistical surveillance. *Statistics in Medicine* **11**, 1489–1502.

Gill, C.A. (1921). The role of meteorology in malaria. *Indian Journal of Medical Research* **8**, 633–693.

Gill, C.A. (1923). The prediction of malaria epidemics. *Indian Journal of Medical Research* **10**, 1136–1143.

Girard, J.F. (1997). Sentinel networks: the National Public Health Network. *Bulletin of the Academy of National Medicine* **179**, 919–925.

Glass, G.E., Check, J.E., Patz, J.A., Shields, T.M., Doyle, T.J., Thoroughman, D.A., Hunt, D.K., Enscore, R.E., Gage, K.L., Irland, C., Peters, C.J. and Bryan, R. (2000). Using remotely sensed data to identify areas at risk for hantavirus pulmonary syndrome. *Emerging Infectious Diseases* **6**, 238–247.

Greenwood, B. (1998). Emerging infectious diseases: setting the research and public health agenda. In: *New and Resurgent Infections, Prediction, Detection and Management of Tomorrow's Epidemics* (B. Greenwood and K. De Cock, eds). New York: John Wiley.

Gunakasem, P., Chantrasri, C., Chaiyanun, S., Simasathien, P., Jatanasen, S. and Sangpetchsong, V. (1981). Surveillance of dengue hemorrhagic fever cases in Thailand. *Southeast Asian Journal of Tropical Medicine and Public Health* **12**, 338–343.

Gylys, L, Chomel, B.B. and Gardner, I.A. (1998). Epidemiological surveillance of rabies in Lithuania from 1986 to 1996. *Revue Scientifique et Technique de L'Office International des Epizooties* **17**, 691–698.

Halperin, W., Baker, E.L. and Monson, R.R. (1992). *Public Health Surveillance*. New York: Wiley & Sons.

Hay, S.I. (1997). Remote sensing and disease control: past, present and future. *Transactions of the Royal Society of Tropical Medicine and Hygiene* **91**, 105–106.

Hay, S.I. and Lennon, J.J. (1999). Deriving meteorological variables across Africa for the study and control of vector-borne disease: a comparison of remote sensing and spatial interpolation of climate. *Tropical Medicine and International Health* **4**, 58–71.

Hay, S.I., Tucker, C.J., Rogers, D.J. and Packer, M.J. (1996). Remotely sensed surrogates of meteorological data for the study of the distribution and abundance of arthropod vectors of disease. *Annals of Tropical Medicine and Parasitology* **90**, 1–19.

Hay, S.I., Packer, M.J. and Rogers, D.J. (1997). The impact of remote sensing on the study and control of invertebrate intermediate hosts and vectors for disease. *International Journal of Remote Sensing* **18**, 2899–2930.

Hay, S.I., Snow, R.W. and Rogers, D.J. (1998). From predicting mosquito habitat to malaria seasons using remotely sensed data: practice, problems and perspectives. *Parasitology Today* **14**, 306–313.

Hay, S.I., Rogers, D.J., Myers, M.F. and Snow, R.W. (2000a). Malaria early warning in Kenya. *Parasitology Today*, in press.

Hay, S.I., Myers, M.F., Burke, D.B. *et al.* (2000b). Etiology of inter-epidemic periods of mosquito-borne disease. *Proceedings of the National Academy of Sciences of the USA*, in press.

HealthMap (2000). URL: http://www.who.int/emc/healthmap/healthmap.html.

Heymann, D.L. and Rodier, G.R. (1998). Global surveillance of communicable diseases. *Emerging Infectious Diseases* **4**, 362–365.

Hielkema, J.U. and Snijders, F.L. (1994). Operational use of environmental satellite remote sensing and satellite communications technology for global food security and locust control by FAO – the ARTEMIS and DIANA Systems. *Acta Astronautica* **32**, 603–616.

Hutchinson, C.F. (1991). Use of satellite data for famine early warning in sub-Saharan Africa. *International Journal of Remote Sensing* **12**, 1405–1421.

Hutchinson, M.F., Nix, H.A., McMahon, J.P. and Ord, K. D. (1995). *Africa—A Topographic and Climatic Database*. Canberra: Centre for Resource and Environmental Studies, Australia National University.

Kafadar, K. and Stroup, D.F. (1992). Analysis of aberrations in public health surveillance data: estimating variances on correlated samples. *Statistics in Medicine* **11**, 1551–1568.

LaPorte, R.E. (1993). Needed - universal monitoring of all serious diseases of global importance. *American Journal of Public Health* **83**, 941–943.

Linthicum, K.J., Bailey, C.L., Glyn Davies, F. and Tucker, C.J. (1987). Detection of Rift Valley fever viral activity in Kenya by satellite remote sensing imagery. *Science* **235**, 1656–1659.

Linthicum, K.J., Bailey, C.L., Tucker, C.J. *et al.* (1990). Application of polar-orbiting, meteorological satellite data to detect flooding in Rift Valley Fever virus vector mosquito habitats in Kenya. *Medical and Veterinary Entomology* **4**, 433–438.

Linthicum, K.J., Anyamba, A., Tucker, C.J., Kelley, P.W., Myers, M.F. and Peters, C.J. (1999). Climate and satellite indicators to forecast Rift Valley Fever epidemics in Kenya. *Science* **285**, 397–400.

Maelzer, D., Hales, S., Weinstein, P., Zalucki, M. and Woodward, A. (1999). El Niño and arboviral disease prediction. *Environmental Health Perspectives* **107**, 817–818.

Nabarro, D.N. and Tayler, E.M. (1998). The roll back malaria campaign. *Science* **280**, 2067–2068.

Nicholls, N. (1993). El Niño-southern oscillation and vector-borne disease. *Lancet* **342**, 1284–1285.

Nobre, F.F. and Stroup, D.F. (1994). A monitoring system to detect changes in public health surveillance data. *International Journal of Epidemiology* **23**, 408–418.

Parsons, D.F., Garnerin, P., Flahault, A. and Gotham, I. (1996). The status of electronic reporting of notifiable conditions in USA and France. *Telemedicine Journal* **2**, 273–284.

Pratt, D.J. and Gwynne, M.D. (1977). *Rangeland Management and Ecology in East Africa*. London: Hodder and Stoughton.

Rogers, L. (1923). The world incidence of leprosy in relation to meteorological conditions and its bearing on the probable mode of transmission. *Transactions of the Royal Society Tropical Medicine and Hygiene* **16**, 440–464.

Rogers, L. (1925a). Climate and disease incidence in India with special reference to leprosy, phthisis, pneumonia and smallpox. *Journal of State Medicine* **33**, 501–510.

Rogers, L. (1925b). Relationship between pneumonia incidence and climate in India. *Lancet* **256**, 1173–1177.

Rogers, L. (1926). Small-pox and climate in India: forecasting of epidemics. *Medical Research Council Reports*, **101**, 2–22.

Shalala, D.E. (1998). Collaboration in the fight against infectious diseases. *Emerging Infectious Diseases* **4**, 15–18.

Shanks, G.D., Biomondo, K., Hay, S.I. and Snow, R.W. (2000). Changing patterns of clinical malaria since 1965 among a tea estate population located in the Kenyan highlands. *Transactions of the Royal Society of Tropical Medicine and Hygiene* **94**, 253–255.

Snacken, R., Lion, J., Van Casteren, V. *et al.* (1992). Five years of sentinel surveillance of acute respiratory infections (1985–1990): the benefits of an influenza early warning system. *European Journal of Epidemiology* **9**, 485–490.

Snacken, R., Manuguerra, J.C. and Taylor, P. (1998). European influenza surveillance: scheme on the internet. *Methods of Information in Medicine* **37**, 266–270.

Stern, L. and Lightfood, D. (1999). Automated outbreak detection: a quantitative retrospective analysis. *Epidemiology and Infections* **122**, 103–110.

Stroup, D.F., Williamson, G.D., Herndon, J.L. and Karon, J.M. (1989). Detection of aberrations in the occurrence of notifiable diseases surveillance data. *Statistics in Medicine* **8**, 323–329.

Susser, M. and Susser, J. (1996). Choosing a future for epidemiology. 1. Eras and paradigms. *American Journal of Public Health* **86**, 668–673.

Swaroop, S. (1949). Forecasting of epidemic malaria in the Punjab, India. *American Journal of Tropical Medicine* **29**, 1–17.

Szecsenyi, J., Uphoff, H. Ley, S. and Brede, H.D. (1995). Influenza surveillance: experiences from establishing a sentinel surveillance system in Germany. *Journal of Epidemiology and Community Health* **49**, 9–13.

Valleron, A.J., Bouvet, E., Garnerin, P.H. *et al.* (1986). A computer network for the surveillance of communicable diseases: the French experiment. *American Journal of Public Health* **76**, 1289–1292.

Van Brackle, L. and Williamson, G.D. (1999). A study of the average run length characteristics of the National Notifiable Diseases Surveillance System. *Statistics in Medicine* **18**, 3309–3319.

Walsh, J. (1988). Famine early warning system wins its spurs. *Science* **239**, 249–250.

Watier, L., Richardson, S. and Hubert, B. (1991). A time series construction of an alert threshold with application to *S. bovismorbificans* in France. *Statistics in Medicine* **10**, 1493–1509.

Wilson, M.E. (1994). Detection, surveillance and response to emerging diseases. *Annals of the New York Academy of Science* **740**, 336–340.

Zeng, G., Thacker, S.B., Hu, Z., Lai, X. and Wu, G. (1988). An assessment of the use of Bayes' theorem for forecasting in public health: the case of epidemic meningitis in China. *International Journal of Epidemiology* **17**, 673–679.

Education, Outreach and the Future of Remote Sensing in Human Health

B.L. Wood,[1] L.R. Beck,[2] B.M. Lobitz[3] and M.R. Bobo[3]

[1] *NASA Ames Research Center, Moffett Field, CA 94035, USA;*
[2] *California State University, Monterey Bay, NASA Ames Research Center, Moffett Field, CA 94035, USA;*
[3] *Johnson Controls World Services (JCWS), NASA Ames Research Center, Moffett Field, CA 94035, USA*

ABSTRACT

The human health community has been slow to adopt remote sensing technology for research, surveillance, or control activities. This chapter presents a brief history of the National Aeronautics and Space Administration's experiences in the use of remotely sensed data for health applications, and explores some of the obstacles, both real and perceived, that have slowed the transfer of this technology to the health community. These obstacles include the lack of awareness, which must be overcome through outreach and proper training in remote sensing, and inadequate spatial, spectral and temporal data resolutions, which are being addressed as new sensor systems are launched and currently overlooked (and underutilized) sensors are newly discovered by the health community. A basic training outline is presented, along with general considerations for selecting training candidates. The chapter concludes with a brief discussion of some current and future sensors that show promise for health applications.

1. INTRODUCTION

Since 1972, when the first earth resource satellite was launched, remote sensing technology has been adopted and used operationally by a number of applications communities, including agriculture, mineral exploration, forestry, oceanography, and climatology. The health community, on the other hand, has been slow to accept this technology, even though the potential of using remotely sensed data for studying disease vector habitats was demonstrated in the 1970s by the Health Applications Office (HAO), established by the National Aeronautics and Space Administration (NASA) at the Johnson Space Center in the USA. During its 6 year life, the HAO organized numerous lectures, workshops, and seminars to acquaint scientists with the potential health applications of remote sensing (RS) technologies. In 1972, the University of Texas School of Public Health at Houston published a report for the HAO entitled *Public Health Implications of Remote Sensing* (Fuller, 1972), which included detailed tables of diseases and their associated environmental variables that might be addressed using remote sensing systems available at that time. Two years prior to the Fuller report, a paper entitled '*New eyes for epidemiologists: aerial photography and other remote sensing techniques*' was published in the *American Journal of Epidemiology* (Cline, 1970). In his paper, Cline summarized how remote sensing might be used to study the associations between indices of disease and characteristics of humans and their environment.

Overall, NASA considered the HAO programme a success as it resulted in the publication of nearly 100 papers and reports, as well as the development of a remote sensing-based model to predict the distribution of potential screwworm fly habitat along the United States and Mexico border (Barnes, 1991). However, the HAO programme was terminated in 1976 because NASA felt that it had successfully demonstrated the utility of remote sensing in health-related applications, and that it was the responsibility of the user community to implement the technology. Unfortunately, the health community failed to translate the potential of remote sensing into operational disease surveillance and control programmes as the HAO programme scientists and managers had anticipated.

Two reasons have been cited for the failure of the user community to adopt remote sensing in the 1970s. First, senior officials within the appropriate health agencies were not familiar with remote sensing and remained sceptical of its potential (Bos, 1990; Barnes, 1991). Second, and perhaps more fundamentally, the education and technical training provided to the user community was limited. With the termination of the HAO, NASA's involvement in the application of remote sensing to health-related issues ended for nearly a decade. During this period, NASA focused its remote sensing and geographical information system (GIS) outreach activities on developing operational application programmes in such fields as agriculture, forestry, water resources, and geology, and the potential health applications of remote sensing and GIS were not addressed. This raises the questions of why, despite the early successes of the HAO programme and the operational use of remote sensing in other fields, the health community was still reluctant to adopt this tool.

2. THE 'OBSTACLES'

In general, there are a number of obstacles (both perceived and real) that have impeded the transfer of remote sensing technology to any user community. These include low general awareness, lack of knowledge of the technology, data and systems support costs, and lack of expertise. Perhaps the biggest obstacle, however, is the fact that health applications are unique; unlike agriculture, mineral exploration, forestry, and other fields, the 'ecology of health' is extremely complex as it includes components of both the biotic and abiotic systems at a diversity of scales. It is further complicated by the uncertainty of the human element. For example, a remote sensing-based environmental model for malaria risk might require information about vector habitat (e.g. vegetation, resting sites), vector survival (e.g. temperature, precipitation, humidity), and breeding sites (e.g. water, vegetation type and condition), as well as information about human activities (e.g. settlement patterns, migration, land-use activities, proximity to vector habitat). To add to the complexity, these factors can be very dynamic, both temporally and spatially.

An additional problem is the matching, in time and space, of available environmental, epidemiological, entomological, and remote sensing data sets necessary to study a specific disease issue. Figure 1 provides an example of the data sets available for Bancoumana, Mali, where malaria is a serious health problem and investigators wanted to use RS/GIS to study the patterns of disease transmission risk. In an ideal situation, the environmental, epidemiological, entomological, and remotely sensed data sets would be acquired at the same time over the same area, thus allowing their inter-relationships to be explored. However, as can be seen in Figure 1, the data for Bancoumana is far from ideal, as they are either discontinuous or missing entirely. This situation is not limited to developing countries, but is a universal problem in health applications. Given this situation, one can see that developing a model of malaria transmission risk could be much more problematic than mapping forests, agricultural crops, or mineral deposits, although significant efforts are underway to address this complexity (Hay *et al.*, this volume).

A further complication is that the health community itself is extremely diverse in that it is composed of a number of disciplines. Continuing with the malaria example, the development of a remote sensing-based model of malaria transmission risk could require input from four major areas: human, environmental, epidemiological, and entomological. Each area involves specialists in different fields, including epidemiology, landscape ecology, vector ecology, entomology, and medical, to name a few. If one wished to include parameters associated with *Plasmodium* survival, then parasitological input would also be required. Clearly, the scales at which these fields operate are extremely diverse and difficult to integrate into a unified study.

The relatively recent development of remote sensing systems and the exploration of the utility of remotely sensed data for health applications initially resulted in a 'retrofitting' of conventional epidemiological models with remotely sensed data; that is, a substitution of classical model parameters with remotely sensed approximations of those parameters. There are problems using this approach; as stated earlier, the supporting data (e.g. vector data, case data, habitat characteristics) are rarely collected at the same time, and seldom at spatial scales appropriate for use with

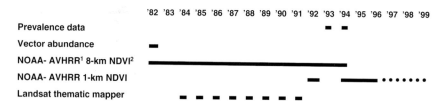

¹ National Oceanic and Atmospheric Administration's Advanced Very High Resolution Radiometer

² Normalized Difference Vegetation Index, a measure of greenness

Figure 1 Available data sets for Bancoumana, Mali, where malaria transmission is a serious health issue. The lack of coincident data sets makes it difficult to investigate the relationships between malaria prevalence, mosquito abundance, and environmental parameters.

Plate 14 (a) Average global air temperature (T_a) estimated from Advanced Very High Resolution Radiometer (AVHRR) imagery using the TvX technique. A total of 288 separate T_a maps (8 years at 10-day intervals) went into the mean calculation. See Goetz *et al.* (this volume).

(b)

Celsius

0   7

Plate 14 (Contd) (b) Coefficient of variation in global air temperature of Plate 14(a). See Goetz *et al.* (this volume).

Plate 15 (a) Vapour pressure deficit (D), estimated from AVHRR imaging using the method described by Goetz *et al.* (this volume).

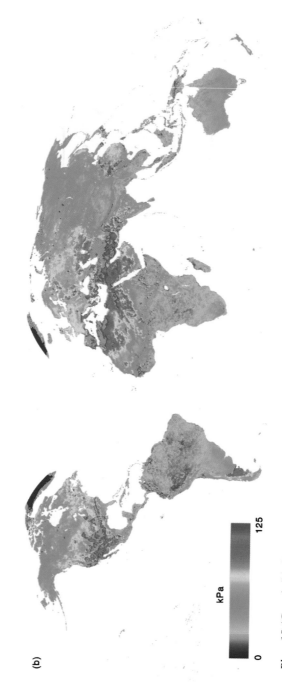

Plate 15 (Contd (b) Coefficient of variation in vapour pressure deficit of Plate 15(a). See Goetz *et al.* (this volume).

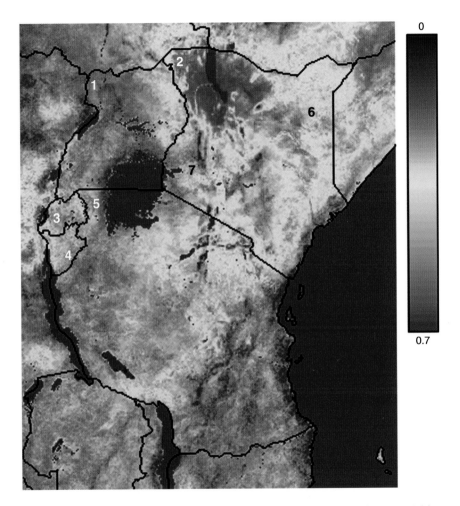

Plate 16 Average Normalized Difference Vegetation Index (NDVI) of East Africa during 1993 at a 1 × 1 km spatial resolution. The image is based on daily retrievals from channels 1 and 2 of the AVHRR on the National Oceanic and Atmospheric Administration (NOAA) satellites. The NDVI is scaled linearly between brown (0) and green (0.7). Blue indicates permanent water, and north is to the top of the page. The white numbers represent the following countries: 1, Uganda; 2, Kenya; 3, Rwanda; 4, Burundi; 5, Tanzania. The black numbers 6 and 7 indicate the location of Wajir and Kericho, respectively, whose malaria epidemiology is contrasted in the text. See Myers *et al.* (this volume).

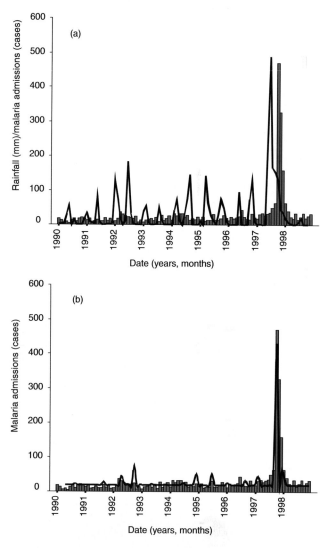

Plate 17 (a) Malaria cases and rainfall by month (1991–1998) for Wajir, Northern Kenya. Red bars indicate malaria cases, and black lines rainfall totals. (b) Observed (red bars) and predicted (black line) malaria cases in Wajir, Northern Kenya. The prediction is based on a simple quadratic relationship between present cases (x) and rainfall (y) 3 months previously; where $x = 19.9635 - 0.0399y + 0.0018y^2$. The 20-case baseline is a statistical artefact, probably resulting from the background of imported malaria cases. See Myers *et al.* (this volume).

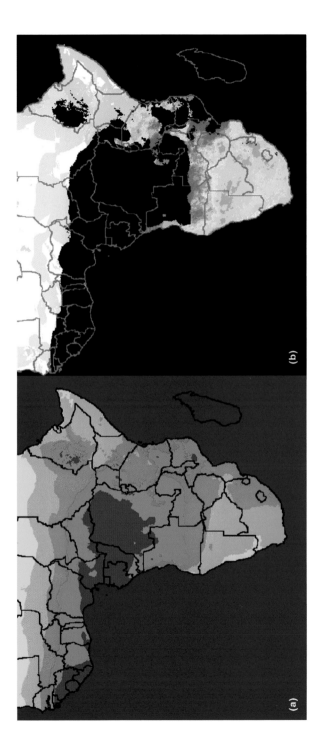

Plate 18 Rainfall anomalies in the arid areas of sub-Saharan Africa. (a) An ecological classification for Africa based on rainfall amount (Pratt and Gwynne, 1977) using long-term climate data (Hutchinson *et al.*, 1995). Yellow corresponds to deserts, and light and dark orange to arid and semi-arid regions, respectively. The light- and dark-green areas are sub-humid and humid zones, which are masked out of the other maps in this plate. (b) Rainfall anomaly maps for January 1999 calculated as deviations from the long-term average. Red areas indicate negative deviations, white no change and green positive deviations, and the data are scaled linearly between the darkest red (−200 mm) and the darkest green (+200 mm) areas. In all maps, north is to the top of the page, and data for Madagascar are not shown. See Myers *et al.* (this volume).

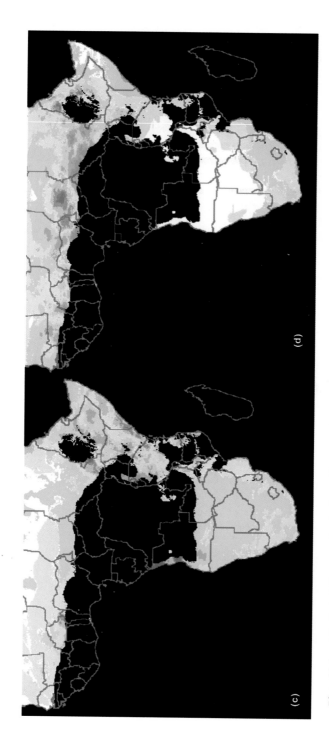

(c)

(d)

Plate 18 (contd) Rainfall anomalies in the arid areas of sub-Saharan Africa. (c)–(d) Rainfall anomaly maps for April and July 1999, respectively, calculated as deviations from the long-term average. Red areas indicate negative deviations, white no change and green positive deviations, and the data are scaled linearly between the darkest red (–200 mm) and the darkest green (+200 mm) areas. In all maps, north is to the top of the page, and data for Madagascar are not shown. See Myers *et al.* (this volume).

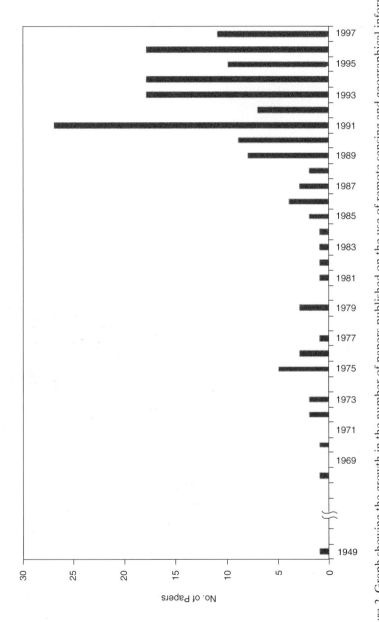

Figure 2 Graph showing the growth in the number of papers published on the use of remote sensing and geographical information system technologies in human health applications.

remotely sensing data. Similarly, remote sensed data are seldom acquired at spatial scales that match the field data.

Since the publication of the Fuller report in 1972, there have been similar papers (Jovanović, 1987a,b; Hugh-Jones, 1989; Washino and Wood, 1994; Hay *et al*., 1996, 1997; Thomson *et al*., 1996; Connor *et al*., 1997), reviewing the potential application of remote sensing to issues of human health. This period has also seen a steady increase in the number and sophistication of remote sensing systems (Rogers, this volume; Hay *et al*., this volume; Randolph, this volume; Brooker and Michael, this volume), as well as a gradual but steady increase in research projects that have explored the utility of remotely sensed data to address a variety of health issues (Figure 2). Unfortunately, little of this work has focused on the issues of technology transfer and operational uses of the technology.

3. TRAINING AND OUTREACH

A key element in achieving the successful acceptance of remote sensing and GIS technologies in the health community is through focused training and outreach. General training in the fundamentals of these RS/GIS technologies can be obtained through course work at most universities, centres such as the International Institute for Aerospace Survey and Earth Sciences (ITC) in The Netherlands, and the Thailand Remote Sensing Centre (TRSC), as well as from commercial data providers and software vendors. Short-term training that is focused on specific topics such as agriculture, forestry, or water resources is also available from the groups such as ITC, TRSC, the United Nations Food and Agriculture Organization, and the United States Geological Survey. One can also pick up some of the basics of remote sensing using online tutorials available over the Internet (see Table 1 for a partial list of those websites). However, none of the groups mentioned above currently provides training focused specifically on the health applications of remote sensing and GIS.

In 1984, NASA's Life Sciences Division initiated a new programme known as the Global Monitoring and Human Health (GMHH) programme at NASA's Ames Research Center in the USA. The focus of the GMHH programme was on basic research on such vector-borne diseases as malaria and Lyme disease, with no emphasis on training or technology transfer. During the course of the GMHH programme, NASA sponsored a series of workshops to solicit input and user requirements from the health community. At each of the workshops, participants expressed a strong desire for education and training in remote sensing and GIS technologies. In response, the Life Sciences Division initiated the Center for Health

Table 1 Some useful universal resource locators (URLs) for remote sensing tutorials available online over the Internet. Note: URLs are subject to change.

Fundamentals of Remote Sensing	www.ccrs.nrcan.gc.ca/ccrs/eduref/tutorial/tutore.html
Geographer's Craft	www.colorado.edu/geography/gcraft/contents.html
Introduction to Digital Processing	www.ccrs.nrcan.gc.ca/ccrs/eduref/exercise/digexece.html#English
Introduction to Radar	satftp.soest.hawaii.edu/space/hawaii/vfts/kilauea/radar_ex/page1.html
Introduction to Remote Sensing	satftp.soest.hawaii.edu/space/hawaii/vfts/oahu/rem_sens_ex/rsex.spectral.1.html
Introduction to Remote Sensing	seaspace.esa.int:8000/exercises/default/index.html.en
NASA Remote Sensing Tutorial	rst.gsfc.nasa.gov
Remote Sensing Core Curriculum	www.research.umbc.edu/~tbenja1/
The Science of Remote Sensing	ceos.cnes.fr:8100/cdrom-98/ceos1/science/science.htm
Utah State Class Exercises	www.nr.usu.edu/Geography-Department/rsgis/rsexer.html

Applications of Aerospace Related Technologies (CHAART) at Ames Research Center in 1995. The goal of the CHAART programme is to provide education, training, and outreach to investigators from the human health community.

Since its inception, CHAART has trained foreign and domestic scientists and health specialists to use remote sensing and GIS in their work. CHAART has also collaborated in training courses with the US Centers for Disease Control and Prevention (CDC), the World Health Organization (WHO), the Liverpool School of Tropical Medicine (LSTM) in the UK, and the International Research Institute for Climate Prediction (IRI). These collaborations have largely been short-term workshops, focused on providing potential users with an overview of the capabilities of remote sensing and GIS and the use of these technologies in health research. The CDC and WHO collaborations were centred on a specific geographical region, such as Latin America, or specific disease topic, such as schistosomiasis in China. In each case, in-country participants in the training course were actively involved in ongoing field studies. The LSTM workshop brought investigators from Africa and Asia to Liverpool to learn how to integrate remotely sensed data at a variety of temporal and spatial scales in a stratified approach to modelling disease transmission risk. The focus of the IRI collaboration was on integrating remote sensing data with long-term climate forecasts to understand the spatial and temporal patterns of disease risk associated with climate variability.

In addition to these collaborative workshops, CHAART has provided longer-term training to individual scientists; this training generally lasts 2 months. Participants have included physicians, epidemiologists, entomologists, ecologists, and public health officials. Because of this diversity of background, as well as the range of health issues studied, each participant has received tailored training based on the specific goals and objectives of that individual. Before training, the needs of each student are assessed and matched with his/her technological abilities. Individuals are also evaluated in terms of the support they receive from their home institution and the technical capabilities of that institution; both these factors are critical if technology transfer and capacity-building are to occur. Goals of the student might include one or more of the following: developing a risk model, acquiring sufficient knowledge to create an RS/GIS laboratory, developing a research proposal, learning how to design a study based on remotely sensed data, or acquiring skills to analyse spatially data already collected. The results of this assessment and the specific goals and objectives of each student help to target training in areas such as the fundamentals of remote sensing and GIS, integration of these technologies into ongoing field studies, study design and field sampling, and capacity-building and sustainability.

The following is a typical training outline for a student with little or no background in remote sensing and GIS:

(1) What is remote sensing? What is GIS? This includes an introduction to the electromagnetic spectrum, spatial/spectral/temporal resolutions, map projections, and scale.
(2) How can remote sensing and GIS be integrated into specific projects? This involves the use of case studies, with an emphasis on issues of temporal and spatial scale.
(3) What remotely sensed and ancillary data would be appropriate for a specific project?
(4) What remotely sensed and ancillary data are available? This includes a comprehensive look at various data sources, including governmental, commercial, and the Internet.
(5) What processing is required to derive useful products? This involves an introduction to basic image-processing techniques, with a subsequent focus on the steps required for each student's specific needs.
(6) How can these skills be transferred to the home institution? This encompasses transferring the skills learned in the training centre to the

hardware and software to be used when the student returns to his/her home institution, as well as guidance on developing a functional RS/GIS laboratory based on cost and the technical support at the student's home institution.

CHAART's training experience has identified some critical issues that must be addressed for training to be successful. Identification of appropriate candidates for training is one major concern. Candidates should submit a short proposal outlining the goals and objectives of their research and how they believe remote sensing and GIS will contribute to it. Each candidate should also produce evidence that his/her home institution will provide the necessary resources to maintain a remote sensing and GIS capability after the training period has been completed; in this way, the trainee can then become the trainer. Successful applications for training should be based on the likelihood of success of each proposal, the availability of supporting ancillary (i.e. epidemiological, entomological, and environmental) data to support the training and long-term investigation, and its consistency with the goals and objectives of the training programme.

Other obstacles, such as the cost of data collection and inadequate spatial/spectral/temporal resolutions, have also been identified as barriers to the transfer of remote sensing technology to the health community. The next section will briefly discuss how data from future sensor systems might help to resolve these remaining obstacles.

4. FUTURE SENSOR SYSTEMS

Since the launch of the first Landsat earth observation satellite in 1972, there has been a steady improvement in the spatial, spectral, and temporal resolution of remote sensing capabilities. This period has also seen an increase in the number of systems that acquire commercial remote sensing data as well as the availability of derived and archived products available over the Internet at little or no cost. The Committee on Earth Observation Satellites estimates that international space agencies are planning more than 80 new missions before the year 2010 (CEOS, 1995). These missions will carry over 200 different instruments, which will provide measurements of many environmental parameters at a variety of spatial, spectral, and temporal resolutions that were previously unavailable. The commercial sector is also planning to launch several systems during this period that should provide complementary data (Stoney, 1996).

Nearly all the studies indicated in Figure 2 used data from one of three major sensor systems: Landsat, the French Satellite pour l'Observation de la

Table 2 The spectral, spatial and temporal resolutions of some potentially useful current and future sensor systems. See text for details.

	Resolution		
Satellite sensor system	Spectral[a]	Spatial[b] (m)	Temporal (days)
Advanced Along Track Scanning Radiometer (AATSR)	Ch 1–7 (0.555–12.00)	1000	3
Advanced Spaceborne Thermal Emission and Reflection Radiometer (ASTER)	Ch 1–14 (0.52–11.65)	15, 30, 90	16
Australian Resource Information and Environmental Satellite (ARIES)	Ch 1–32 (0.4–1.1)	10, 30	7
China-Brazil Earth Resources Satellite (CBERS) CCD IR-MSS	Ch 1–5 (0.51–0.83) Ch 1–4 (0.50–11.45)	20 80	3–26 26
Global Imager (GLI)	Ch 1–34 (NA)	250, 1000	4
Ikonos	Ch 1–5 (0.45–0.90)	1, 4	11
Linear Imaging Self Scanner (LISS III)	Ch 1–3 (0.52–0.86)	23	24
Moderate Resolution Spectroradiometer (MODIS)	Ch 1–36 (0.62–14.35)	250, 500, 1000	2
Sea-viewing Wide Field-of-view Sensor (SeaWiFS)	Ch 1–8 (0.402–0.885)	1100–4500	1–2

[a] The spectral resolutions are the electromagnetic wavelength range in µm where 0.3 is at the visible and 14 at the thermal infrared end of the spectrum (Plate 1a).
[b] The spatial resolution is given as the diameter of the viewing area of the sensor, D, at nadir.

Terre (SPOT), and the National Oceanic and Atmospheric Administration's Advanced Very High Resolution Radiometer (AVHRR). Most of these studies focused on vector-borne diseases, such as malaria, trypanosomiasis, and Lyme disease (Beck *et al.*, 2000). In an attempt to familiarize the health community with the new capabilities offered by future sensors, as well as to identify some current sensors that have been thus far overlooked, the CHAART programme conducted an inventory of current and future satellite sensor systems and an evaluation of their possible utility in human health applications. The results of this evaluation are available at the CHAART Web site (http://geo.arc.nasa.gov/sge/health/sensor/sensor.html).

The selection of an appropriate sensor will depend, of course, on the specific health application. For example, developing a model of Lyme disease transmission risk might require information regarding tick habitat or the proximity of deer habitat with human settlements. Dister *et al.* (1993, 1997) used Landsat Thematic Mapper (TM) data to map forest patch size (related to deer habitat), indices of greenness and wetness (tick habitat), and forest/urban ecotones (human-deer contact risk). Other current/future sensors that could provide information about forests, ecotones, or human settlement patterns include ARIES-1, an Australian satellite scheduled for launch in 2001; the CCD and IR-MSS sensors on board CBERS, recently launched by China and Brazil; Ikonos, a commercial satellite with 4-m spatial resolution; LISS III, on board the orbiting Indian IRS-1C and -1D satellites; and ASTER, on board the recently launched Terra satellite. Investigators doing regional studies using patterns of vegetation green-up derived from NOAA's AVHRR sensor might also consider data from SPOT's Vegetation, launched by the French in 1998; ADEOS-II GLI, to be launched by late 2001 by the Japanese; ENVISAT's AATSR, scheduled for launch by the European Space Agency in 2001; or Terra's MODIS. An example of a non-terrestrial application is cholera. Lobitz *et al.* (2000) used sea surface temperature data, derived from NOAA-AVHRR, and sea surface height anomaly data, derived from TOPEX/Poseidon, to search for temporal patterns in the Bay of Bengal associated with cholera outbreaks in Bangladesh. Until the recent launch of SeaWiFS, the Sea-viewing Wide Field-of-view Sensor, there was no orbiting satellite capable of mapping ocean colour. Data from this new sensor are already helping cholera investigators to map chlorophyll concentration in the Bay of Bengal, an indicator of phytoplankton abundance. (In the marine environment, *Vibrio cholerae* preferentially attach to zooplankton; zooplankton, in turn, are closely associated with phytoplankton, both spatially and temporally.) See Table 2 for the specifications of the sensor systems mentioned above.

The new capabilities offered by future satellite systems will bring significant improvements in spatial, spectral, and temporal resolutions, making it possible to address health issues previously thought to be beyond the capabilities of remote sensing. In addition, advances in the understanding of the ecology of disease organisms, vectors, and their reservoirs and hosts have directed researchers to assess a greater range of environmental factors that promote disease transmission, disease vector abundance, and the emergence and maintenance of disease foci. Advances in computer processing, GIS, and global positioning systems now make it easier to integrate epidemiological, environmental, entomological, and remote sensing data for the purpose of developing models of disease transmission risk that can be used in operational surveillance and control programmes. These developments have also led to the initiation of an increasing number of interdisciplinary investigations, thereby resulting in the need for training in remote sensing and GIS.

5. CONCLUSIONS

Remote sensing technology has played an important role in the biological and Earth sciences by enabling researchers to study the Earth's biotic and abiotic components in ways not possible before. The extent to which this technology has been used to study the spatial and temporal patterns of disease has depended on a number of obstacles, including lack of familiarity with the technology. The main goal of outreach and training is to expose health scientists to the potential utility of remotely sensed data, thus creating the interest and desire for acquiring the skills with which to process, interpret, and analyse these data. The ultimate goal is to transfer the proper use of the technology to health investigators so that they, not remote sensing technologists, can continue to establish the extent to which remote sensing technology can contribute to science and its applications to health. This chapter briefly presented some of the considerations and issues that should be addressed to enable this transfer of knowledge to occur; these include the selection of the appropriate person to train, which involves assessing his/her goals, the home institution's hardware/software capabilities, and the commitment of long-term continuing support for that trainee so that he/she can then become a trainer.

Other obstacles, such as the cost of data and inadequate spatial/spectral/temporal resolutions, have also limited the use of remote sensing within the health community. Fortunately, many of these obstacles will be addressed by the array of sensor systems scheduled for launch within the 5 years up to 2005. With the development of so many new systems and processing capabilities, however, there will be an increasing need within the health community to link basic research with outreach and training. This has already resulted in the initiation of several new collaborations with national and international organizations, which have begun to link basic research and training with the goal of developing operational programmes. As with the HAO programme in the early 1970s, the authors believe that the potential application of remote sensing and GIS has been demonstrated and the time is right to transfer these tools to operational research, surveillance, and control programmes. This will be achieved only through intensive focused training, outreach, technology transfer, and long-term technical support.

ACKNOWLEDGEMENTS

The Center for Health Applications of Aerospace Related Technologies (CHAART) was established at Ames Research Center by NASA's Life Sciences Division, within the Office of Life & Microgravity Sciences & Applications, to make remote sensing, geographic information systems, global positioning systems, and computer modeling available to investigators in the human health community.

REFERENCES

Barnes, C.M. (1991). An historical perspective on the applications of remote sensing to public health. *Preventative Veterinary Medicine* **11**, 163–166.

Beck, L.R., Lobitz, B.M. and Wood, B.L. (2000). Remote sensing and human health: new sensors and new opportunities. *Emerging Infectious Diseases* **6**, 217–227.

Bos, R. (1990). Application of remote-sensing. *Parasitology Today* **6**, 39.

Cline, B.L. (1970). New eyes for epidemiologists: aerial photography and other remote sensing techniques. *American Journal of Epidemiology* **92**, 85–89.

Committee on Earth Observation Satellites (CEOS) (1995). *Coordination for the Next Decade (1995 CEOS Yearbook). European Space Agency.* Smith System Engineering, UK.

Connor, S.J., Flasse, S.P., Perryman, A.H. and Thomson, M.C. (1997). *The Contribution of Satellite Derived Information to Malaria Stratification, Monitoring and Early Warning.* Geneva: World Health Organization, WHO/MAL/97.1079.

Dister, S.W., Beck, L.R., Wood, B.L., Falco, R. and Fish, D. (1993). The use of GIS and remote sensing technologies in a landscape approach to the study of Lyme disease transmission risk. In: *Proceedings of GIS '93: Geographic Information Systems in Forestry, Environmental and Natural Resource Management.* Vancouver, BC.

Dister, S.W., Fish, D., Bros, S., Frank, D.H. and Wood, B.L. (1997). Landscape characterization of peridomestic risk for Lyme disease using satellite imagery. *American Journal of Tropical Medicine and Hygiene* **57**, 687–692.

Fuller, C.E. (1972). *Public Health Implications of Remote Sensing.* Final Report, NASA contract NAS-9–11522. Houston, TX: University of Texas, School of Public Health.

Hay, S.I., Tucker, C.J., Rogers, D.J. and Packer, M.J. (1996). Remotely sensed surrogates of meteorological data for the study of the distribution and abundance of arthropod vectors of disease. *Annals of Tropical Medicine and Parasitology* **90**, 1–19.

Hay, S.I., Packer, M.J. and Rogers, D.J. (1997). The impact of remote sensing on the study and control of invertebrate intermediate host and vectors for disease. *International Journal of Remote Sensing* **18**, 2899–2930.

Hugh-Jones, M. (1989). Applications of remote sensing to the identification of the habitats of parasites and disease vectors. *Parasitology Today* **5**, 244–251.

Jovanović, P. (1987a). Satellite medicine. *World Health*, January–February, pp. 18–19.

Jovanović, P. (1987b). Satellite remote sensing imagery in public health. *Acta Astronautica* **15**, 951–953.

Lobitz, B., Beck, L., Huq, A., Wood, B. *et al.* (2000). Climate and infectious disease: the cholera paradigm. *Proceedings of the National Academy Sciences of the USA* **97**, 1438–1443.

Stoney, W.E. (1996). *The Pecora Legacy – Land Observation Satellites in the Next Century*. Paper, Pecora 13 Symposium, Sioux Falls, SD, 22 August 1996. (The complete paper can be found on-line at http://geo.arc.nasa.gov/esdstaff/landsat/wes.html.)

Thomson, M.C., Connor, S.J., Milligan, P.J.M. and Flasse, S.P. (1996). The ecology of malaria as seen from Earth-observation satellites. *Annals of Tropical Medicine and Parasitology* **90**, 243–264.

Washino, R. K. and Wood, B.L. (1994). Application of remote sensing to vector arthropod surveillance and control. *American Journal of Tropical Medicine and Hygiene* **50**, 134–144.

Index

Page numbers in *italic* indicate illustrations and tables.

Contents of Volumes in This Series

Volume 47